EUROPEAN POLITICAL, ECONOMIC, AND SECURITY ISSUES

A MODERN HISTORY OF SOCIOLOGY IN ITALY AND THE VARIOUS PATTERNS OF ITS EPISTEMOLOGICAL DEVELOPMENT

EUROPEAN POLITICAL, ECONOMIC, AND SECURITY ISSUES

Additional books and e-books in this series can be found on Nova's website under the Series tab.

EUROPEAN POLITICAL, ECONOMIC, AND SECURITY ISSUES

A MODERN HISTORY OF SOCIOLOGY IN ITALY AND THE VARIOUS PATTERNS OF ITS EPISTEMOLOGICAL DEVELOPMENT

GUGLIELMO RINZIVILLO

nova
science publishers
New York

Library of Congress Cataloging-in-Publication Data

ISBN: 978-1-53614-761-2

Published by Nova Science Publishers, Inc. † New York

CONTENTS

PREFACE

This work aims at fostering interest in the links between a particular theoretical and conceptual development of sociological science in Italy and the debate surrounding the history of scientific subjects, here called the 'epistemological history' of various disciplines. The author sets out to trace the 'points of view' emerging from Italian epistemological sociology between the early twentieth and twenty-first centuries and related to the debate on the historical and philosophical sciences. The intention is that of revealing the distinctive characteristics of the latter, their 'internal' (logical and methodological) rather than 'external' (historical and social) factors, while investigating the task of the new historians of science confronted with the changes caused by various 'knowledges' and forms of technological 'know-how'. The text provides the contemporary scientific community with a new route to follow when seeking to rationally establish and write the 'history of sociology'. The addressees are, more than anything else, the new generation of scholars of the history of science.

ACKNOWLEDGMENT

I wish to thank Kay McCarthy for her help with this extended English-language version of my book.

Introduction: The History of Sociology and the History of Science: New Epistemological Issues, Ambits and Perspectives

I believe that one of the most important issues twenty-first-century historians of sociology need to address is that of relocating the discipline in an age of scientific knowledge, thus renouncing the historical-cultural and academic literature which represented the evolution of sociological studies within a framework of the paradigmatic dualism existing traditionally between the history of sociological thinking and scientific sociology. This dualism still persists, conditioning relationships between accredited academic historiography and results obtained in the field of science. This dichotomy has also succeeded in generating the idea of scientific sociology as a separate discipline, a science endowed with an ascertained historical-epistemological and methodological autonomy all its own within the field of the 'sciences'. In short, sociology as a science has always appeared as a history in its own right and not as the product of interaction with other disciplines belonging to the field of science. The painstaking quest for a distinctive character within the sphere of historical studies of science, above all during the nineteenth and twentieth centuries, diffused the idea that scientific sociology derived its truths and foundational topic-problems from itself and that, for precisely that reason, it did not need to be located within the history of the most accredited scientific subjects, at a time when those same disciplines were opening up to the notion of considering science a social issue, that is, as a society-science product, thanks to the close links between scientific issues, structures and historical-social developments in a number of areas etc. The greatest results to this regard were obtained in Europe starting from the late 1970's, with the opening up of the social sciences, sociology in particular, to the history and philosophy of science. This is what we shall discuss here.

A history of scientific sociology devoid of the 'internal' results of the epistemological debate which arose in Europe during the second half of the twentieth century would have no chance of surviving in the world of modern disciplines. It would simply fail to exist, at least as the history of a modern scientific discipline among its contemporary peers.

The truth is that this theoretical subject does not as yet exist or remains in a state of perennial logical-empirical becoming, as is the wont with sociology: a should-be science, valid in some developmental areas, not in others. The situation is, at present, highly sectorialised and proceeds alongside an increment in specialization.

In my opinion it differs from the concept of interest in the field of the sciences of culture and religion, or in that of the historiography of civilization. In those ambits sociology asserts itself alongside social history, the history of peoples and cultural anthropology, thus deriving its consistency only from confrontation with them. This holds similarly for the economic and political sociological disciplines which, historically speaking, expand the juridical factors of the existence of grand issues concerning social organisation etc.

It is in the field of the history of science, therefore, that sociology has not as yet acquired those requisites which might make it part of a consistent and accredited development of the scientific disciplines capable of handing down 'strong' historiographical models as an expression of their epistemological maturity and the progress made with regard to method. Serious consideration of the transmission of their scientific solidity as an attribute indispensable to the progress of knowledge and the discipline itself is lacking among modern sociologists, more so than among experts in other fields of specialist knowledge. This also holds for the social sciences. Let us take social psychology for example. This remains perennially grappled to experimental or political-economy models, which find it difficult to free themselves from the economic-historical or cognitive models developed in a given historical period.

But how can we judge these issues today? What has the total dissolution of scientism, beginning with neo-positivism, bequeathed to contemporaries? And what may be said of the results of already obsolete neo-positivism? We shall try to respond to these questions by examining the 'period-bound' production of a number of historians and scholars of scientific matters as well as the history of sociology, to provide answers of an epistemological nature which the debate of the 1970's left unresolved.

To this regard, it is worthwhile noticing how long it took historians of sociology and/or historians of sociological thought to formulate perspective interest in the matter. Very often, in the face of various interrogatives, these scholars of science came to realise that they had no precise answer or, more generally speaking, that the questions raised did not yield the answers hoped for and that, in given periods of the development of scientific sociology, these had been archived and/or failed to appear in debates relevant to matters concerning the foundation of/in science.

In our opinion, the serious error committed by historians of sociology has been and continues to be that of having abandoned certain crucial paradigmatic turning points in the development of sociological science. This means admitting that some issues should, by now, be considered as surpassed and that it is necessary to move forward. This way it seems that positivism and scientism have vanished from the horizon of sociology, like evolutionism and other currents of scientific thought of the past centuries. This seems to hold too with regard to cognitive-sociological perspectives, the social-action theory and American functionalism, critical theory and even methodological individualism, the orphaned brainchild of Raymon Boudon. And what can we say of relations between sociology and Marxism?

At this rate sociology would end up by doubting its very own existence, finding no connection between its historians of thought and the historians of science of the twentieth and twenty-first centuries, or unable to count on discussions within the ambit of other disciplines which began in the eighteenth and nineteenth centuries and continued up to the present. This is why we believe that it is still well worth our while to discuss issues of interest to sociology from the beginning, that is, confrontations between natural philosophy, rationalism, nineteenth-century positivism and the dissolution of scientism into Ernst Mach, Jules Henri Poincaré and Pierre Maurice Marie Duhem's conventionalist epistemology and into twentieth-century constructivism. Again, we might also discuss the debate surrounding Marxism, were it not for a certain kind of demobilization undertaken about twenty years ago, in Europe at least. We shall speak of this too.

Now, let us deal with the thesis whereby sociology has belonged to the history of science since the reform of natural philosophy, inaugurated in the seventeenth century by Francis Bacon, and which began to involve disciplines called *encyclopaedic knowledge* in the period between the Enlightenment and nineteenth-century positivism. As we know, with Auguste Comte, sociology becomes the *scientia scientiarum* of reputed modernity, a science of sociality and of social or socio-physical organisation. Until then, the experimental scientific method had not been applied as yet to political issues, as Henry Saint Simon stated in his *Réorganitation de la societé europenne*, 1814.

During this phase of the nineteenth century, encyclopaedic knowledge gave way to a revolution begun by the science of society, where, as Saint Simon demonstrated, the experimental philosophical model managed to circulate among the various knowledges. Science then seemed to be its own master, as reason had been for the Enlightenment and scientific reason for the natural world.

From the point of view of the history of science, investigation by the social human being of nature and of the empirical world had been conducted down through the ages according to the modalities of natural philosophy, where the ancient and mediaeval traditions had left a number of questions of a finalistic and vitalistic nature unresolved. In the sixteenth century, but in the seventeenth in particular, there was a certain return of topics inspired by atomism and mechanicism when Aristotelian physics began to decline. In the century of Descartes, which refuted atomism, a quest for a theory of matter took centre stage alongside the idea of uniform linear movement as the determinant of the resting state of bodies, as we read in *Principia philosophiae*, 1644 (published in French in 1647). The principle whereby knowledge derived solely from the interpretation of and commentaries on ancient texts, as a given form of *principium potestatibus*, was gradually replaced by a philosophy of the history of progress embodied by the experimental method in the ambit of scientific knowledge. The scientific revolution culminated in the application of this methodology to natural philosophy, to the sciences of life (biology, anatomy, zoology, botany), to medicine, chemistry, physics (optics, magnetism, thermodynamics, electricity) as well as to William Harvey's discovery of blood pressure, Evangelista Torricelli's of atmospheric pressure, Francesco Redi's regarding living organisms and Isaac Newton's concerning the nature of light. The history of the sciences was based prevalently on the above-illustrated classification expressed by Baconian science, with the help of mathematical investigation and demonstration of the world which Galileo Galilei extended to include astronomy, cosmology, theoretical physics, mechanics and dynamics. The history of the sciences thus tended to establish itself as a *notum facere* capable of interesting various fields of scientific knowledge and most of that of the philosophy of nature. The axiom of this period, which went well beyond the sixteenth century, was to fight against the obscurantism which had endured in time, ensnaring projections of human perceptions of nature, anchoring them to the prejudices of ancient philosophical doctrines which simply diffused the use of a terminology unsuited to descriptions of the natural and cultural phenomena of the time.

As we know, the slow decline of nineteenth-century positivism coincided in Europe with the advent of a logical-empiricist revolution which found correspondence in an examination of scientific language, known as 'neo-positivism' which took place within the so-called *Wiener Kreis* [Viennese Circle], which included in its ranks economists, logicians, philosophers of science, sociologists, etc. Alongside its ambitious project to unify the sciences, clearly set down in the *Wiener Kreis's* programmatic manifesto, and expressed by its generally anti-metaphysical attitude -two pillars of the neo-empiristic theory- one particular feature around which the Circle's theses rotated was the great attention it paid to logical analyses of language. As we know, it was by means of this method, more or less, that it was possible to demonstrate the uselessness of assertions, that is, that metaphysical statements were devoid of content.

The introduction to the Vienna manifesto, *Wissenchaftliche weltauffassung. Der Wiener Kreis* considerable importance was dedicated to the historical-social and intellectual context which made the development of this new movement possible, that is, the Vienna of the early twentieth century. Towards the end of the nineteenth century, as we know, the political and social unrest which had been quelled after the 1848 uprisings, came to the fore again in Vienna. Here the Christian socialists clashed with the German

National Party and with Viktor Adleris Social Democrats. Besides historical-sociological matters, the development of a new form of modern scientism was the only other issue which actually prevailed. Vienna became the focus of an empiricist attitude destined to give rise to a new scientific revolution in a developmental context which found an alternative to Kantianism, thanks to its anti-metaphysical, positivist, utilitarian, pragmatic, materialistic attitude.

The neo-positivists began to discuss general problems strictly related to empirical science. These were mainly epistemological and methodological issues regarding physics. We might mention, by way of example, Jules Henri Poincare's conventionalism, Pierre Duhem's conception of the purpose and structure of scientific theories, as well as matters concerning the fundamentals of mathematics, axiomatics and logistics. Vienna, in the early decades of the twentieth century gave rise not only to neo-empiricism but also to psychoanalysis, juridical normativism, dodecaphonic music, modern architecture and abstract painting. Neo-positivism set out to establish a theory of a scientific 'view' of the world, which was obliged to provide a tool and/or set of tools capable of promoting serious social and economic transformation, as Otto Neurath, Hans Hahn and Rudolf Carnap proposed.

One of the cornerstones of the scientific conception of the world consisted in rigorous criticism of any metaphysical pollution of scientific work. The primary objective of scientific neo-empiricism was to provide a scientific language capable of expressing expectations and conditions of control availing of so-called 'observational statements'. The object of the polemic was not metaphysics only, that is, its coexistence with idealistic or rationalistic topics, but, in particular what the neo-empiricists deemed to be the metaphysics intrinsic to Kantian apriorism. As mentioned above, neo-positivism did not stem from the problem of knowledge, but from an examination of the structure of science. In short, we can say that the scientific conception of the world recognized only empirical propositions formulated with regard to objects of all kinds and the analytical statements of logic and mathematics.

The fact is that the anti-metaphysics of twentieth-century neo-empiricism also performed an anti-philosophical role, where science acted as the yardstick of philosophy, not vice versa. The same may be said of the relations that prompted sociological research to redefine the discipline. The historical debate regarding science thus took place within disciplines, where analyses revealed the limits of not having to resort to the historicity of statements but to a kind of extended naturalism with a minimum of logical analyses of language. Sociology in particular tended to deal with its own existence with more or less verifiable utterances, where the neo-empiricist theory of knowledge concerned comparisons between protocols and experiences, between facts and propositions, which, ultimately, opened the way to every possible kind of dualism and a return to traditional issues.

The concept of sociology as a science of the observed behaviour of human groups remained the core of the discipline, at least for some of the members of the *Wiener Kreis* like Otto Neurath, who spoke of propositions detectable in the study of groups. He extended the concept of sociology with a view to reducing scientific descriptions of the various disciplines in terms of space-time and observable predicates. In this sense, Otto Neurath's positivism, in particular, outlined a larger-group type of sociology which ceased to be a specific science on a par with others, considering it merely as part of a sole, all-embracing science, grounded in original physicalist assumptions and belonging, therefore, to so-called 'cosmic history'. The latter because it availed of a universal language, that is, of a physicalist language, contained the same propositions as those of a unified science. This way the unified science of the Encyclopaedia might be understood as a language typical of (cosmic) history.

Thus, the outcomes of scientific sociology were once more commensurate with the outcomes of an accommodation of physicalist terminology, where some programmes indicating a plurality of possible 'histories' appeared. This, even if neo-positivism failed to collocate sociology within a historical perspective of epistemology, that is, within an epistemological history capable of reaching us. The efforts of the neo-empiricists seemed to be encapsulated in an anti-metaphysical tendency of its own, aimed at reducing science to a play on language consisting in verifiable statements, with a view, also, to re-establishing that demarcation between science and philosophy achieved later by Karl Popper in particular.

The test of Popperian philosophy was to be Marxism, something of which Otto Neurath was not unaware. We can say that many British and US historians of science, thanks to Popper's teaching (1935-40), trod the pathway of mere criticism of totalitarianism and historiographical determinism, which sought, once and for all, to act as the hub of history and the social sciences, without recourse to any real epistemological criticism. Thus, for example, Thomas Samuel Kuhn's well-known discussion regarding the scientific revolution took place without Karl Marx and grew up around a dominant paradigm which did not seek historicity in the practice of scientists and/or the social sciences. In our opinion, this had an impact on the formation of the modern and contemporary generation of historians of science greater than that of positivism and evolutionism on the development of sociological knowledge in the course of the nineteenth century. The fact is that sociology took shape more and more within western culture without paying sufficient attention to its concordances and dissonances with Marxism, which had always experienced a kind of self-criticism of the knowledge produced, as the chief among the social sciences.

Thus, excess criticism was harmful to sociology, even to areas of research outside its own domain. In part, transcendentalism in the sociological sciences (favoured by critics like the Austrian Max Adler of the neo-Kantian school of Baden which gave life to the famous *Methodenstreit*) helped make the debate misleading according to a more 'open' Marxist perspective.

With regard to this point, it is interesting to read Max Adler's criticism of Heinrich Rickert, Wilhelm Windelband, Rudolph Stammler and Hugo Munstenberg (ignored, for example, by 'official' historiography of sociology), as well as strictly Weberian positions, or better still, those attributable to Max Weber's scholarly production and books. Not only, but there are many, more or less well-known examples of those who sought in Marxism only sociological tools to avail of to explain the genesis and development of ideology, ideas and even political institutions (the Pole, Ludwick Krzywicki, by way of example). Alongside these, we find also examples of those who sought the genesis of sciences like mathematics and physics by examining structure and ideology. Their theories regarding the formation and production of knowledge were based on highly different scientific grounds. In short, the basic assumption was that an objective and empirical science of social relations, in particular, ran the risk of belonging to ideology.

The problem was also that of considering the history of science as more or less essential to the development of scientific sociology. It was not a 'matter of words', as it had been at the time of positivism and Italian evolutionism within a context of ideas where a dispute arose between those Marxists who, inspired by the works of Antonio Gramsci, sought the practical usefulness of the *philosophy of praxis,* and the positivists whose research and predictions regarding the origins of social problems were scientific and, therefore, a-critically separate from the practice of history – something that went 'beyond' the serious intentions of the logical positivists of Vienna. It was a modern problem, raised by those who had already crossed the threshold of epistemological relativism and, therefore, of those who, on the other hand, were attempting to do so. Was it the case to proceed rationally making one way more feasible than another? The question was: which way was the most convenient?

If we refer to research into the genetic blueprint of scientific procedure, we can assume that Marxist sociologists lost and, maybe rightly so, the methodological key to the problem. We might say the same of some cultural historians or historians of ideas known as 'social scientists' belonging to the Marxist area and, therefore, considered as belonging to the Marxist tradition of studies regarding the production of science. Furthermore, this may be true for those who renounced 'historical' studies, preferring to avail of them only as 'subjects' required to foster interest in the sociology of knowledge and the sociology of scientific knowledge. The sociologists of knowledge referred, in particular, to historians of physics and epistemologists who were not strictly Marxists (like the aforementioned Thomas Samuel Kuhn, Popper and Bachelard) to describe changes and developments in science with a view to coming up - eventually- with a critical, conflictual theory of the changes themselves. On the whole, it is not easy to study and understand the nature and the origin of these problems or carry out a systematic examination of that assimilation of a number of sources and categories of Marxism by contemporary sociology which took place in order to have people believe that the genetic explanation of causal social phenomena was quite different from the production of scientific material capable of explaining the genesis of science. Until proven otherwise, it was sociology which indulged in Marxism, not the opposite.

Conversely, what we do find is a quantity of books and essays which discussed sociology's systematic vocational failure to seek within its own scientific history a consistent development of rationally conceived methods. Furthermore, there was also the confusion generated by the terms 'systemic' and 'systematic', caused, at times, by their translation from German into English, starting with the anti-dogmatic lessons held by Karl Mannheim at the London School of Economics and which Marxists considered as critiques of the 'real issues' addressed by Karl Marx. In reality, many of these 'issues' historically concerned sociologists and their particular theoretical-intellectual approach. Many of them, as we know, were clarified thanks to the mediation of some sociologists of knowledge or simply vanished when confronted with the sociology of scientific knowledge.

What more modern historians of sociology found on the horizon was a simple attempt at exploring and tracing the contributions of scholars belonging to that approach to sociological studies of science known as *the sociology of scientific knowledge* (abbreviated SSK) developed in England in the early 1970's, and which referred in particular to its first phase (1970 to the early 1980's). During this period, in Europe, between sociologists, philosophers and historians of science there was a keen debate surrounding Merton's paradigm and rejection of the regulatory model by some scholars who deemed it inadequate to explain the social dynamics of science and account for its importance as a social problem This referred to its then clear role as a social institution capable of influencing other political, economic, etc. institutions while, in turn, being affected by them. This was the topic American sociologist Robert King Merton brought to the attention of sociology (especially the sociology of knowledge) when he formulated the paradox destined to bear his name. These scholars who belonged to what was first called the 'Science studies unit' were informed by the debate which arose concerning the reception of Kuhn's work by sociologists, philosophers and historians of science who examined the matter of the social dimension of science from their various perspectives.

This reference led analysts to reflect critically on the topics advanced by Thomas Samuel Kuhn and Robert King Merton as a filter at their disposal in order to narrow the scope of the discussion while providing them with the most concise, best-oriented view possible of a specialised area of social science, otherwise so full of sundry contributions by authors and scientists both from the social and other scientific areas. The work of the historian of science Thomas Samuel Kuhn, especially after the critical discussion surrounding *The Structure of Scientific Revolutions,* may be considered as a watershed, a point after which sociological reflection on science came up with something different. What was required here to guide epistemological discussion in the field was a choice of authors and aspects of their writings, that is, a rational choice which many historians of sociological thinking and science might as yet fulfil.

After the criticism received by *The Structure of Scientific Revolutions*, many scholars, historians and philosophers of science were obliged to address issues of a more specifically sociological nature. To a certain extent, this led some to gradually consider the model proposed by Robert King Merton no longer adequate and shift their attention towards the heterogeneous materials produced by the history of science which also entered a dialogue with the philosophy of science as an essential point to ponder. This, also because the members of the SSK themselves came from different disciplinary backgrounds.

Following this line, consisting in a somewhat general introduction to the sociology of knowledge, with some concession to its macro and micro dimensions, the many fields of interest which then arose, expanded and affected the new-found task of twenty-first-century historians of science, that of pursuing epistemological history. These historians were required to narrow increasingly, with comparison to the scope of the discipline itself, the range of their own particular interest in the sociology of science and scientific knowledge. With this in mind, historians of sociology and / or historians of sociological were, first of all, able to address the paradox of Robert King Merton, the first to take an interest in scientific institutions as a social phenomenon and relationships between science and society. Secondly, they were able to describe his functionalist and legislative-type approach to the sociology of science, formulated as a veritable paradigm, to the point of dealing with its slow decline. The paradigm itself became part of a historiographical reconstruction of science, included in a serious debate concerning the history of science. Having discussed Merton's model, the historians of sociology were able to throw further light on important issues regarding the *ethos* of science, deal with scientific change and the remoteness of the reality's prevalent regulatory system, while dealing also with the theoretical and empirical criticism advanced by American and European scholars of science and culture.

It then became opportune to present the SSK in handbooks of sociology and / or history of sociological thought, with relevant historical references and the change of the department's name from 'Science studies unit' to 'Sociology of scientific knowledge'. It was also possible to provide a list of its main exponents and the works they published from between 1971 and the early 1980's: the so-called "strong programme." This led to serious consideration of the exponents and important, seminal works by authors like Barnes, Bloor, Mackenzie, in particular, but also contributions by Dolby, Whitley, Mulkay and Shinn. For each of them it was possible to browse one or more of their works, chosen from among those regarding science and the relationship between sociology and science, with a view to highlighting essential topics. Further articles were also included to provide an idea of the real problems which the sociology of scientific knowledge was addressing at the time.

At this point, historians of sociology and / or history of sociological thought, after the "strong programme," began to explore the approach to the sociology of scientific knowledge which had spread beyond the confines of Britain and illustrate the works of some of the authors who had contributed to its development in the early 1980's, while referring also to the constructivist theory.

The challenge to modern scientism occurred with the emergence of conventionalism, i.e., the philosophical conception whereby principles or propositions, both cognitive and ethical-political, were conventions based on an agreement or a choice (even implicit), and therefore not evaluable in terms of truth or falsity. The appearance of conventionalism established new standards for research and an epistemological vision of the history of science. In antiquity, conventionalism had been supported by the sophists on moral and political grounds. Much of the legal and political thought of the seventeenth and eighteenth centuries supported the conventional nature of the fundamental rules of the social order, based, as they were, on agreements between human beings, grounded in criteria of general well-being and utility. Towards the end of the nineteenth century, conventionalism was extended to the field of formal languages and therefore modern conventionalism became a school of thought which emphasised the conventional nature of a number of fundamental concepts of mathematics and physics. It arose from the discovery of so-called non-Euclidean geometries and the relativisation of space and time, determined by early twentieth-century investigations of physics.

In short, non-Euclidean geometry questioned notions of space based on common sense and psychological inquiry and used as the foundations of geometry. It was also possible to demonstrate the legitimacy of geometries which did not recur to this principle or analysis in terms of psychological, perceptual or physical space. As to physics, the objections raised regarding the absoluteness of space (the basis of Newtonian physics) especially by Ernst Mach and Henry Poincaré, came to fruition with Einstein's theory of relativity. In this context, where neither a privileged geometry nor an absolute space was provided, it was conventionalism which reiterated the fact that, in the long run, privileging one kind of geometry or a system of spatial references depended exclusively on conventions established by the operator and based on scientific criteria of opportunity and convenience. The problem of the affirmation of scientific orientation ceased to exist.

So, the need to seek an objective, univocal criterion was rejected as a false problem associated with a traditional, metaphysical way of thinking.

This notion, extended first to the fundamentals of mathematics, and then to all mathematics, transformed this science into a set of concepts and theories, the adoption of which was bound only to conditions of non-contradiction and manageability. Thus, the existence in mathematics of *a priori* knowledge and recourse to evidence, sustained by theory was denied due to the fact that specification of the meaning of the terms used in this science depended solely on conventions that were gradually stipulated In the early twentieth century, this position was advocated not only by Henry Poincaré, but also by

Maurice Pierre Duhem. The vast repercussions of the philosophical works of the former meant that conventionalism served as a substratum to numerous and sundry philosophical positions, like that of pragmatism or French spiritualism. It should be noted, however, that as far as the fundaments of mathematics and logic were concerned, the conventionalism of the early decades of the twentieth century soon lost ground in the face of a greater wealth and variety of currents like formalism, logicism and intuitionism. Theses like conventionalism emerged from works produced during a certain period by Rudolf Carnap, more generally, from logical neo-positivism. However, the fate of conventionalism applied to studies of the fundaments of physics (the focus of recent, lively debate among epistemologists) has been somewhat different.

PART 1

Chapter 1

PRELIMINARY CONSIDERATIONS

The late twentieth-century Italian debate concerning the conceptual history of Sociology (here considered differently from '*Begriffsgeschichte*' and its methodological applications), its epistemological production and/or the historiography of scientific sociology, seems to encounter an enormous number of difficulties. One such difficulty regards so-called *formal matters* which respond to the interests of some who aspire, more or less individually, to consideration as historians of scientific subjects and historians of science, following the 'internal' debate during the second half of the twentieth century between Karl Popper and Thomas Samuel Kuhn with his considerations regarding the issue of paradigmatic revolutions within the social and the empirical sciences. We refer, above all, to those who should by now reason in terms of given conceptions of knowledge, that is, in terms of hypothetical-deductive concepts and models as well as theories of mathematical language, etc. From our 'point of view' it is pretty clear that they do not aim in the least at bringing their epistemological research within the fold of traditional historiography. They seek, on the contrary, to study the contemporary development of sociological science availing of the changing tools belonging to the rational domain they possess. Their task seems to be that of revealing the lacunae of 'official' historiography as well as reading the history of science using the most recent tools provided by the debate concerning scientific knowledge. Moreover, in Italy to be exact, an experimental philosopher of knowledge like Antonio Aliotta, following in Ernst Mach's genial footsteps, claimed that there were three main periods in the historical evolution of science: the first '*experimental*' or based on contact with the concrete becoming of reality, the second '*deductive*' and the third, '*formal*' or a summing up of scientific results. The dominant motif of his 'experimentalism' was to become assertion of the theoretical value of experimental science as against the denials of pragmatism, intuitionism and neo-Hegelianism.

Is Mach and Aliotta's prevision destined to come true in the twenty-first century or has it already done so?

One hundred years ago exactly, some of the questions considered fundamental might have been identified. This could have occurred initially among sociologists and/or positivist social writers, but more so among post-Kantian historicists, both those favouring the methodological legitimisation of the historiographical work carried out within the ambit of the science of values and culture (something which occurred only in the west) and with the diffusion (in the 1930's) of physicalist theories which sought to put a definitive end to the distinction between '*Naturwissenschaften*' and '*Geistwissenschaften*' while leading to a strategic and more pragmatic assertion of the inexorable organizational work required to organise the sociological theories produced – besides those most readily accredited (American structural-functionalism, 1950) – as well as the sources (those regarding Italy since 1970, for example). Nowadays, we ought to reflect, instead, on the lack of awareness of the need – pointed out by those engaged in the 'debate' of the late 1990's - to review and check the consistency of the development of the scientific knowledge produced continually, in the light of various disciplines which, as time passed, should have been able to assume an independent epistemological position. There is no doubt that this observation regards the social sciences from the mid-1900's on.

In actual fact, sociology in Italy at the beginning of the twenty-first century does not present an acceptable profile of the type mentioned above. It does not seem to stimulate research either within its own ambit or into its historical methodological and, above all, epistemological unity. Very often it presents itself as it is within a temporal dimension: that is, as the increasingly fragmented and specialist, though more or less paradigmatic outcome of the relationship between the production and reception of a certain theory and empirical research. A striking example of the fragmentation of sociology in Italy is that of sociological specialization, which expects epistemology to define its nature as a social and academic practice. This is true of many of the specialist branches of sociology that came to the fore in Italy up to the first decade of the twenty-first century.

A certain perspective which sought to draw on part of the epistemological views concerning the fate of industrial society central to sociology in Italy failed to arouse a true debate stemming from the interests of the sociologists' organization led by scholars like Federico Butera, Enzo Bartocci, Fabrizio Battistelli, by sociologists of i-work such as Domenico De Masi, Paolo Zurla and Aris Accornero (dead in 2018) or other authors like Giuseppe Bonazzi, who was the secretary of the Economics Section of the Labour Organization of the Italian Association of Sociology during the 1990's. It must be said that in various ways the reports produced are well suited to outlining the present sociological discipline in its various historical, practical and academic contexts but without highlighting a unified point of view capable of expanding and / or representing the scientific production of a certain type of innovative trend. Many of these theorists incorporate the Mittle-European theories of globalization and technological transformation of the world of work and production. The same applies to the phenomenon of migration and population studies, analysed by labour sociologists such as Enrico Pugliese and Maria Immacolata Macioti. As regards these interests, see also Italians studies of economic sociology by authors like Arnaldo Bagnasco, Massimo Paci, Carlo Triglia, Marino Regini and, in general, those published in the journal "State and Market."

Moreover, pondering on research might mean following ways of implementing the purpose itself, to such an extent as to conduct the problem back to an epistemology of research considered in the light of its own structural characteristics and, therefore, directed by surveys on human cognition, leading to the creation of a relationship between the researcher (the operating subject) and his/her field of research (the object with which to become acquainted). Resuming the discussion and, to be more practical, it must be said that in Italy, sociology seems to have not as yet confronted the theoretical problem producing an epistemological history of its own, which, moreover, cannot be identified even by its use within an operational framework of research. Such use is not practicable, simply because the logic of research stems from a different source.

The discipline's epistemological history seems to be, instead, the creation and development of a 'point of view' within the historiography of scientific subject-matters; and here we hold that there is no trace of that 'point of view' in any production up to the end of the previous century.

One needs to say, furthermore, that for many contemporary sociologists the issue of the *historicity* of science is a veritable 'crossing of the Red Sea': it remains motionless, entangled within a formula of relational 'circularity' between theory and empirical research, encapsulated in an area of genesis and development of generality which, most of the time, risks closing in upon itself; for others, on the contrary, the quest is identified with '*nomological*' prospects or certain 'programmes', as indicated by the dictates of the philosophy of science, in antithesis to a merely descriptive and historicist definition.

However, it seems that the so-called research 'programmes' need to be centred around the grand theme of changing ideas in science and, consequently, implying an analysis of the social changes surrounding the historical development of scientific ideas. We can also affirm that, from a *formal* 'point of view', models and theories might well replace 'history', if only a new generation of competent researchers and historians capable of managing and producing it were born, one capable of renouncing the attraction of yet another investigation of context-text within which to develop ahistorical prospective of cultural models.

Obviously, one crucial matter might be the non-affirmation among Italian sociologists between the late twentieth and early twenty-first centuries, of a practice of research capable of producing concepts of the *historicity* of science, a theme generated both by the specialization of scientific work in recent decades and by certain structures of knowledge or 'knowledges' and the consequent amplification of the social frontiers of knowledge itself leading -as an ultimate evaluation- to the discovery of various kinds of sources and the gradual abandonment of others. The fact is that an entire generation of researchers does not seem to have sufficiently stimulated younger researchers to take an interest in continuous exploration of similar issues and carry out in-depth analyses of 'materials' dispersed and/or buried by academic and/or 'official' historiography.

Another important point is that a certain reflection on this type of issue might well have generated a whole series of considerations concerning epistemological problems '*strictu sensu'* and claims to validity of sociological knowledge in our country. The principal 'important matter' surrounding Italian sociology is trapped within the renouncement on the part of contemporary researchers of scientific development to discuss again the theory of the sources and materials of its foundation (which was not the case with Italian philosophy during the whole of the twentieth century). Moreover, they tend to spontaneously cite the title of one of Shakespeare's famous comedies "Much ado about nothing" when responding to objections regarding their recent ambition to produce a 'sociology of sociology'.

In reality, in Italy, a genuine need to conduct research into the sources leading to the foundation of sociology (around the time when the " La Rivista italiana di sociologis" - Italian Journal of Sociology- appeared in 1897) began to emerge simultaneously with critiques of Positivism prompted also by theories of scientific socialism and the 'experts' of the works of Karl Marx and Friedrich Engels. In particular, Antonio Labriola's theory of 'epigenetics' may be considered a striking example of developments in thinking regarding the empirical social sciences, in that, starting with him, it became difficult to resort to the use of traditional sources of knowledge like those belonging to the sphere of philosophical speculation. A large sector of so-called 'official' history has spoken erroneously of a 'philosophical renewal', in another sense. It needs to be mentioned that even if certain Italian historians of sociology recognize some of the limitations found in 'single' contributions to the systematic development of the discipline, they can disavow neither the existence of rational sources of production (let alone empirical sociology's substantial research investigation of the historicity of science) nor a rather 'conventional' interpretation of these e sources by 'official' history. To be clear, this is no mere fluke that some of the clichés surrounding Italian positivism are free to circulate without specific investigation and re-definition- conducted by 'sociologists of history' or sociologists of science, etc. and not by historians '*strictu sensu'*- of the genetic links existing, for example, between sociology and positivism in the logical-historical sense.

Sociology was not entirely positivistic from the onset; it is probable that the persistence of the need to recognise, even of only in general terms, a continuous genetic self-definition may lead to a new exigency to sound sources of production availing of research 'pathways' capable of explaining the development of a kind of processuality and founding theories in the name of the logical unity of knowledge, without expunging historiographical work from scientific texts.

To this regard, one of the most relevant issues is that in Italy the development of an interest in scientific subjects seems to assume, in time, a peculiar stance when it comes to the genesis of the different philosophical activities, for example, theorisation or the processes by which theories are elaborated and from which fundamental matters of ontology or the theory of knowledge stem and lead to distinctions between theoretical and moral philosophy, aesthetics, the philosophy of history, the philosophy of science, etc.

It was no coincidence that until well into the twentieth century that, in their books, Italian philosophers availed of the Aristotelian differentiation alone between the theoretical sciences (Physics, Mathematics, Metaphysics) and the practical sciences (Ethics and Politics). Moreover, an absolute novelty perhaps, is that our country lacks analytical works of reconstruction, targeted towards the rational development of the sociology of knowledge, considered the methodological basis of rigorous scientific discussion, derived from, for example, an elaboration on the Marxist approach.

It would be useful, at this point, to present a brief overview. The eclecticism of the authors, practically those whom Lenin attacked in his book '*Materialism and Empirio-Criticism'*, professed with a touch of neo-Kantian philosophy and two-fold reception of sources belonging to evolutionistic positivism and the historical materialism of nineteenth-century Europe, is the likely cause of the historicist assertion of criticism of the tradition and, moreover, of a way of presenting the empirical social sciences required by 'official history' (not that derived from the Marxist matrix).

Labriola's work itself is usually portrayed as 'philosophical as well as theoretical renewal', within a context of criticism of the philosophy of history and of studies regarding Bertrando Spaventa, Antonio Tari and the Neapolitan Augusto Vera's Hegelian school, until writings concerning the materialistic conception of history and more were published.

Moreover, placing Antonio Labriola within the history of socialism has long been a matter of controversy, something discussed quite overtly by philosophers ranging from Benedetto Croce to Antonio Gramsci until sixty years after his death (1904).

In the 1970's, and in part of the 1980's, the issue of the pre-Marxist Labriola began to be interpreted more clearly, also as an intellectual intermediary in favour of detaching scientific reflection concerning historical-social phenomena from idealism, as confirmation of a national tradition of more or less systematic theoretical studies of knowledge. It is precisely the relationship between Antonio Labriola and the problem-topic of the foundation of knowledge which appears interesting, even though, it appears not to apply his relationship 'of the time' with sociology, kept at an appropriate distance but only in his well-known critiques of a more naive and destructive positivism. In actual fact Labriola's theoretical reflections on Marxism and knowledge, reveal a marked inclination towards a quest for logical-empirical correspondences between the historical phenomena of thought, the reality and social analysis. The basic hypothesis is also that whereby in Italy, thanks to Labriola himself, not only was it plausible to found a methodological approach to epistemology derived from post-Kantian Herbartism filtered, in some instances, by the debate on the 'sciences of spirit', but it was also possible to arrive at an explicative definition of the historical-social phenomena according to canons typical of analyses of processuality (or of 'relationships' between phenomena according to empirical psychology).

The genetic method (as posited in "*Genesi sociale del fenomeno scientifico*" by the Cremonese Alessandro Groppali, annulled by 'official' historiographers as a sociologist of science in an age of production-foundation of 'materials' of interest to sociology as an empirical science) authentically reflects the scientific nature of the theoretical approach even if in contrast with a certain conception of 'dialectics' which was subjected – according to Labriola himself- to a semantic 'overload' ("*La concezione materialistica della storia*" - The materialistic conception of history). The same seems to apply not only to metaphysics, as discussed as a crucial matter in "*Discorrendo,*" but also for the series of epistemological issues regarding the relationship between Marxism and science, which is quite another matter, but which - at the same time - witnesses consideration, on the part of scientists of knowledge, of the methodological outcome of a mode of reasoning about the need to investigate the logical structure of fundamental scientific results to the point of reproducing their concreteness as something mentally concrete: the enormous problem of the formal worth and the relationships existing between logic and history.

Undoubtedly, many other issues regarding the foundation of scientific knowledge are actually constitutive of epistemology, and, for this reason, closely linked to the development of epistemological history.

For historians of science, this may mean that even the philosophical themes of the foundation of so-called 'certain' knowledge, belong fully to are fully-fledged members of the epistemological history of the various disciplines and that scientific work may identify itself with the quest for a whole series of authentic 'alternatives' to the philosophical theory of knowledge or epistemology. We know that, in some cases, the justification of knowledge has become the task of those philosophers who have sought, above all, the derivability of principles applied to modern science; these scholars, including many positivists of the late 1800's and also some philosophers of the 1900's, have failed to solidly anchor their premises in such a manner as to prove useful to future reflection concerning the genesis of a new historical view of scientific subjects.

Epistemological history does not seem capable, actually, of identifying with the justifications of the entire patrimony of knowledge of one century or another. In this sense, it seems to have been born neither with the Enlightenment and rationalism nor with positivism. In some ways, it does not seem to have arisen even from Kantian criticism, the direct heir of rationalism and empiricism which held that the preliminary tasks of philosophy were simply those of examining the possibilities and limitations of human knowledge, so that, if on the one hand, it recognised mathematical and physical knowledge, on the other, it decreed the impossibility of recognizing metaphysics as a science, perhaps for the very first time.

It is worth our while to highlight, for a moment, what some historians of ideas or historians of the development of knowledge now know, and recall that for Immanuel Kant the content of metaphysics, that is, God, the immortality of the 'soul', the freedom of man, etc., may be redeemed only at the level of rational faith, that is, reason acting practically as a source of 'morality'. In this sense, it is useful to recall that criticism arises historically as a doctrine of the limitations of human knowledge subject to boundaries, up until the advent of idealism which seeks to transcend those limitations and, by stating the absolute also states that scientific knowledge is abstract and partial and may be overcome by philosophical knowledge, the only one capable of yielding the reality in all its concreteness and totality. It is 'Hegelian' idealism that begins to hold that higher spiritual activities like art and religion, may find their full realization in philosophy, as absolute knowledge, meaning that, it is philosophy that, in a conceptual sense, possesses that truth which art and religion possess only in a sentimental and mythical form.

At this point, a series of further analyses of the course of Hegelian idealism, lead to the pervasive atmosphere of positivism, which for many was supposed to represent its antithesis, which, it was to some extent, except in its affirmation of 'epistemological Romance' as a true alternative to philosophy, intended as the methodological conscience and the principle capable of unifying the various sciences. Not even the 'Romanticism of science', as the Italian pragmatist Nicola Abbagnano defined it in the twentieth century, seemed to represent the actual beginning of the historical establishment of epistemology. This did not occur automatically as positivism proclaimed its unlimited exaltation of science as the only true source of authentic knowledge, capable of ensuring a harmonious development of humanity in coincidence with the development of the various sciences. As we have already mentioned above, philosophy represented its methodological 'conscience', capable of allowing man's full dominion over nature, but also of establishing a new religion, new ethics and politics to satisfy all human needs. This philosophy genuinely influenced and accompanied a revolution greater than that occurring in the field of studies constituting an epistemology that was more modern and closer to its contemporaries: it took a hand in forming the industrial society of the West and shared in its initial optimism.

This set of ideas accompanied the rise of sociology as a science and the crisis of positivism (preceded, however, by Romantic idealism) and witnessed the birth of contemporary philosophy, amid alternating returns to idealism –as an anti-positivistic function- and to spiritualism. When addressing this complex pattern of developments in contemporary philosophy, it is possible trace the extreme usefulness of the slow work to which the historians of epistemology resorted during the various stages of its later definition, in particular, when faced with calls for the scientific redefinition and renewal of philosophy and, ultimately, in the presence of a new and more mature attempt to do away definitively with metaphysics. But we shall return to this argument later when dealing with the main problems of the theory of knowledge, of logical empiricism, and, in particular, of the reception of analytical philosophy.

For the moment, we need to say that, as far as the context of limited issues is concerned that the *ex-ante* work itself of the historians of sociologic scientific thought should have grasped the strategic aspect of confrontation with the philosophy at the time of those last evolutions of its.

Moreover, that reflection should have taken place *ex post* and ought to have regarded not only positivism as an expression of its reduced capacity to generate a historical perspective and a modern historical epistemology within the confrontation which had occurred concerning sociology's claims to autonomous foundation of, but also regarding the subsequent and partial reception in various contexts of both neo-Positivism and Popperism, often reconsidered, even by authors contemporary to us, as prospects exclusive and unsurpassed within the panorama of twentieth-century ideas and scientific studies. In particular, it is these authors and philosophers of science closer to us in time, appeal to a similar diffusion causing it to produce an entire series of unrepeatable *exploits*.

To the contrary, a historical perspective of scientific sociological knowledge might indicate the need for an even epistemological reconsideration of the problem of the quest for the foundations of sociology, especially as far as its relationship with 'traditional philosophy' is concerned. We shall start from this consideration and proceed by degrees. The problem seems serious and should not be neglected: as far as possible, it needs to be led back to the exercise of the twentieth-century profession historians of science.

Chapter 2

FROM A CRITIQUE OF METAPHYSICS TO SELF-CRITICISM IN SCIENCE

For us, the use of the term 'traditional philosophy' refers almost necessarily to the utilitarian concepts of religion and to the Hegelian and neo-Hegelian stances against science; according to us, a lack of reference to the potential arising from deliberation of this dual 'retarding' characterization, places the issue of the foundation of scientific sociological knowledge in Italy on a different plane than that from which new investigations seek to move.

This seems to apply also to individual claims regarding the foundation we find in authors bent on seeking an auspicated character of autonomy. In addition, it needs to be said, that a number of the self-based assumptions found in the notes and ideas of Comte and Spencer regarding modern science, are unable to lexit the orbit of philosophical decisions imposed upon scientifically designed theories the fate of which seems to be that of maintaining that eternal coexistence which Comte was the first to seek to destroy. The fact is that, starting with this very point, several twentieth-century philosophers and spiritualists have been pressingly critical.

These latter have been, ironically, avenged of those postulates which, during the decade following the death of Georg Wilhelm Friedrich Hegel originally pervaded European thought, advocating the reinforcement of a *social theory* which evaded the constraints of rationalism to reunite under the aegis of the study of reality from the naturalistic 'point of view' of indisputable objective necessity. Later, under the onslaught of mere 'empirical certainty', sociology's attempt at free its scientific foundation from dogmatic beliefs no less short-sighted that those of which spiritualistic metaphysics was accused, proved a failure.

At the end of the nineteenth and beginning of the twentieth century, quite a few assertions stemming from this attempt, progressively enter the ambit of Italian sociology where the sociologists do not even remotely doubt that the positivist mission to kill the 'dragon of Hegelism' might actually produce those very adverse effects that, under pressing problems arising in the future, everyone sought to avoid.

In the light of a more 'modern' accommodation of events this failed several times. First, it drove criticism to question science's alleged all-inclusiveness due to the lack of a hypothetical-deductive research phase, so that Auguste Comte, Herbert Spencer and others were now obliged to defend themselves from the charge of having pursued limited inductive methods *à la* Bacon. It was they, after all, and not others, who generated the rigid reason/experience opposition which 'modernity' avoids admitting. Secondly, it claimed that with Comte it established 'sociology' as a '*scientia scientiarum'* based on a least common multiple shared experientially by all the sciences (*sociality*), entrusting the latter with the illusion of escaping the formulations of traditional morality. Thirdly, it did not foresee that the pursuit of naturalistic, physical-biological and experimentally common models, might conceal a Romantic mystique of unification of knowledge from the sciences of society, so that a critical review of the procedures and results of the natural sciences capable of denying it might arise, as it did, in the pragmatic currents in Germany (1843-1896 Richard Avenarius and Ernst Mach 1838-1916), in the Anglo-American world (William James 1842-1910) or in the movements of ideas that occurred during the final decades of the nineteenth century and the beginning of the twentieth century in Germany with Rudolph Hermann Lotze (1817-1881), Wilhelm Wundt (1832-1920) and the *philosophy of values* characteristic of Kantian neo-criticism (1848-1915 Windelband Wilhelm and Heinrich Rickert 1863 1936), as well as in England, in Italy and in America with neo-Hegelian tendencies, and finally in France, with a number of philosophers ranging from Secrétan Charles (1815-1895) to Émile Boutroux (1845-1921) and Maurice Blondel (1861-1949).

Within this scenario one can already see Comte and Spencer's joint ambition of designing a new universal form of the knowledge of things based on a systematic coordination of the natural relationships between phenomena, in such a way as to permit the human *facts* suppressed in history only to subsist. This suspicion, triggered by nature conceived as self-sufficient by positive science, becomes more and more evident: while the nascent sociological determinism that stemmed from is clear to all, the notion of its sociologically justified historical evolution and the verification of its organic implementation in the course of things become questionable.

Sociological organicism, its mechanistic guidelines nurtured, besides, because it excluded humans from all claims to the supra-sensory (whereby, according to the principles of evolution, the very faculties for conceiving supra-phenomenal realities were destined to decline until they vanished completely, being non-functional organs) suddenly saw the compromission of its ability to believe: a) in the infallible determination in objectivity of the ultimate form of the human spirit; b) in a tendency to define the historical advancement as an obligation for the human community of individuals to move all in the same direction as those of these assumptions; c) in an evolutionary non-ambiguous, abstract and/or artificial history of development or development of history; d) in an explanation of social, moral and religious facts, automated by judgment, introspection, and consciousness.

In practice, the assumptions advanced to support the legality of an exclusive submission of *science* to *science* –as always in the name of the mechanistics of nature– do not escape the temptation of being discussed as the outcome of the schematised work of intellects. It is sociology, in fact, as conceived by Comte as a science of the intellect, that opens up the way towards a re-evaluation of the cognitive processes leading to the *spirit*: no longer reached, therefore, by means of observation of the activities and efforts of social individuals in history, as already demonstrated by illuminist interpretations of the models of progress in science drawn up by the philosophies of history prior to evolutionary determinism (Condorcet), but through recourse to *will* or *intuition*.

When generalized spiritualism assumed the task of appealing to the *conscience* as a necessary condition of any kind whatsoever of cognitive or moral acts, the possibility that sociological positivism might become simultaneously philosophy and sociology, dwindled. The contradictions contained in claims to posit an a-critical philosophy as the synthetic point of arrival of the various sciences, directed and achieved by humanity independently of the ability of reason to reflect upon the conditions and results of science, seems to emerge more evidently.

Science itself, which had assumed the onus of revealing the mysteries of the universal reality within a scenario - guaranteed by 'the empirically operative' – where the absolute was replaced by the relative (true *positivity*) accused itself of being an artificial representation and/or simplification of reality, with exclusively practical and utilitarian tasks.

The positivist *chances* of seeing, for example, psychology return within the ranks of biology in order to address the physiological manifestations of psychic facts, or ethics reduced to the level of scientific sociology, come to grief against a deeper, more mature, sociological too, form of awareness, which accuses the doctrine of universal determinism of 'forcing' the concrete possibilities of nature, in the sense: a) that it stifles everything that, inside and outside of the human dimension, is contingent; b) that it admits that the discovery of clear results in the rigid schemata of the various *laws,* means having discovered a precise correspondence between theoretical predictions and the reality. So, philosophy is still able to exert its domination over science.

Availing of the advantages of a more objectified adjustment of sources, it is possible get the moment when claims of the affirmation of sociology in Italy tend to attenuate coincide with the year 1905 when Benedetto Croce published his *Logica come scienza del concetto puro* (further elaborated on, in 1908), provisionally concluding his early studies of the history of art, conducted in hopes of reaching a less abstract definition of the peculiar characteristics of Italian philosophy.

Already in *Lineamenti di una logica del concetto* (1904-1905) we find the first instance of true awareness of the difference between philosophical and scientific concepts and of nature as a mere pseudo-concept. Science- from a practical point of view - may be seen as having been expelled from the sphere of knowledge which remains that of art, philosophy and history and as already on its way towards that of logic. The project of self-review proposed by Croce may –on the contrary- be carried out, undisturbed, upon two planes of negation: the first aimed at rejecting the theoretical validity of the natural-empirical sciences, which due to the confusion of a material reality of positivist origin, need to be seen as belonging to the conventional, utilitarian and practice ambit; the second aimed at reducing all religious transcendence to the ranks of mythology. In one foul swoop, Croce judged religion and natural, 'special' or 'particular' sciences as a historical anachronism and placed –more generally speaking- philosophy and science upon two completely separate planes, causing science's claim to autonomy to derive from an exclusively philosophical finding. In Italy, those who during the preceding period, had tried not only to reinstate the moral and social sciences as well as their will concerning the foundation, upon the same plane as that of the sciences as '*Naturforschung' (*meaning research into the natural sciences), had to admit defeat. Furthermore, in 1904-1905, particularly important events should be remembered taking into account that in other areas of scientific European thinking, this date coincided with the publication of Albert Einstein's theory of relativity and Max Weber's essays *Protestant ethics and the spirit of capitalism* and *Objectivity in social science and social policy.*

On the contrary, in view of Croce's treatment of problems regarding social (scientific-philosophical) knowledge, sociology, in particular, risked losing momentarily losing the possibility of having its self-foundation derive seriously from confirmation of valid independence stemming from its own enunciations, which, on the contrary is triggered elsewhere; it can rise to autonomy only through philosophy.

All this seemed to justify redefinitions of the object of study assumed by historians of scientific subjects as far as the topic of foundation was concerned. The operation ended up by postponing even to the moment of its possible discovery- the marked urgency to resort to a preliminary confrontation with 'traditional philosophy' (previously defined) in order to justify the birth of some unprecedented scientific, revisionist and autonomous potential capable of establishing sociology not only as: a) an antithetical reaction to the centuries-old alliances stipulated between religion and philosophy to reconfirm the function of speculation, but also as b) knowledge, qualified for *permanent reinterpretation* of various forms of individual modern life and associated with its own capacity for anti-dogmatic self-revision.

It is in the light of this two-fold meaning that we intend speaking again today of claims to autonomous foundation. Furthermore, analysis of an object assumed within a likely area of foundation, might reveal today, in actual fact, a marked fragmentation of perspectives, attributable above all a) to practical-historical need to see Italian society and public opinion reformed after unification, functional to scientific progress but weakly antithetical to tradition; b) to pressures brought to bear by multi-revisionist trends arising historically from Christian spiritualism and historicist idealism in terms of 'speculative reconfirmation'; c) by obligatory demands for autonomisation that emphatically vein the social sciences, which, moving away from analyses of real historical processes, can expect to witness the decline of the nineteenth and early twentieth centuries, bringing to a close all positivist influences over general criticism of metaphysics. In fact, if these three conditions are allowed to act simultaneously, is this is, in particular to the partial need established by the sciences to discuss those themes contained in traditional philosophy', which due to Benedetto Croce's processes of renewal, have come, little by little, to establish themselves as indisputable methodology, or better still, as a reduction of philosophy to the rank of 'methodology', thus declaring the momentary failure of sociology's requests of autonomous foundation.

This was intended by some as instrumental (positivist and evolutionist) metaphysics-science and/or specialist knowledge on the eve of structural economic and political upheaval of twentieth-century society; but- by others- it was established originally as a possible secular, scientific reinterpretation of Catholic morality and speculative philosophy, and, if assumed as a possible philosophical system, likely to be considered an ideal restoration of reformist schemata, confirmed *in toto* by allies of the Christian tradition along lines of thinking historically subject to theology.

It is not difficult to demonstrate (after a century) how attempts made by some authors may have centralised the problem of the foundation of sociology in Italy, by leading the treatment of morality back to a certain kind of traditional historicity to subvert totally at antithetical level and to demands of integration with philosophy. Some, like Roberto Ardigo, did not deny the possibility of reaching a critical philosophy-science compromise; he denied that religion be considered a traditional ally of philosophy. For him, this alliance had devastating effects on science. This problem cannot be avoided unless one wants to risk the illusion of pursuing a non-existent autonomous route, and discussing science as a figment of the imagination, or, a '*fisciones mentales*' unpurged from theological ghosts established by history: the sole-history claim 'sung' by those 'perennial philosophers' who, as interpreters, often appear in Italy as the pontiffs of culture replacing the Catholic Popes.

Furthermore, entire series of studies show how the grafting of critical potential upon scientific reflection, aimed at self-revision of the dogmatic principles present in the positivism, can turn against science in a counterproductive way, whenever the umpteenth separation between science and philosophy takes place in order to draw up a possible definition of the concept of autonomy. In actual fact, Vilfredo Pareto's subtended renunciation, in his '*Trattato di Sociologia Generale*', to discuss whether sociology might be considered or not an autonomous science is rather significant. It is probably associated with a desire to avoid topics established in 'traditional philosophy' and, favour a specialist assertion of social science that may also turn into factualism, while undertaking a dispute with itself such as to shift interest towards irrationalism. There are no doubts regarding the merits of Pareto who sought to have science follow 'the world of things' as an antithesis to metaphysics and, maybe as a legacy inherited from Galileo Galilei, who claimed that the problems of science ought to focus on the 'sensible world' and shrink from 'pieces of paper'.

In reality, the choice of a rigorous pathway would have obliged him to compete with 'traditional philosophy', had he not believed that once metaphysics had vanished from the scene sociology might be set up undisturbed, in such a way as to avoid 'speculative returns' that might compete with its acquired self-reviewing logical capabilities, now established as a privileged theoretical platform in keeping with the validity of the natural sciences. It is a fact that Croce's '*La Logica*' –Logic- was able to deny this kind of theoretical validity and, unlike Pareto, addressed the problem of autonomy in science by having it derive from philosophy; so much so that Pareto, in 1916, had to abandon the field of the dispute logic regarding the autonomous foundation of sociology and base – instead- all the strength of his *'Trattato'* on that.

He simply followed a different course, the same one which sought supposedly to lead scientific reflection to bring philosophers face to face with problems of how philosophy itself may avail or not of its thrust towards totality or, when all comes to all, renounce it definitely.

This is the same route followed by modern sociology and later pursued by contemporaries, which based on various categories of *action* the possible denial of a supremacy which found it convenient to admit that each foundation has already occurred *ipso facto* in the spirit and/or intuition. Already, the exponents of the 'Vienna Circle' had become the interpreters of the need to lead all mediation back to the intellect, towards rationality. In all cases, it is maybe when science too –like philosophy- demonstrates the will to discuss itself, presumably a separation, that it is possible to have autonomy derive from scientific discourse.

In this sense, some authors founded a scientific way of considering sociology in Italy until it dissolved into the criticism it managed to arouse.

Pareto, on the contrary, eluded the temptation of discussing sociology philosophically and replaced a desire for possible integration with another aim, that of 'achieved separation', derived, practically, from a lack of confrontation. Most probably Pareto availed of this to leave the tradition definitively and abandon terms belonging to so-called 'traditional philosophy' which, if not acknowledged, might have decreed the creation of an independent scientific tradition. Here Pareto risked failing because his thinking resembled that of the oldest of positivists.

They tried to have the self-foundation of sociology as a science stem from a multiplicity of appositive concepts which had already proved compromised from the out start, so that a contrived quest for autonomy from philosophy became its ineluctable destiny. Pareto by denying his desire to address the issue of the autonomy of social science, abandoned the initial assumptions of Comte and Spencer but found himself at a loss when seeking recourse to a scientific tradition which he himself had forsaken, laying himself open to an entire series of topics which –without being either attacked or contradicted- were inexorably thrown at him, with all the speculative force of a philosophical and traditional self-interpretation in action. Given this, only the strengths of the anti-dogmatic self-review he himself unleashed in his scientific reflections on sociology, may act permanently to officialise the passage from a critique of generalised metaphysics to a self-criticism of science and to admitting, as a result, that sociology may be founded.

This admission may also lead acknowledgements by a number of twentieth-century historians of science rotating around the disciplinary spheres that characterized the development in Europe and in Italy of sociological thinking with important investigations into *experimental philosophy*, which operated simultaneously with a more general need for an anti-dogmatic interpretation of the need itself for scientific self-revision. It is a fact that criticism of metaphysics by science, did not suffice to overcome that very criticism arising from philosophy and which has succeeded in spreading new religions and conceptions of the world in a country historically educated by religion and 'institutionalized philosophers' until the 1950's.

From the precise moment, scientific thinking regarding the foundations of sociology, has been able to interact differently towards the speculative potential at work in the tradition, without being able either to impose on the choice an essential integration with criticism of the topics contained in 'traditional philosophy' or dwell completely outside of it. The 1950's seem to mark the simple rebirth of a science scarcely purged from a quest of the dogmas of assertion of certain theoretical meanings, which occurred with regard to the relationship between 'traditional philosophy' and the unleashing of subversive forces peculiar to a really revolutionary historical age like that of the foundation of sociology.

The truth is that among modern and contemporary historians of sociological science a void would have surrounded the topic of the theoretical assertion of sociology in Italy if, as is already the case here with the development of scientific culture and ideas, philosophers had not experimented certain topics, for example, on the same grounds as those of the so-called 'idealistic reaction' against science, intended –long before the well-known historian of physics Thomas Samuel Kuhn- as a veritable operative model for a 'scientific revolution'. It is no coincidence that some contemporary scholars have referred in their analyses to particular moments regarding the debate surrounding early sociology in Italy, starting from the outcomes of certain stances on the part of philosophers regarding the issue, such as the proposal to resume the discussion begun in 1967 in the *Ressegna Italiana di Sociologia* between philosophers Ugo Spirito (who as we know was a disciple of Giovanni Gentile's), Camillo Pellizzi, Agostiono Palazzo, Vittorio

Castellano and Gianfranco Morra (first Italian Professor of the Sociology of Knowledge at the University of Bologna in 1974) as well as Franco Ferrarotti in "La Critica Sociologica," [Sociological criticism], etc. as a response to a certain type of stimulus and to criticism of sociology and the 'history of sociology' with further interventions by Franco Crespi in the "Rivista di Sociologia" and by Vincenzo Thaulero with his *Metafisica della scienze e sociologia* in the same year.

In all these cases and, apart from examples due to handovers between different generations of sociologists, the assimilation of urgent self-criticism accomplished by many epigones of *fin de siècle* positivism concerning a variety of issues and topics regarding alleged theoretical significances of an epistemological nature, appears to be a topic that should have gone on urging Italian historians of sociology longer, as it was already considered a subject to be investigated by specifically scientific subjects during the early decades of the second half of the twentieth century.

Before then, self-criticism of science was brought to maturity and was involved in a confrontation springing from 'traditional philosophy' which, in Italy, stemmed from a permanent sphere of speculative interest which re-proposed assimilation and also -in the long run-the debasement of Hegelian philosophy.

Let us say that, it would be well worth historians of sociology's while to follow this pathway whenever they seek to undertake a realistic discussion of the completeness or not of their investigation of claims of the theoretical validity of sociological science in Italy, that is, of the assertion of a historiography of sociological science, deprived, when all comes to all, of an essential comparison at a 'crucial' moment, as we tried to discuss above. Furthermore, from certain points of view, the problem of the foundation of the sociological sciences may also be taken for granted, that is, considered as an indisputable principle, an irrefutable truth: for some, a true dogma. This is not so, however, from the 'point of view' of the historiography of scientific knowledge. But, let us now depart again from Kant.

Chapter 3

ADDRESSING SEMANTICS AND EPISTEMOLOGY

It seems evidently clear for late twentieth-century historians of science, both the metaphysical issue and the foundation itself of problems-matters concerning the epistemological status of a variety of scientific subjects, could have developed only in the direction of recognition of the destitution of the logical validity claimed by a vast range of extra-scientific concerns. One way or another, all of us recall Kant's assertion that "Concepts without intuitions are empty, intuitions without concepts are blind."

It is with Kant - not Comte- that both philosophers and scientists began to question themselves concerning the meaning and value of the human knowledge of things as well as the use of particular terms.

As is more or less well known, Kant's criticism of Metaphysics referred to those 'fundamental propositions' advanced in his *Kritik der reinen Vernunft* [Critique of Pure Reason], backed by a use which might also surpass the boundaries of experience. In fact, metaphysics as gnoseology was Kant's specific field of research, although, the concept had already been considered by Bacon a universal science, the mother of all others and the common route all doctrines ought to follow. Kant shared with Francis Bacon, his insistence on the principles of science rather than its purpose. Kant considered metaphysis a study of cognitive principles, constituents of all finite reason and, therefore, of human reason.

In Kantian thinking, both ontology and criticism coincide as the sole manner by which to confer scientific value on Metaphysics. Kantian metaphysics is, therefore, a science of pure concepts, so that the knowledge derived from them, is alien to all experience. In other terms, metaphysics (from the Greek *τὰ μετὰ τὰ φυσικά* pronounced *ta metà ta physikà*, that is: the books following those written by Aristotle concerning physics) was believed to represent an indeterminate field which sensory experience was unable to access and where, reality finally coincides with reason.

It may be useful to recall, from the 'point of view' of the historiography of knowledge, that the historical continuity of Kantian metaphysics found later in Edmund Husserl, did not refer back to general principals, but to those which constituted the fundaments of a science or group of sciences.

Metaphysics has still got an active role to play within cognitive investigation: it has to do with the issue of the meaning (or meanings) to attribute to existence, according to the language of the various sciences while also taking part in the issue of relationships between the different sciences and the quest for purpose, placed at the meeting-point between one science and another.

Moreover, again from the same view point, we should not forget that, as regards the meaning of existence, Willard van Orman Quine accepted ontology almost as if it were a scientific theory, which places data within a normally disorderly, simple conceptual scheme of raw experience. Rudolph Carnap, from his point of view, was later led to accept Quine's statement in substance, even though; he was reluctant to accept the term 'ontology', where he detected some links with metaphysics. Only some of these topics were developed in a specific text by Quine published in 1969, translated in Italy, in 1986.

Here, however, it is particularly useful to recall that as far as the problem of relationships between the various sciences is concerned, conventional metaphysics will be replaced by methodology which seeks to bestow order on the 'conceptual universe' so that communication between one science and another may be favoured, without impairing the autonomy of either; to that end, at each stage of the research, the relations between the various sciences are problematized to the advantage of each and for the sake of the clear use human beings can make of them.

These are topics of likely interest to historians of twentieth-century methodology of the human and social sciences. Analytical and logical neo-positivist philosophy will endeavour, in fact, to blend philosophy into the methodology of science. It is opportune to remember first of all, that, within the ambit of analytic philosophy, there is considerable efforts have been made in favour of the recovery of metaphysical-theological discourse.

It is in this sense that, due to steps taken gradually in Europe and within the so-called historiography of science, which owes its origins to the British Isles and the USA, some of the most significant reflections have emerged not only as regards a critique of metaphysics but also concerning semantic analyses of scientific languages, to the point of determining a kind of epistemological, reference: a broad field of historical research where even the issue of the foundation of the science of sociology has begun to be placed.

On pain of dismissal as partial identification with a perspective of the history of sociological thought destined to re-propose supporting some vetoes, paradigmatic of a dichotomous order, already circulating within a historical, philosophical and scientific context of a traditional kind such as ours. This is, in fact, what happened.

It might seem to be a novelty that, at the end of twentieth century, many Italian historians of social thinking chose a route totally divergent from that proposed by epistemologically-informed nascent historiography, anchoring their discourse to a pure re-proposal of positivism-idealism the dualism which, in reality, and in close relationship with reality, did not lead to any result, other than that which, as we have seen, deprived possible scientific-philosophical comparison of theoretical-perspective analysis and a historical-conceptual outlook of a matrix other than mere historicity. As far as it is concerns itself, philosophy has always emerged unscathed, as if science were the result of its permanent renewal, of its transfiguration. And so, the definition of 'historicity', from the idealist philosophers' point of view, has sought to preserve a vast heritage as 'presentialisations of the past'. Italy failed to promptly receive European anti-philosophy beginning from the issues discussed.

It had to wait until well into the second half of the twentieth century, therefore, for the spread of neo-positivistic themes to the launch a view of studies of the philosophy of science all its own.

From another perspective– from the 1930's to the early 1950's – it became clear to some important exponents of neo-positivism, how at the root of the issue of the birth of an in-depth scientific reflection between the modern and the contemporary, the theory of knowledge might find a place with its themes-problems regarding meaning and verification. We refer, in particular, to Rudolph Carnap and, in all cases, to that logical-analytical linguistic approach which characterized the efforts of philosophers from the 'Vienna Circle' which sought to dethrone metaphysics once and for all.

For a period, the problem for Rudolf Carnap, was also that of contesting the presumed objective validity of values or standards not subject to empirical verification or deducible from empirical propositions, to the point of stating that the assertion of 'philosophy of values' themselves had no meaning at all, in the light of the construction of propositions expressing valuable judgments; for Maurizio Schlick too, metaphysical emptiness was representable in the demonstration of knowledge of the essence of things which, as a quality, might appear only in experience etc.

From our 'point of view' one may say that the results of Carnap's research lead to an important assertion, that is, that metaphysics has no theoretical value and that therefore, its historiography is capable of proposing nothing other than non-sense.

As we can see, then, in the eyes of twentieth-century historians of science, the problem of its foundation may certainly regard attempts to give birth to a new tradition of studies regarding the logical validity of its discourse, in the light of transformations induced by the so-called 'philosophy of science', as shown in accounts by authors like Hans Reichenbach and Philip Frank, for whom it is philosophy that continues to remain anchored in the revolution of science during the nineteenth century and not the contrary (as was the case with Italy and imported positivism).

It is probable that, in a similar direction, epistemological history might well represent the philosophical progress of science which, from a perspective considered ‘satisfactory’, may combine logic with the sociology of science, although there are not many modern historians of scientific subjects who avail of this interpretational position; and this is the point.

One of the reasons for this choice, might be that an understanding of philosophy as logical empiricism leads not only to a rejection of statements that are not empirically verifiable but also to an assertion that philosophy itself would be left the task of analysing scientific language and thus clarify the methods and structures of science (neo-positivism) or analyse the language shared, in order to clarify the linguistic expressions commonly used as in the case of analytical philosophy.

Indeed, it is logical empiricism the proposes opposition to the traditional notion of philosophy conceived not so much as a doctrine as an activity aimed at freeing oneself from conceptions of language that generate the pseudo-problems of metaphysics.

In this sense, epistemological historiography might be taking shape thanks to admitting a function of ‘traditional philosophy’ which turns out to perform a function of a therapeutic type only.

In actual fact, epistemology would rely above all on philosophical analyses of language and methods, internal organization and the results of the various sciences in order to define the nature and value of scientific knowledge.

It would appear more as the discipline that studies the formal structure of science critically, born, in the strict sense, to create a demarcation between what is and what is not science, like logical positivism which attributed meaning only to scientific discourse, leaving philosophy the exclusive task of clarifying specific aspects of science up to the point of identifying philosophy with epistemology.

Down over the years, many of those theses were abandoned, even if the broad and complex field of analysis that was investigated by the neo-positivists constitutes the reference point for a large part of contemporary epistemology, within which sphere, some theses which assert the need for an operative reduction of all scientific (or physicalist) concepts have come to grief; this is the case with the positions assumed by Carl Hempel and Ernst Nagel that, even while moving in the conceptual environment typical of the philosophers of Vienna, made a number of choices regarding so-called ‘liberalised neo-empiricism’. And not only.

It is with a view to obtaining a historical reconstruction of the destiny of epistemology of empiricist origin in the twentieth century, that one should recall -but only for the sake of convenient exposition- that Karl Raimund Popper went further and replaced the concept of verification with that of ‘falsification’; furthermore, one also needs to recollect the controversial ‘critical theory’ theses of Max Horkeheimer, Theodor Adorno and Jurgen Habermas as well as the total separation from neo-positivism of Jean Piaget, founder of ‘genetic epistemology’, etc.

On the whole, much of the genesis of a historical view of contemporary epistemology is taking shape in the presence of these topics while also incorporating its dogmas and the negations of contemporary philosophy - to continue an overview- leading to a certain kind of pragmatism and Anglo-American neo-realism to the point of veritable 'returns' of metaphysical speculation which reside in existentialism or, rather, return, simply as ghosts, in view of requests to 'return to the things themselves' of phenomenology, re-proposing the theme of the foundation of science not only by trying to find its root in the complex 'world of life' but also by opening up to recognition of the historical-social dimension of man, and providing, ultimately, a new method for the study of the activities through which it expresses itself.

The foundation of scientific discourse may become –and have every right to- the answer to some of the issues brought into play by neo-empiricist epistemological and may also place at the centre of epistemological investigation the relationship between scientific and social organization, between the knowing subject and society, placing within a heritage now common to historians of the development of modern and contemporary scientific subjects, topics and problems of the theory of knowledge (including linguistic analysis), the same ones which as far back as 1953 Carnap summed up in two: the first, concerned the conditions under which prepositions possess meaning or not in the cognitive or factual sense; the second, posed the question of how we can know something, of how can we establish whether a given proposition is true or false?

In actual fact, if the most incisive assertion of the impossibility of producing a historiography of metaphysics takes shape simultaneously with the birth of studies of linguistic analyses, it is the development of epistemological historiography itself that will be marked by the event. It is metaphysics that is will be gradually reduced to seek its own impossibility as far as the obligatory course of twentieth-century scientific studies is concerned.

The route seems to have been plotted to some extent due to the self-maturation of sociology as a science caused by the resolution of serious and still poorly concealed 'theoretical questions' and when faced with the need felt more and more by historians of scientific development to imbue the various contributions made by social scientists operating in different concrete-historical environments, not only with acceptance of analytical philosophical studies but also with the contributions of philosophers of language like Friedrich Gottlob Frege, Bertrand Russell and Ludwig Wittgenstein who, as we know, drew closer to logic again thanks to the birth of mathematical logic to the beginning of the 1900's.

Here, both semantics and epistemology played an extremely important historiographical role as regards the future redefinition of a framework for twentieth-century scientific developments, capable of relating to the research and different contributions made by those authors who remain–when all is said and done- a distinctive trace of the accomplishment of separation from problems affecting the philosophical bases of scientific disciplines, let us say, from Kant to Carnap, roughly.

It is probably along this route that some interests of a considerable number of historians of logic and of scientific knowledge took shape; in Europe, those who followed the ups and downs of philosophy and the *Wiener Kries*. Frege's semantics, for example, or 'logics', as it began to be defined, which resumed – in part- considerations advanced by a precursor, the mathematician and philosopher Benhard Bolzano, who questioned epistemological idealism in open contrast with Kant.

Frege and Kant agreed only that logic is based on the intellect and a field where neither the sensitivity nor any of its various forms, play a role. The Frege's study of arithmetic etc. led to other important results. Russel too rebelled against idealism at a time when he was most affected by the studies of Giuseppe Peano who linked logic and mathematics. In fact, the attempt to reduce mathematics to logic is associated with the issue of fundamentals.

In general, semantics questioned logical knowledge and different kinds of *a priori* knowledge; that is, it began to address the problem of fundaments which Frege associated with 'sources' of knowledge, leaving the task of continuing the discussion regarding both the metaphysical use of the 'quality' of things and alleged logical 'truth' to authors like Russell and Wittgenstein. This way, the birth of a history of epistemology was made possible, even in the face of themes-problems which were to prove indefinable, like those identified by Russell, when reflecting on what for him was fundamental, and entering, moreover into contrast with Henry Poincaré.

As with analytical philosophy, semantic research became central to logic, sociology and science during the whole twentieth century. Very few historians of European sociology have understood this. The philosophical movement, as we know, was able to interest contemporaries and achieved its greatest historical development in Great Britain, in particular, after The Second World War.

On one side we find Frege himself, the 'Vienna Kries', Reichenbach, Carnap and Wittgenstein; on the other Georg Edward Moore and Russell. The peculiar character of the movement was rejection of a systematic conception of philosophy and language as foci of investigation.

In the end, this turned out to be central to all cognitive processes and was not considered in a general, abstract sense only, but also on the basis of its everyday function and in particular those on that of specific linguistic uses capable of generating possible philosophical issues. In practice, linguistic research is considered a fundamental requisite for philosophical investigation of any kind; a sort of philosophical equivalent of the experimental method. This, also to avoid every possible kind of linguistic, conceptual confusion and of pseudo problems.

In actual fact, it needs to be said that the 'philosopher analysts' by harking back both to the teachings of the English philosopher Moore and to the thinking of the later

Wittgenstein, attributed a task of clarification linguistic analysis, which, on the one hand, might eliminate the obscurities and misunderstandings underlying false problems while, on the other, facilitating the practical use of propositions.

One needs to recall, in particular, two representative believers in this trend: the English thinkers Alfred Jules Ayer and Gilbert Ryle, editor of 'Mind' magazine. One should not fail to remember, besides, the appearance of new trends and developments in thinking which proved capable of surpassing purely formal analyses of language and address the function of meaning; one regarded Charles Morris an ex-student of George Herbert Mead's at the University of Chicago, who sought to build a general theory of signs or 'semeiotics', by means of which to attribute a certain degree of importance to the pragmatic dimension of language, and, therefore, to its psychological, sociological and pedagogical implications (*Foundations of the Theory of Signs*). Analyses of meaning or semantics was to become the object of further research by Carnap in his '*Introduction to Semantics'*, 1942, leading to a particular flourishing of particular trends in general semantics in the United States (Alfred A. Korzybsky).

Semantic research has undoubtedly generated a number of different approaches and given birth to a vast variety of investigative perspectives: analyses of the common, everyday use of language (Gilbert Ryle and John Langshaw Austin) by what came to be called 'linguistic phenomenology'; analyses of the formalized languages by Peter Fredrick Strawson of the Oxford school of analysis, with his references to the later Carnap and also to the exact definition of the fundamental notions of science and of philosophy: or, the analytical investigations, extended to aesthetics (Ronald W. Hepburn), to theology, -as well as to philosophy and to logic-(Frank Plumpton Ramsey, Hepburn himself, Antony Kenny), to ethics and 'meta-ethics' (Richard Mervyn Hare, Alan Montefiore) as well as to political philosophy (Thomas Dewar Weldon, Richard A. Wollheim).

All this, without taking into consideration the increasingly strong interest accorded to the issue of language in different contexts and which led to the integration and broadening of several perspectives, frequently coincident with trends in thinking (phenomenology, historical materialism, philosophical structuralism and the hermeneutic currents of the second half of the twentieth century).

We should not, therefore, underestimate the fact that particular relevance is now attributed by historians of contemporary epistemology, as far as the development of similar scientific studies is concerned, to that particular period which saw the resizing of philosophy with regard to the production of research by authors who sought to provide the 'science of language' with a definition which, as Austin held, was obliged to clash with 'traditional philosophy', having accounted for studies of logic and grammar.

Austin, for example, identified the problem of Kant with that of 'traditional philosophy', and, later, he addressed questions shared, in turn, by Wittgenstein's *Philosophical Investigations'*.

It was the Viennese logician and philosopher who in his '*Tractatus*' defined thinking as a 'logical image of facts'. The text revealed the influence of George Edward Moore and Russel which were Wittgenstein had acquired from grey Cambridge and which led, on one hand, to all logical inference being reduced reduction to pure tautology, at the other, the reduction of science to contingency and factuality. This was reached by means of clarification of the nature of propositions. In practice, outside of the two above-mentioned contexts, there would be no room for assertions of any kind, like those which are mainly the stuff of philosophy.

It is important to remember, as many *excursi* by contemporary historians and philosophers of science highlight, that the latter analytical trend, and, in general, the entire anti-metaphysical structure of Wittgenstein's thinking, was destined to exercise enormous influence over neo-positivism, in particular over the Viennese School. It is also important to remind these historians that in the period that followed, in particular during the one influenced by the intuitionism of Luitzen Jan Brouwer, author of studies concerning the foundation of mathematics (1925-1927), Wittgenstein abandoned the conception of logic primarily as ideally 'perfectible' language and that established that the purpose of philosophy should investigation of the languages as they effectively are.

It may be relevant that Wittgenstein's studies prove useful to the de-composition of language as a given, that is, as an institution created over time thanks to connections deriving from practical aims, from the 'linguistic play' comprising it.

So, philosophy becomes mere description of language, with an essentially therapeutic purpose, that of banishing the illusions that arise from the ambiguous character of common language, from which the problems of 'traditional philosophy' stem; the main task of the philosophers, therefore, was that of ridding themselves of the problems that preoccupy them.

More precisely, we need to remember that for Wittgenstein thinking meant the logical image of facts and making possible a pragmatic structure capable of addressing in depth all that may be thought.

The gifts of thought and language were meant to guarantee not only the ability to build languages with which all meanings might be expressed, but also to provide 'common language', capable of concealing logical relations like the well-known known ones of proposition-fact. Philosophy is an instrument that permits criticism of language: that activity that can free us from error due to badly concealed comprehension of linguistic rules and confused use of terms and concepts. Wittgenstein established this proposition as an image of reality and, as such, it was taken as a model of reality which we consider it; this, in order to be an image of reality, should be comprehensible thanks to the existence of 'forms of logic, capable, therefore, of manifesting themselves. A proposition can reveal -in the long run- the 'logical form' of reality. Pursuing the non-sense of philosophical propositions, seems to become the task of Wittgenstein's philosophy, assuming, in the eyes of twentieth-century historians of science a precise function of clarification.

The question inherited from Wittegenstein which catches the attention of historians of science, seems, on the contrary, to be that of prompting philosophers to seek the possibility of irrefutable truth and clarification of principles of given data; this obliges them to address those who seek to oppose the ineffable, the unutterable and the enigma: everything that seems to remain within modern contemporary thought as a domain of the indescribable.

Philosophy may be able finally to investigate its own dogmas though only after acknowledging the errors of its propositions and, therefore, after revaluating and admitting that natural-scientific propositions should be considered the only ones capable of expressing meaning. It is science that can become meaning and the new enigma of its own course: authentically philosophical propositions, simply do not exist.

From a historical-scientific stance, one should note that this logical-philosophical conclusion of Wittgenstein's underwent many changes; the same ones which affected the period between the two well-known developmental 'phases' of his thinking: the first, that of the *1914-1916 Notebooks* and the *Tractatus*, where the form of language depended that of the world; and the second 'phase', which included his *Philosophical Observations, The Blue Book* and *The Brown Book* as well as his *Philosophical Research*, where language was structured by means use of 'linguistic play', world 'sociality'. There are many authors, including some Italians, who now believe that the decisive moment in Wittgenstein's philosophical development occurred roughly around 1929 and may be represented by the substitution of the 'phenomenological language' of the *Tractatus* with the 'physicalist language' of the 'everyday' as the 'operative' language of twentieth-century philosophy.

What seems important is that *The Philosophical Research,* unlike the *'Tractatus'*, finally founded -its semantics on the social dimension of language: the use of enunciations or of an entire language might appear as 'linguistic play' and, therefore, as a kind of 'social container' of the reality and also of its ontological nature.

In actual fact, this does not seem to have meant full assimilation of Ludwig Wittgenstein's 'point of view' by the social sciences, especially in some historical ambits of the development of the scientific disciplines such as ours; in the end, the Viennese logician and t philosopher and many of his Europeans epigones, seem to have maintained a certain epistemological neutrality that, in some instances, was as yet to come, lacking a solid approach to studies of the main issues of science.

In all cases, it is semantics and epistemology that seem to have established the bases upon which to construct the essentials of contemporary thinking regarding the relationship between philosophy and science.

As Bertrand Russel wrote in 1945: during the twentieth century philosophy was able, at last, to renounce its dogmas (*History of Western Philosophy*). From his point of view, epistemological historiography stemmed undoubtedly from the results of a grand confrontation with logic and the debate concerning the rudiments of logic; another source of his is the analysis of language and, as mentioned above, acceptance of analytical philosophy. A third, though no less relevant feature, is the philosophy of science, embracing the great theoretical debate regarding the sphere of physics. For this reason, to depart from Kant would prove useless to all.

Chapter 4

The Dogma of the Philosophy of Science

The quest for schemata and conceptual changes related to scientific activities seem to coincide with the work of historians of science who -during the twentieth century- managed to associate their studies with the epistemological genesis itself of some issues or subjects, like physics, to the point of extending a number of the results obtained to the human and social sciences.

The most indicative example seems to be that of Thomas Samuel Kuhn, who succeeded in developing (at least from his *Copernican Revolution* to *The Structure of the Scientific Revolutions*) a historical and sociological perspective for analyses of 'compound of events' which led to major transformations in astronomy and mathematical astronomy, even in the fields of cosmology, physics, philosophy and religion.

In reality, Kuhn's problem consisted, mostly, in seeking to violate the boundaries between the field of science and those of history and philosophy. The starting point of a similar 'brainteaser' was physics, above all, the attitude of philosophers towards attempts made to interpret the main processes of the beginning of the century. Today we might tell the historians of scientific sociology that, from the point of view of historians of knowledge, and, little by little, from the privileged observation point of contemporary historians of science, the study of a number of developments, constitutes an important example from which to begin to pose problems of an epistemological nature. By way of further example, many philosophers consider relativity from a positivist angle which in itself would raise problems of a mathematical kind in particular. Yet, this position has permitted us to delimit the epistemological aspect of the field of physics better, providing it with greater freedom in the formulation and experimentation of its theories without entertaining worries of a metaphysical nature. It would be useful to mention that in the field of knowledge, the theory of relativity, leads, on the one hand to subjectivism, on the other to a quest for expressions that are independent of all subjective elements and of the laws of nature themselves (an absolute datum, for example, like the speed of light), always adherent to experience and free from mental habits.

Historically, it was, thus, possible to reach a sort of 'critical realism', which revised Isaac Newton's and Kant's notions of time and space. The 'fusion of space and time' brought about by general reality (drawn up by Albert Einstein in 1915) made it possible for philosophers to recognize the distinction existing between the concepts of space and time to such a degree as to argue that the four-dimensional model of space was useful to research, although still only a model.

Also from a conceptual 'point of view', the fact that philosophers admitted that unity of physical phenomena - like that pursued by Einstein- was possible, paved the way towards a whole series of further scientific interpretations. It is a more or less accepted fact that, to the very end, Einstein pursued the ideal of classical physics: that is, that of providing reality- as existence independent of our observation- with a conceptually simple, substantially deterministic description, where it was possible to establish a clear relationship between cause and effect for each phenomenon. Moreover, this attitude caused him to reject generalizations of the quantum theory of which he was one of the founders. Similarly, we should not forget, for example, that the epistemological analysis of the theory of relativity by Hans Reichenbach -author of the *Philosophical Foundations of Quantum Mechanics*, 1944- provided proof of the inadequacy of Kant's *a priori* conception of space and time following the studies he carried out to systemise quantum physics as well as those regarding randomness and induction- finalized, more than anything else by claims of their empirical and probabilistic origin. Reichenbach's experience as director with Carnap from 1930 to 1940 of the neo-positivist magazine *Erkentnis*- (Knowledge) should not be underestimated; neither should his '*The Rise of Scientific Philosophy'*, a more accessible expression of his research on fundamental topics.

The same applied to a number of authors and producers of 'points of view'. We might say, rather, that the more or less staunch acceptance of the perspectives of the philosophy of science within the ambit of the development of scientific ideas during the twentieth century depended on the ability – or lack of ability- to produce, within existing philosophical perspectives, a sector whose task it was to study method and the meaning of the sciences within the ambit of each one's specific problems and connections and as related to a general view of the natural reality; in actual fact, it depended on their ability to produce a specific 'point of view' capable of explaining that self-same reality.

To this regard, as connoisseurs and historians of European positivistic sociology know, at the end of nineteenth century, due, in particular, to an increase in the interest shown, criticism was also voiced against the theoretical lacunae present in the mechanism, which, meanwhile, was brought into question by the results and discoveries achieved in the field of thermodynamics and electromagnetism.

The mechanism had always based the whole of natural science on principles which, as time passed, proved inadequate in the face of Darwinian developments, of physiology and psychology. For many positivists of the second half of the 1800's, these principles had been the 'certainty' whereby open affirmation of the autonomy of science from metaphysics, as promulgated by Comte, might be achieved.

When sociology was founded in Italy, positivist sociologists, social philosophers and those who wrote about sociological issues, were able to consider much of what has been said until then, as a veritable perspective of the philosophy of science, as the only one practicable one in existence as the automatic consequence of the development of knowledge. In actual fact, this mode of reasoning came to establish itself in Italy along with some undisputed truths.

If on the one hand, the philosophy of science seemed to end up among the dogmas which social scientists might apply, on the other, poor theoretical processes led philosophers –also for ideological and political reasons- to diffuse a refusal of materialistic and positivistic views in academic circles, to the point of accepting and encouraging subjectivist, even irrational, philosophies.

Therefore, in Italy, the philosophy of science ceded to others the task of producing a 'point of view' concerning the physical and natural sciences useful to the formation of sociology as a science and allow the social sciences to assert themselves, within the developmental framework of the history of science and epistemology, naturally. Within other historical contexts regarding the development of the philosophy of science the old Protestant motto *sola scriptura*, began to be applied.

In Italy, the historiography of scientific subjects emerged severely penalized and was unable to take shape for several decades afterwards. In any case, it did not succeed in arousing the interest of Italian sociology or in imposing its own specificity, except for scholarly developments formulated elsewhere and without taking into consideration the importance of some historical moments which affected other disciplines like Ernst Mach's European formulation of philosophy of science which held that science was human adjustment to reality and therefore an economic ordering of sensorial data which did not permit comprehension of the essence of nature and that all claims to absoluteness relating to mechanical concepts like space, time, force, mass, etc. were to be refuted.

Within the ambit of his phenomenological physics and radical phenomenism, Mach also rejected atomism and professed a strongly anti-dogmatic approach called 'empirical criticism'.

According Mach's 'principle of economy, science may develop along lines which make various disciplines congregate around simpler and simpler principles whose application grows vaster and vaster.

Besides acceptance of similar principals by the social sciences, it ought to be said that Mach's endeavour was most relevant especially because he grafted onto scientific thinking the possibility of verifying the conceptual constructs of the scientist while also promoting a whole series of procedures useful to the establishment of a historiographical epistemological of physics. We should remember that Pierre Duhem too was hostile towards mechanicism and that he expressed his rejection by opposing it with energetism considered the fundamental theory of physics, but, unlike Mach, he recognized formal hypothetical-deductive procedure as essential for science, which would, eventually, link empirical laws to general theoretical principles.

The birth of non-Euclidean geometry brought the assumption of a sole applicative model of reality to a critical head and thus questioned the need to follow one of the original perspectives of philosophy and science only. A number of postulates were criticized precisely because of their dogmatic nature so that the absolute truth might finally replace a system of hypothetic-deductive reasoning. In fact, many Euclidean truisms had endured for centuries, conveying with them the irrefutable gnoseological foundations of thought regarding the realities of geometry, mathematics and physical space.

Later, Henry Poincaré attributed a conventional character to conventional geometrical axioms and considered all scientific theories conventional, arguing that, when all comes to all, they are neither true nor false; most epistemological implications are contained in the phrase 'one geometry cannot be truer than another; it may only be more convenient'. These interpretations of phenomena, which claimed their rightful place in the history of science alongside cautious methodological reflections regarding the human and social sciences, were opposed by the Austrian physicist Ludwig Boltzmann, who defended atomism and declared himself favourable to science's fundamental need to go beyond experience by availing of theories unbound by empirical laws but aimed at establishing connections between phenomena; one should remember that for Boltzmann, science should proceed through crises and reconstructions, and, therefore depend on a process we might call dialectic.

Thanks to the prevalent approach outlined by Mach, Duhem and Poincaré, the philosophy of science was able to establish itself in the twentieth century, and therefore thanks to new-positivism or logical positivism; the general theory of science was developed due to the influence of Mach and Wittgenstein's '*Tractatus Logico-Philosophicus*' (1921) and, also to the extraordinary links existing between the members of that group of philosophers who, starting from 1923, sprung up around Sclick Moritz, professor of the philosophy of the inductive sciences at the University of Vienna; as we know this group included Rudolph Carnap, Friedrich Waismann, Otto Neurath, Kurt Godel, Karl Popper, Hans Kelsen and several others mentioned above. departing from developments in logic (Georg Boole, Giuseppe Peano, Gottlob Frege and Russel) the historians of contemporary science consider it essential that this group of philosophers was able – back then-to restore to science a rigorous formal language endowed with empirical content and thus introduce criteria by which to determine the truth or falsity of various elements. In view of a series of retrospective analyses, there are interpretations which lead back to a rediscovery of the initial issue of the 'verifiability' of 'factual experiential' propositions advanced by Schlick, a topic, which with Neurath, became a linguistic theme-problem, as logical operations can only be achieved in language and make the verification of propositions significant.

Of interest is the history of Neurath's studies (introduced to Italy by sociologists only at the end of the 1960's) with his theory of 'physicalism' and his general interest in: a) the development of diverse principles according to different contingencies (ethical studies); b) applications regarding effective conditions (studies of sociology); c) and, finally, Marx's theories (economic studies); however, in addition to the translations of Neurath's writings by Gianni Statera (1967 and 1968) we need to mention studies favouring a new season for the 'philosophy of science', like those, for example, by Francesco Fistetti *(Neurath contro Popper. Otto Neurath riscoperto* published by Dedalo, Bari in 1985) or those by Danilo Zolo (*Scienza e Politica in Otto Neurath. Una prospettiva post-empiricista*, Milan, Feltrinelli, 1986) and others, on which Paolo Parrini commented a few decades later (see *L'empirismo logico. Aspetti storici a prospettive teoriche,* Rome Carocci, 2002). But it was with Carnap that science was to become 'a logical construction of the world' and philosophy to run itself out in analyses of scientific language destined –in turn- to assume the traits of radical 'formalism', as established in *Logische Syntax der Sprache (1934*) where, among other things, it dissolved into pure observation of the symbols used without any reference to real data. We need to recall as far as this is concerned, the development of Carnap's thinking overcame early 'restricted empiricism', according to which a proposition had to concern real objects; this conception meant rejecting the possibility of scientific laws seeing that a law may apply to an infinite number of cases that cannot ever be analysed.

We need to say, that Carnap was able to renounce complete verification by admitting the possibility that a sole proof might establish the truth of a statement or provide confirmation such as to establish the possibility of asserting it. From this position it was possible to highlight the problem of the relationship between theoretical affirmation and empirical statements; if a theoretical system consists in a calculation, that is, of signs connected by means of syntactic rules, these signs lack meaning; it will be necessary, therefore, to specify a semantic interpretation capable of providing the signs with meaning, that is, rules that may relate or link the signs with observable properties or objects.

Again, semantics reveals its relevance in the face of the historical formation of contemporary epistemology and, the development of the social sciences and empirical research methods. This is even more evident when reviewing the contribution made by Carnap's philosophy to contemporary science and the enormous importance of his research into inductive logic, in particular the theories expressed in his *Logical Foundations of Probability (1950).*

Carnap departed from a distinction between probability as statistical frequency of an empirical kind, and probability as a 'degree of confirmation'. As regards the concept of probability he sustained that between inductive logic and inductive processes there had to be the same relationship as between deductive logic and deductive processes. In other words, as far as the introduction of a criterion by which to determine the truth or falsity of elements is concerned and considering that no general statement may be verified definitively by means of a single empirical proof, it has to be admitted that a similar criterion should be that of its 'degree of confirmation' that is, its probability.

Historically speaking, it was Karl Raimund Popper, however, who held that scientific statements could never be verifiable but considered only falsifiable.

It should also be remembered that the most radical position which denied that science was capable of knowing the reality beyond phenomenal experience appeared with the Percy Williams Bridgman's American operationalism. He assumed that the meaning of a scientific concept was nothing other than the set of empirical operations it entailed; the idea that science acknowledged as pure knowledge of phenomena might be overridden emerged, within positivism itself when analysing the relationship between experimental theories and laws. According to some authors theoretical terms (like atom, electron, etc.) were not attributable to purely experimental or observational data but ultimately indicated a reality that transcended them.

Furthermore, this represented, at least for a period, a historical return to realistic positions, similar to those sustained by Boltzmann and other approaches such as dialectical materialism. The latter was probably due to historical analyses of concrete developments in contemporary science (theory of relativity, quantum theory, Heisenberg's uncertainty principle, etc.) This change proved rather important for philosophers of science and historians of science during the second half of the twentieth century. Epistemology obtained a field in which to express itself historically: besides acceptance of analytic philosophy by some areas of the development of scientific ideas, which extended the problem of scientific language to ordinary varieties of human languages; besides the possibility of conceiving the greater relevance to the philosophy of science of the historical method to the point of revealing the need that it may, against increasing specialization, unify various scientific results and provide a global view of natural processes.

Chapter 5

The Historical Method and Issues of Knowledge: The New-Rationalistic Turning Point in Italy

The introduction of studies regarding the historical progress of scientific knowledge and its growth within the context of the development of ideas during the 1900's and, until the end of the century, represented an authentic advancement of perspective for the philosophy of science that attracted the interest of the social sciences in the years following acceptance of topics relevant to the study of scientific hypotheses and laws, the logic of explanation and prediction, as well as the better-known (at least for the purposes of historical debate) demarcation between science and metaphysics and, finally, the formulation of concepts and theories within the empirical disciplines. In general, if it is true that the history of science is full of *experimentum crucis*, it may also be true that the acceptance of the debate concerning fundamental issues constitutes irrefutable proof of the historical genesis of epistemology in various theoretical-practical developmental contexts. This seems to be confirmed by the fact that starting from a number of themes-issues, an interest in the historical method was coming to maturity also within Italian epistemological culture as well as within the French circles of the 1900's, (with authors like Pierre Duhem, the Russian Alexandre Koyre, Edouard Le Roy, Emile Meyerson and Gaston Bachelard) and those of mid-twentieth-century England which owed their origin to Popper (as well as to Imre Lakatos, Thomas Samuel Kuhn and Paul K. Feyerabend). We need to add that basic problems had always aroused the interest of scholars in a sort of historical genesis of scientific concepts. To this regard we might mention the anti-dogmatic attitude of Heinrich Rickert (a pupil of Wilhelm Windelband's and Max Weber's teacher) towards the foundation of the cultural sciences, especially when he tried to show how reference to the conceptual dimension was always accompanied by theoretical operations which made knowledge possible.

However, as far as a historical reconstruction of crucial events for contemporaries was concerned, epistemological research was able to pursue, especially within the French and English contexts, a veritable autonomisation from the themes imposed by gnoseology, and, already beginning with J. S. Mill's logic, or the ideas of W. Whewell, Mach and Poincaré to arrive at the development of an authentically European kind of logical empiricism, at Popperist falsificationism (that of John Watkins, Alan Musgrave and Joseph Agassi) and, finally, at the critical current of the new philosophers of science (Nortwood Russell Hanson, Stephen Toulmin Eldelston, Hilary Putnam). On the other hand, it must be said that in the face of the development of the historiography of science and the historiography of the social sciences, the traditional 'point of view' imposed by the neo-positivist, Popperist myth, did not appear to have exhausted its 'discovery' of a range of philosophical contributions, within which, in the light of much production by many areas of research, the epistemology of the 1990's assumed a different connotation, especially in view of a future generation of historians of scientific subjects and, consequently, of new –much longed-for- twenty-first-century historians of social sciences.

In this sense, it is useful to examine studies regarding the philosophy of science by writers like Evandro Agazzi, Fabio Minazzi and Ludovico Geymonat - availed of little by Italian social historians of the second half of the twentieth century - which recall not only contributions by Federigo Enriques regarding the European nature of the birth of a new form of critical rationalism, but also those of Ludwick Fleck (whom many sociologists associate closely with the development of Kuhn's thinking) or of Ferdinand Gonseth and Bacherlard, of Jean Piaget and Hugo Dingler, Hermann Weyl, Émile Meyerson, Mario Bunge in opposition to the prevailing neo-positivist tradition; not to mention early-twentieth-century studies concerning the fundamental issue of relationships between science and philosophy (Giuseppe Peano, Giovanni Vailati and Enriques himself) or the philosophy of mathematics etc. It is interesting to note, however, that these observations date back to the last decade of the twentieth century, when the prospect of a conceptual revolution within twentieth-century epistemology closely linked to a more complete perspective of science history studies arose. The idea that conceptual history and epistemological history –when all was said and done- could not be separated from the creation of a new kind of historian within various philosophical and empirical fields, like sociology, came to the fore. The umpteenth turning point in the philosophy of science and in how it might be used concretely in work by scholars (initially 'scholars' in general, later 'historians of science', in an increasingly autonomous and specialised sense) appear clear enough as does the conception –which existed already in the early 1950's- whereby the increase in knowledge has in time come to pay its debt to existing links between the philosophy of science and its historiography.

As we know, Popper used strictly logical criteria to address the problem of the historical sciences, traditionally excluded from neo-positivist analyses as irrelevant to the scientific field. In some ways, we should recall that the attitude of Popper's followers towards the methodological differences between 'natural science' and the 'science of the spirit' that had established themselves during the development of German historicism, among moderns and contemporaries and a number of even more recent hermeneutic currents, proved totally adverse. In other ways, we need to clarify whether the origins of Popper's critical rationalism induce one -eventually- to dissent more or less from the foundation of a '*Methodenstreit'* historiography, exclusive to the historical and social sciences, especially in view of Weber's liquidation of the historiography/sociology dispute within the ambit of his 'cultural relativism'.

The fact is that Popper's positions succeeded in influencing - much more than Weber's- historians of scientific knowledge in contexts where epistemology has developed in the light of studies of its genesis and complex, establishing itself as a historical and rational subject. This occurred, roughly, between the mid-1960's and the 1970's, especially with acceptance of the debate by the English-speaking world, although it is also mentioned by the Congress of the German Society of Sociology, held in Tubingen in 1961, where a definitive maturation of a number of epistemological issues into a profoundly anti-dogmatic vision of the 'logic of the social sciences' was highlighted and owes much more to Max Weber than English historians of science are inclined to admit.

Hans Albert, for example, held that Weber, as author of historical and social sciences texts, was a true precursor of Popper's, although the Popper school of epistemology later embarked on a pathway of anarchism, summed up later in Paul K. Feyerabend's counter-induction methodological proposals to the point of recognising the ups and downs of an epistemology bearer to the debate of *ex novo* arguments recollective of those to which the logical positivists had subjected physicalist and conventionalist postulates.

Contestation of the new trends - which no longer took place in the presence of the ghost of anti-metaphysics- did not fail to trigger a continuous, recurring debate concerning 'formal' issues like the relationship between historiography and epistemology. This debate involved historians of the empirical social sciences and of the cognitive and linguistic sciences, etc., who seemed, little by little, to have become absorbed by the grand issue of criticism and the growth of knowledge within their own increasingly more restricted and specialized spheres of interest. In this sense, the epistemology of the 1900's and its historiography tended to make the interests of historians of science converge on many of the issues the philosophy of science proposed to those interested in its development. A number of these may be found in historiographical investigations of the growth of knowledge or changes in scientific ideas. We can also find them in studies which followed the route of scientific revolutions often in conflict with the very history of epistemological itself, which provided an entire range of approaches different from those that it was logical – useful in any case - for science to discard all-inclusive interpretations of scientific knowledge. This was the case of studies inspired by Thomas Kuhn which defined the well-known criteria of consensual acceptance of theories and paradigms within scientific communities. In actual fact, epistemological history does not seem to appear as a variety of sociology of scientific knowledge if it is unable, moreover, to produce a solid historiographical 'point of view', preceding or containing an all-inclusive theory. The historiographical production of science and epistemology itself seems to belong to essential moments of the constitution of paradigms. This was, perhaps, the reason why the social sciences might consider themselves at a 'schools' stage, in a paradigmatic phase, as Kuhn wrote.

It appears that Kuhn spoke about the lack of a model for the historiography of the empirical social sciences or a historiography of scientific sociological knowledge in the making. In any case he did not intend his observations to refer only to the acceptance by his contemporaries of the results of the debate surrounding the philosophy of science.

Kuhn raised a number of questions for the historians of science, ones which may be deemed valid from his epistemological point of view, which rests on the concept of the incommensurability of theories and paradigms and refers, in a sense, to the impossibility of reducing the various 'worlds' of scientists to a sole set of rules. He believed that epistemological history should seek a common ground for scientists within various contexts of historical development, in antithesis, in any case to opposite visions which it had managed to generate during the second half of the twentieth century. One of these is undoubtedly hermeneutics, that is, the alleged, widespread belief that we can do without epistemology when trying to create a rational discourse concerning the history of science. Another example might stem from the anti-foundational and pragmatic vision of knowledge and meaning sustained by authors (some already mentioned here) like J. Dewey, Wittgenstein, Quine, Wilfrid Sellars, and Donald Davidson which rotate around holistic theories which often lead to the abandonment of epistemology. In fact, some of Kuhn's findings favoured a purely subjective and structurally 'internal' view (defined later) of the historiography of science which began to assert itself, not only among the usual philosophers of science, but rather by among social scientists too, to the point of affecting methodology and the validity of some of the procedures to be adopted by a scientific community as yet to be created. The various attempts made by prominent European scholars, to assimilate the ideas of Kuhn regarding the paradigm and/or the research programmes of Lakatos, etc. have led to just as many results. On the whole, this occurred wherever the debate on the history of science reached a certain level of maturity and also claimed some of the discontinuities present in the historical development of science in the west, which, according to Popper's legacy, would have endorsed the affirmation that scientific knowledge – ultimately- can be considered a conquest of rationality and, therefore, something to be defended.

The same may hold, at this point, for the philosophy of science, which, deprived of its historiography, appeared just as blind or empty (depending on the 'point of view') and, in all cases, lacking the sole possibility of rationally envisioning the conceptual history of the various disciplines as a history of epistemology produced within theoretical and practical developmental contexts. Furthermore, this may prove useful when outlining a series of non-competing links between epistemological history and the historiography of the empirical social sciences. In this sense, and especially in some developmental contexts, reference may be made again to the legacy of the philosophy of science and the failure to accept some of its goals by historians of the social sciences. In Italy, for example, this outlook would have made it possible for all those seeking investigate, when the early works Ludwig Geymonat appeared. He was the 'natural' heir of the mathematician Federigo Enriques. In his studies, besides attributing relevance to the historical dimension of epistemology, he claimed his rationalism and criticised neo-positivism for not having grasped some of the scientific results associated with history.

In around the 1930's and after 1945, a new kind of methodological rationalism saw the light in Italy, thanks to re-evaluation by the founders of modern logic, to Dewey and to various programmes of a neo-illuminist nature drawn up by that 'inconvenient philosopher' Nicola Abbagnano who, in 1934, published '*La fisica nuova. Fondamenti di una teoria della conoscenza* where he advocated neo-Kantism in the face of scientific revolutions; an auchthonous rationalism, antithetical to that of metaphysics but, above all, endowed with a versatility based on investigation of the new problems that science and practice might have to address and see imposed upon philosophical research. Positivism and neo-positivism were thus embraced by different scholars in Italy and from different angles, of little relevance to the academic philosophical culture of the time and conditioned, above all, by other studies in philosophy of science like those regarding mathematics, born in Turin thanks to Geymonat's frequentation of Peano's theories. In addition, the ripening of a similar interest in mathematics emerged as a proposition alongside reception of some of the main theses circulating among members of the Wiener Kreis which Geymonat had met around the time when the group was set up and which was always esteemed by twentieth-century Italian philosophical and scientific culture. Moritz Schlik's 'realist' teachings, where philosophical inquiry follows but does not precede that of science, was always in the foreground. In 1959, Geymonat wrote '*Peano e la sorte della logica in Italia*' [Peano and the fate of logic and in Italy] and the following year he published '*Filosofia e filosofia della scienza*' [Philosophy and the philosophy of science], where he advocated the pursuit of 'historical science' and that scientific theories should open up toward themselves and towards problems related to formal recognition of experience. It became clear to Geymonat how, from Galileo on, the history of science had to come to terms with the major problem of determining a methodology based on a combination of induction and deduction, mathematics and experience. To be precise, Geymonat is obliged to discuss the fact that formalized theories might be reconciled- in their various axiomatic and conventionalist aspects- with an 'opening up' to experience. For him, the history of science was capable of proving that mathematics had contributed most to the construction of the rational structures of scientific theories, especially physics, so that, in the long run, a theoretical flexibility of theories (with regard to each other as well as a common language and experience) might lead to a clear definition of the use of the concept of the 'historicity of science', investigated above and beyond all philosophical immanentism.

The problem of 'historicity' was thus able to become s source of continuous development of epistemological thinking within areas of interest where the philosophy of science had the task of investigating the specific characteristics of scientific knowledge, avoiding general conceptions of the world. This way the philosophy of science was able to provide help to disciplines seeking to define their own historical development by tracing their history of logic through methodology and epistemology in order to bestow upon it a neo-rationalistic type of direction attributable then to a veritable methodological 'reform' of contemporary philosophy. We shall return to this later on. It is opportune to note that the relationship between science and philosophy was re-proposed as re-evaluation of a rational way of structuring scientific theories implying that changes and transformations had taken place in the history directly investigated in the light of a pre-defined or definable 'philosophy of science'.

Ludovico Geymonat, seemed, instead, to address some of the epistemological themes-issues contained in the ideas of Enriques who sought to conjugate philosophy with science following some of the transformations which had taken place within the sphere of scientific knowledge, especially the debate surrounding the 'reform of rationalism' as well as that around the up and coming philosophy of pragmatism. We should also keep in mind that Enriques in his '*Nei problem della scienza science' (1906)* already asserted the theory that scientific knowledge occurred through concepts capable of representing facts. Furthermore, we must not overlook a number of further 'openings' taking place already within early twentieth-century Italian culture thanks mainly to Giovanni Vailati who attempted to define the tasks of philosophy in the face of excessive specialization within the sciences. This way, even the issue of the multiplication and differentiation of the various fields of knowledge was able to become the dominant feature of a grand transformation. Vailati considered philosophy as a discipline capable, from a linguistic, logical and conceptual point of view of critically analysing the methods of science, of grasping its similarities and differences, that is, its fundamental concepts. This also meant that in Italy thinking regarding so-called 'scientific philosophy' might have succeeded into stimulating the growth of a conceptual history concerned with philosophy itself and with a re-evaluation of research into the history of science, as we read in one of the Torinese '*Prelusioni'* dating from 1897. Vailati, following in Peano's footsteps, in his '*Formulario'* had taken an interest in critical notes regarding the development of mathematics. He intended to highlight the interest he took in the historical dimension of scientific activity. For him the history of mathematics meant a continual development and growth of knowledge where the concepts advanced (mechanics, mathematics) might become the tools by means of which to achieve a purpose that presupposes the method to adopt. Even if we refer to logic and to the interest Vailati shows in 'logical' tools, it is possible to return to the study of the scientific method and to the various processes of investigation and demonstration, as he himself defined them. Logic regards to all the sciences and human activities which aspire to the scientific status typical of mathematics, geometry, analysis, and formal logic. Upon the same level as philosophy (it is important to note!), logic tends to induce scientists to seek unity of method which, with mathematics, should constitute the backbone of a new scientific vision. At this point, it would be profitable for historians of science to acknowledge Vailati's unquestionable tendency to re-evaluate the quest for links within conceptual history, insofar as it is mathematical logic which permits us to speak of the definability or not of a concept after having first determined its terms and specifications. It needs to be said, besides, that contemporary researchers did not dwell very much on the similarities 'of the times' between the thinking of Vailati and that of the English Russel (or even on those between the Italian and Charles S. Peirce or William James) especially as regards issues concerning conceptual constructions of mathematics. The fact is that the Lombard philosopher held that the instruments of logic might become a kind of phenomenology of the deductive structure of science analogous to the deductive structure of mathematics. And this needs to be considered some kind of result.

Moreover, it seems that, according to Vailati, the truth or falseness of a proposition concerns a logical consequentiality between the parts, which on the one hand, establish the conditions by means of which an event occurs, and, on the other, represent the event itself. This means that some of his research succeeded in soliciting a series of results at formal axiomatic level, starting from the application of a hypothetical and deductive model capable from its introduction, to tie in with confirmed historical investigations concerning the development of knowledge. Another kind of result attributable to Vailati is that stemming from his investigation of the origin of language. This kind of analysis became, in the early twentieth century when it appeared, the linguistic method, since Vailati was convinced that linguistic confusion was the cause of many major philosophical problems and controversies. Valiati proposed a pragmatic analysis of philosophical nomenclature, re-evaluating the work of the historian of science compared to that of the historian of philosophy as well as proposing the adoption of the methodical standard of pragmatism.

His research and 'Observations' on the 'issue of words' in the history of science and culture spurred him to investigate variations in the meaning of terms and the assert of the discipline of semantics with considerable emphasis on the methodological aspects and the analysis of language which, in the end, might be interpreted as true critical analysis of knowledge in general. This way, the connections between logic and the history of knowledge became amply detectable even in the rational 'products' of early Italian pragmatism, which pre-announced to historians of science and other disciplines, a view of philosophical studies of science of which the neo-rationalists were to become the bearers, with their 'openness' towards the work of methodologists entrusted with the task of studying the historical achievements of human rationality, concretely.

The methodologist, more than the philosopher, was now given the task of revealing how to investigate scientific theories, of rationally expanding the horizon of knowledge and technique to the point of defining methodology a real operation performed on science and not only within its boundary. What better opportunity for historians of Italian sociology of the period to grasp the meaning and the new direction of a programmatic vision of scientific work capable of impacting upon the broad reality of a major 'philosophical renewal' operating within the scientific sphere, the second probably, after that created by Antonio Labriola's Marxism and his sociology of scientific knowledge. Broader perspectives were sought and found which began with positivist investigations of knowledge availing also of Comptian epistemology and arrived at a discussion of systems of propositions, pragmatism and various 'logical' issues and mathematics. As early as 1945, Geymonat, with his '*Studi per un nuovo razionalismo*, sought to spread the neo-empiricist approach among Italian scholars with a view to introducing novelty into their mind-set, and thus addressing the problem of the logical structure of systems of knowledge. This was extremely relevant because the empiricist approach had to confront the formulation concepts regarding scientific and philosophical research. It invited scholars to face serious investigation of the formulation of these concepts by examining the insufficiencies of classical logic compared to mathematics and modern physics. It meant most probably a serious invitation to view and interpret the history of scientific subjects through new lenses. Furthermore, this new type of rationalism tended (in the light of acceptance of a particular late-nineteenth and early twentieth-century debate) to renew itself even as far as so-called 'logistics was concerned: now philosophy might mean the 'philosophy of language' and its task was to clarify the value of recurrent connections between words, that is, ways in which one word can join another so that a proposition may acquire full meaning, etc. All this caused some scholars to acquire a 'formal logical' perspective (as mentioned above) and assume the task of surpassing the principle of identity (upon which 'classical logic' had always been based), without, however insisting on the claim of determining the fate of external reality, seeing that this can be considered rational only insofar the thought that expresses is rational. All told, for historians of sociology who were not strictly positivist, ideas regarding the history of science seemed tantamount to considering it as being pervaded by the 'rational development' of sources already found within the overall debate concerning the fundaments of knowledge produced by 'materials' rather than by conceptual history. Therefore this meant no mere history of thought capable of imposing itself upon the various scientific disciplines during the course of their specialization. The attention of scholars was able to turn, definitely, to a quest for theories of change. It is understandable how change could take place both at socio-historical and 'intellectual' level as a historiographical process. As it has been possible to sustain, if this were not the case, sociology as a science would emerge from history deprived of history and, in particular, of 'scientific thinking' regarding history. This would make it a science without a 'history of science', something logically impossible, implying the absence of differences attributable to specific moments or concrete historical contexts which trigger concepts and theories unlike previous theories or concepts. There is no doubt that for historians of sociology in Italy (regardless of the fate of individual scholars) the neo-rationalistic turning point was the source of a major shift in perspective which proved capable of imposing, for the first time, perhaps, an epistemological view of empirical social sciences such as to legitimize the conceptual history of scientific disciplines related to traditional and/or 'academic' historiography and methodology, no longer intended (by philosophers of science in particular) as a philosophical system, but as a programmatic research sector. The extraordinary capacity for methodological innovation shown by some philosophers and other scholars was due to the fact that the problem of rationality began to be considered and studied on the basis of its concrete implementations, that is according to the achievements of modern science and not simply its general abstractness or as a problem dependent on its own faculties. The problem of rationality was crucial to all implications referring to the various fields of science and inquiry, including sociological history. In other words, it was scientific methodology that in the various spheres of interest represented that set of studies capable containing the outcomes of the development of knowledge and the various scientific theories antithetical to tradition. By way of example, from the 'point of view' of neo-rationalism, the history of sociology might become a subject to avail of for an in-depth examination of the conceptual history of the development of logical reasoning or empirical observations, not born a-priori within philosophical systems, but within the context of various productions. This way, a new kind of sociological historiography should have emerged given developments within scientific methodology and topics related to the epistemological genesis of the single disciplines. It is precisely these Italian philosophers of science of the second half of the 1950's who deserve credit for the idea of having scientific subjects combine contemporary epistemology and its issues with methodology and matters related to empirical research, considering how assertions like that, might well recall – in paradoxically opposite terms and as far as Italian historians of sociology were concerned – the position sustained by Paul Felix Lazarsfeld in the United States in the early 1960's, when he held that philosophers of science had 'unfairly' dedicated little of their time to empirical social research. Lazarsfeld advocated a whole series of technical-methodological facilities that had actually been attributed conceptually to methodologists by the philosophers of new scientific rationalism a decade previously. Moreover, Lazarsfeld (a mathematics graduate) asked philosophers of science, as far as matters concerning empirical social research were concerned, to collaborate as much as possible to avoid further neglect of the history of sociology seen as an analysis of procedures. This is exactly what (in a different developmental scenario) Italian historians of sociology or historians of sociological thought overlooked doing in their studies dedicated to the 'renaissance' of sociology at the precise moment when Italy began to take an interest in the 'methodological question' –imported from the United States- which outlined horizons quite different from those of certain philosophers of science who, meanwhile, promoted important conferences concerning methodological studies (Turin 1952, 1961 and 1967) or met in centres which gave life, most times, to epistemological traditions (like Bologna and Milan) as well as to groups engaged in intense and original research (Roman centres).

Chapter 6

MARXISM, EPISTEMOLOGY AND JOHN DEWEY'S PRAGMATISM IN ITALY

The horizons available to new historians of science during the second half of the twentieth century were capable of comprising the results of the diffusion of pragmatism in Italy and those due to absorption of that other form of provincialism coming from American scientific culture, which, if on the one hand, it stemmed from Peirce's desire to establish a true theory of meaning, on the other it harked back to William James, who, at that time, claimed the general superiority of practice over theory in order to uphold utilitarian criteria capable of judging the truth of scientific and philosophical propositions. When all was said and done, we can say that American pragmatism has always remained the same; even when present-day researchers seek links between what Dewey and Quine, endeavoured to learn, from James's *Pragmatism*, at the beginning of the century they arrive at Quine, Davidson and Richard Rorty themselves. In Europe, on the contrary, pragmatism fitted into a broader epoch-making perspective which may be defined, *ex post,* as an anti-intellectual derivation due, in France, to Boutroux's 'contingentism', Poincaré's 'conventionalism' and Bergson's 'intuitionism' Bergson, in Germany, to Mach and Avenarius's 'empirio-criticism' and, in England, to F. G. C. Schiller's 'humanism'.

These historians were able to witness a turning point in the new kind of rationalism just mentioned above, saw, in Italy the dissolution of neo-positivism which even contained, in part, results of the establishment of Peirce and John Dewey's pragmatism in the 'positivistic' version provided by Giovanni Vailati and Mario Calderoni antithetic to that professed by Giovanni Papini, during the *Leonardo*. It was Calderoni in *Variazioni sul pragmatism*, a work long forgotten by the historians of sociological positivism, who dealt with the characterizing varieties, connections and 'identities' existing between positivism and pragmatism. However, we should not forget that pragmatism managed to influence, and in a no simple way, Nicola Abbagnano and his existentialism which, in the 1950's, saw Marxism dialectically emptied as American sociology took centre stage.

It was a crucial moment. From 1948 Abbagnano directed his efforts towards the creation of a new Enlightenment, which, subsequently, and in Dewey's wake, was to embody a whole series of inspired choices, from instrumental reason to scientific rationality. The model it gave rise to was clear enough also because of a series of historical and conceptual prospects that might have arisen, but did not as they initially sprung from an exclusive 'point of view' of the history of ideas and of culture etc. encompassing important results of the debate surrounding Italian philosophy. One of these was undoubtedly the development of a positive existentialism capable of paving the way towards the concreteness of the experience that seemed to stem both from Abbagnano's attitude towards neo-positivism and his critical relationship with Marxism when he defended sociology, at the time of the "Quaderni di Sociologia" which he founded with Franco Ferrarotti. This way sociology in Italy suddenly plunged into a quarrel over ideas: between Marxism, pragmatism and Dewey. But let us proceed with order.

We find several signs of this in the writings of many authors who by reflecting deeply on neo-positivism, on the 'formal' aspects of science and on conventionalism etc. managed to identify the matrix of their own intellectual historical-epistemological renewal which took place in a theoretical-practical context, like ours, which lacked an empiricist tradition and was engaged in surpassing naive positivism and achieving the emancipation of philosophy and scientific ideas from the metaphysical speculation conducted, from the beginning, to the detriment of concrete scientific activity based on a particular vision of the theoretical activity. The secular view of the origin of scientific knowledge requires that the scientific tradition be considered as self-creating, derived from itself, therefore, self-generated.

It is beyond dispute that in Italy at a given moment, existentialism, like neo-rationalism or pragmatism, underlined not only that the discussion regarding science needed to raise numerous logic-gnoseological and philosophical issues but that also, by admitting the contrary, it should surely have had counterproductive effects on the possible education of a new generation of researchers and historians of scientific subjects. Many authors, like Franco Lombardi, considered justification of the conceptual validity of its pathway the task which would permit philosophy to reconstitute itself continuously in the face of the advancement of science and the accomplishment of a modern theory of knowledge. According to Lombardi in his *'Il mondo degli uomini'* (first edition 1935, second edition printed in Florence, by Sansoni 1967) a new conceptual definition of critical, philosophical thinking would indeed be necessary, with the addition of an *ex novo* request whereby philosophy might and should intervene to establish the research of the scientific, the historian, the psychologist, etc. without any claims to affecting the results. It became clear that certain questions might now be of interest not so much to historians of philosophy as to historians of scientific thinking and historians of the development of scientific sociology as a subject superimposable upon conceptual investigation during the second half of the twentieth century. These were actually assigned the task of recording not only the entry upon the rational scene of new and different research routes, later exploited not only as far as the foundation and development of an epistemological history of social science was concerned, but to probe deeply into the reasons why the tendencies gradually splintered into myriad and individual positions characterized –in the long run- by independent, autonomous pathways, as in the case of the academic sociology and Italian sociologists of the time.

Each one seemed to go his/her own way. The historians of sociological science might have grasped, at a time of significant development, gradual dissolution and methodological discussion of the destinies of Italian philosophy, its relationship with the new epistemological exigencies emerging within the historiography of scientific disciplines of European - not only Anglo-American- origin, due to 'internal' issues.

The alternative, as we know, meant falling, practically, into the void of Croce's epistemology; but that is another story. It may, however, be more useful to recall that some scholars of the new generation, without practicing sociology, nurtured their interests by passing through Marxism but without investigating it in depth. Others like Antonio Banfi, was able to acknowledge the existence of a 'historical humanism' in Marxism, antithetical to sociological subjects, admitting the possibility for people to alter the historical reality and thus become masters of themselves and their world (*L'uomo Copernicano,* 1950). Again, in 1953, Banfi asserted his staunch trust in socio-historical praxis by availing of the criticism Marxism levelled against sociology, which it considered merely empirical and descriptive, and, in his opinion, responsible for not affecting anything (like its 'official' historians) except what existed already.

For lovers of sociological history, a retrospective view of some stages in the development of Banfi's scientific thinking, -and that of some of his pupils like Giulio Preti- may be somewhat useful as regards the 'breakthrough' that led to Dewey's neo-rationalism, from his rationalism and social pragmatism to his encounter with Marxism. Beginning with his *Principi di una teoria della ragione*, 1926, Banfi outlined a 'critical rationalism' based on an idea of reason as the law by which to organise processes of experience in all its fields. Reason was expected to materialize as scientific thinking whose task it was to use interconnections between the data of experience to create laws and philosophical thought, more problematic and more open to development and, therefore, capable of acting as theory and praxis alike. By acknowledging scientific reasoning Banfi assumed the task of interpreting the theoretical structure of science and philosophy in relation to a rational *a-priori* –of a transcendental nature- to investigate the variegated phenomenology of scientific reason and the execution of scientific concepts and the principle of knowledge. As he himself admitted in 1955, and in his *La ricerca della realtà*, 1959, the problem consisted, above all, in a quest for a synthesis of a theoretical nature which had compelled him to arrive at an arduous acknowledgment of the gnoseology and philosophy of his day and treatment of transcendental rationalism as an integrative system of those philosophical visions capable of producing a number of theoretical models and methodological discussions where many experimental, philosophical sciences and social disciplines, from the psychology to sociology, might acknowledge each other. One of Banfi's decisions meant admitting that science (as a concrete resolution of the data and activities of reason into the multiple truths of experience) created, as it proceeded, its own system of categories. Every one of its results depended on a rational structure of knowledge leading to the admission of a law regarding scientific reasoning itself. Practically speaking, the author's reflections on the meaning and theoretical value of science arrived at considering its assumed rational developments as tantamount to rules of knowledge. He also admitted the need to investigate the dialectic process existing between the pragmatic and theoretical moments, between experience and generalization, induction and deduction by means of which science is compelled to implement itself.

Relating concepts to rationality meant emphasising the importance of devising a way of confining the philosopher's efforts and experiences to a quest for concrete knowledge. It is in this sense that Banfi's entry into the Marxist debate and into later discussions about critical views of the development of sociology as an empirical science in Italy, might have proven truly stimulating. This also in view of the scarce consideration in which some sociological and scientific authors held the transition from investigation of gnoseological, conceptual and theoretical issues to examinations of those regarding epistemological conceptions in science, which both Marxism, pragmatism, and other 'points of view', identified with the praxis, that is, with given practical applications. This occurred during a certain period and, wherever Marxist interpretations were carried out in the light of the relationships with the theory of science, some of the contents proved very interesting to several advocates of the empirical social sciences.

It is, for example, on the issue of relationships between the conception of science and the logic of history, that Italy was able to make a conspicuous contribution, thanks to the work of Galvano Della Volpe who not only considered Marxism a scientific, experimental method applicable to history, but also posited the theory whereby, all told, the sole logic of use to researchers in various spheres, including sociology, was that of the empirical sciences and of which Marxism simply represented its application in the areas of economics and social history.

Della Volpe's thinking regarding Marxism followed a given approach where it was the logic that appeared 'concrete' because of it is fusion with epistemology by admitting of a process of formalization involving the uniqueness of logic when considering science. In short, that meant that the logic-argument might submit to comparison both with the whole of its formal aspects and with those derivable from dialectics and therefore, in perspective, from history too. This was a result that could not be overlooked when undertaking a series of investigations of the conceptual dimension of the development of scientific subjects. It appears, rather, that Della Volpe's interests addressed the problem of the formalization of scientific and historical logic. The epistemological results of this kind of thinking were not long in coming and being accepted.

To counter Hegelian idealism, Marx posed the fulcrum of a logic which, as it evolved, was intended as a historical science, in the wake of the assertion of an epistemological model already experimented by Aristotle and, due to uniformity of viewpoint, by Marx himself. To this regard, Della Volpe underlined the relevant function of Galileo's criticism of scientific and academic Aristotelian deductive claims. Galileo had kept his distance from Baconian individualism and, therefore, his model imposed itself beyond possible redefinitions of Aristotelianism. It is in this sense that Della Volpe followed the genesis of the dissolution of scholastic logic in favour of the assumption of a Marxist methodological 'point of view' capable of establishing a materialist epistemology. It is the history of science which contains the different moments of the development of concepts and all further rationalizations which include, besides, the boundaries of theoretical ambits typical both of science and of logic. In any case, history contains the various moments of its claims to scientificity.

The economic-political method of strategic importance assumed when attempting to determine a strong bond between history and logic, something Marx insisted upon. The specific character of 'material production' justified recourse to the methodological definition and redetermination of possible differences between the political- economic approach and, when was said and done, to processes of 'concrete' production, to the point of appropriation of the 'concrete'. As scholars of Italian Marxism know, genetic investigation of the dialectics of concreteness, became, with Galvano Della Volpe the demonstration that the concrete-abstract-concrete circle might contain the political-economic method. Tracing the genesis of the process may lead to assertion of an epistemological status which a quest of this kind tends to admit, to the point of justifying its exact nature or better still, its exactness, that is, its historicity. The topic of the history of science emerges in Della Volpe as generated by Marx's thinking and his findings of a methodological order.

At the end of the day, the close association of the dialectics of the concrete with recognition of Marxism as a science, that is, with epistemology, was one of Della Volpe's the greatest investigational outcomes and was discussed by critics in the early 1960's, and developed historiographically as a simple philosophical debate during the decade that followed, until it vanished from the scene completely after that. Reception too of similar problems by social scientists and historians of the empirical social sciences was to remain rather circumscribed and weaken with as time passed. In the end, what held for Della Volpe and Marxism held also for historians of social science and meant that epistemological issues favouring science were abandoned upon the terrain of philosophical historiography.

Della Volpe's 'methodologism' and that of his 'school' matured not only in the presence of an antiquated idealistic-historical, anti-sociological-empirical culture of Crocian and Gentilian origin but also in a context where the relevance of epistemological themes was slow to be absorbed within the broader debate regarding the education of intellectuals and political scientists among Italian Marxists. This is a very important aspect which has not been reflected upon sufficiently. Backwardness in this area caused considerable delay in the formulation of possible working hypotheses following reception of the importance of these topics and their expansion to other domains of interest and study.

However, research on methodological issues and everything that stemmed from them continued to be relevant during the years when Della Volpe's 'pupils' had to come to terms with the criticism levelled against their teacher, in a different intellectual climate. Despite having been brought up on Della Volpe's philosophical texts, many of them found themselves embracing 'metaphysical issues' and following Heidegger, Jean Paul Sartre and Wittgenstein, dissolving the epistemological heritage of crucial issues in another sense.

In any case, Della Volpe, in his *Critica dell'ideologia contemporanea* expounded his investigation of the logical -historical method and therefore Marx's dialectics: a kind of verification of the 'theory', that is of the logic of *Das Kapital* by means of the 'history' of the famous *Theorien über den Mehrwert* [Theories of Surplus Value]. As Della Volpe himself held, these were theories capable of identifying the historical and dialectical antecedents of economic categories (like labour productivity) vital in a capitalistic society. Della Volpe added a number of clarifications concerning the method and the use of historically precise concepts like 'social relations of production', all this in opposition to the use of 'indeterminate' concepts formulated by classical economy (for example 'population', 'division of labour', 'exchange value' etc.).

The notion was mentioned by Marx in his 1857 Introduction to *Grundrisse der Kritik der politischen Ökonomie* [Fundamental Criticism of Political Economy] and taken up again by Della Volpe in his *Saggio sulla dialettica* in the appendix of *Libertà comunista*. The problem seemed to be of a methodological order and concerned with the possibility of arriving at a reproduction of the concrete starting from abstract definitions. Practically speaking the thinking is obliged to proceed according to the concrete-abstract-concrete circle and always start from the consequences of historical processes that regard the present or 'the concrete', study its causes and historical antecedents as if they were logical antecedents. This for Della Volpe meant examining the 'historical-present' by means of its logic-historical antecedents as *The Introduction* of 1857 indicated and Louis Althusser contested in his *Lire Le Capital* (*Reading the Capital*), emphasising Della Volpe's interpretation of Marx. It was the so-called 'determined abstractions' that constituted the object of Della Volpe's research on the dialectical method, which he identified with the logical-historical method. This way Della Volpe had already indicated in his work of the early 1950's called *Logica come scienza positiva* (later *Logica come scienza storica)* the epistemological importance of similar aspects as well as the depreciation the work of some authors like Althusser had been submitted to.

On the whole, Della Volpe revealed how the successful transformation of the principles of dialectics in favour of the assertion of a historicity acknowledged by historical dialectical reasoning itself, led to admitting its validity as a cognitive tool. The *ratio major* seemed suited to the development of a way of analysing historical and social facts, as materiality and concreteness; a dialectical analysis of contradictions, capable of linking historical materialism and historical-critical sociology stemming from its logical tools as it emerged as a human science. Above all, this thorny issue ought to have urged reflection on the part of those who, availing of Marxism, might have provided a solid landing-place for the development of a conceptual history of social sciences, following epistemological implications, while bringing dialectical examination and history close with a view to providing the theory of science with new tools.

In a certain sense, if Della Volpe decided to establish Marxism as a methodology of the human sciences, used to create a conceptual, epistemological brand of historiography, the gravest problem remaining seems to be the one which justifies the present essay, that is, that of outlining or not the consistency of certain proposals made by a handful of scholars of the empirical social sciences of the 1950's, where the latter seemed to stem from operations of a profound synthesis of Western thought flowing neither from the 'technicalities' of Italian sociologists nor from 'strictly Marxist' social scientists, but which led most often to syntony with philosophers dealing with issues arising elsewhere, for example, logical positivism.

The example of Giulio Preti, the most anti-metaphysical of Banfi's followers is very emblematic, because he is the representative of logical positivism who made some of the greatest contributions to this field with his studies on semantics and the problem of verification, and not only. In his main work, *Praxis ed empiricismo* (1957), he tried to bring logical empiricism and philosophy closer by discovering the links between the young Marx and John Dewey's pragmatism. By means of this operation, coupled with *Storia del pensiero scientifico*, this philosopher tried, in some way, to restore the array of problems that troubled Ludovido Geymonat when the latter questioned himself on the issue of science and history, gradually changing his 'point of view' regarding the kind of conventionalism derived from neo-positivism. Following this pathway, the historical contingencies signalled by Marxism were expected to acquire decisive importance. In reality, as far as Geymonat is concerned, we can say that it was mostly with the restoration of materialistic historicism that epistemological interests were able to assume a certain physiognomy in Italy. Furthermore, a work which bears testimony to the solidity of the choice he made in the 1960's, was *Scienza e realismo*, published as late as 1977.

In any case, many of the theoretical proposals that appeared on the Italian scene during the second half of the twentieth century and up until the late 1970's, seem to derive from a wide range of historical studies (not only those of Giulio Preti) regarding the most significant moments in the conceptual development of science and philosophy, and intended as syntheses capable of reaching Marxism through issues imposed not simply by neo-rationalism but also by Husserlian phenomenology (like the debate on the foundation of sciences by philosophers like Enzo Paci or Giuseppe Semerari) or neo-empiricism etc.

Moreover, at a time when philosophers like Bruno Widmar (see his *Introduzione alla filosofia della scienza* published in 1970 by Leonardi, Bologna) founded magazines and periodicals of cultural and scientific interest like "Il Pitagora" (1959) etc. that later became 'historic', and a number of new scholars were in the making, like Fabio Minazzi, born then (1955) and influenced by the epistemologist Geymonat and the historian of philosophy Mario Del Pra. This young philosopher studied Giulio Preti's historical epistemology and later the scholarly 'systemiser' of his works thanks to an assignment given him by the University of Milan's Philosophy Faculty in the 1980's (different but very fervid years) within the ambit of initiatives aimed at promoting, in keeping with the assimilation of a certain kind of thinking, a renewed debate on matters concerning the sciences and philosophy[1]. One should also remember some other disciples of Preti's like Alberto Peruzzi and Paolo Perrini, prolific authors of works published in the decades that followed, leading to renewed interest in logical empiricism and in projects favouring coordination between humanistic and scientific disciplines.

In the 1970s, many contributions bore testimony to the reception of topics/problems circulating within the ambit of logic positivism and leading to the creation of a number of synthetic conceptual prospects upon which various scholars and enthusiasts of the sciences (within the empirical social science or among the Marxists and the famous 'Marxisms') were to reflect, in order to direct the future generations to take a greater and deeper interest in the study of the historiography of science within their own fields. The historiography of scientific sociology matured when and where the majority of sociologists began to profess themselves 'historians of sociology' though they brought their thinking to a head only - in Italy during the second post-war period - after the spread of classical works and important translations, without any particular accounts of an epistemological nature from the ambits of positivism and neo-positivism now slowly declining. This means that the 'official' heirs of positivism too should be held responsible for some of the lacunae and delays affecting our epistemology. Despite the presence of Vailati and Enriques (unlike Benedetto Croce and Giovanni Gentile) the debate was able to direct itself, for most part, elsewhere, as far as an approach practicable today is concerned and in the light of the epistemological debate produced in forty years, so

[1] The conferences, symposia and seminars held by The Banfi Institute between March 1979 and the end of 1984, those on the age of positivism and socio-biology; May 1985 on the issue of reason in Antonio Banfi and his school published by Dino Formaggio in "Fenomenologis e scienza del uomo," n.3, April, 1986; between spring 1987 and November 1988, conferences on logic of its historical and conceptual development and the seminary *on Epistemology and Biology* with Vittorio Somenzi, Reggio Emilia, 13th-16th November.

different from that spread by 'official' Italian historians of philosophical events between the end of the nineteenth and the beginning of twentieth century, which often resembled that of those called 'historians of sociology' or 'historians of sociological thinking'.

Returning to the contributions of the 1970's, we should also recall the results of scrutiny of logical positivism that followed on the tardy diffusion of Italian translations of original 'material' produced according to a scientific conception of the world and due consideration of the relationships existing between epistemology and critical analyses of science, seen as an important acquisition for the work of future historians of science and not only for philosophers of science as such. Formulations ' of the time' of this kind bring us, for example, to Alberto Pasquinelli (a student of Felice Battaglia's and also of Carnap's in Chicago) and to a number of other scholars interested in examining relationships between scientific subjects and philosophy, like Evandro Agazzi and Vittorio Somenzi. For some of them, including Pasquinelli, the basic task was to 'take stock' of the over-thirty-year-old work of some epistemologists like Enriques, Geymonat and Preti, while asserting the validity of a quest for an 'Italian route' to epistemology and not just a simple debate on 'fundamentals' in traditional philosophical style.

Furthermore, the exponents of the *Wiener Kreis* had, in their day, indicated the importance of knowing how to keep 'philosophical' or 'fundamental' investigations within rational lines of development, submitting them to the scientific conception of the world, that is, to examinations or logical clarifications of concepts, propositions, scientific methods to the point of asserting a range of 'formal possibilities', detectable in logical and gnoseological analyses rooted in human experience such as to provide choices best suited to the needs of scientists.

Historically, the terrain of confrontation of the scientific conception of the world as coincident with the empirical sciences extended also to the world of production had to be accepted wherever modern empiricism entered into inevitable collision with materialistic instances, as intuited by Otto Neurath, who, due to his contorted personal situation, maintained his original label as 'Marxist' (acquired in Europe in the 1920's). Moreover, the relationships between neo-empiricism and Marxist formulations, physicalism and the rational organization of principles based on the social and economic sciences aroused little interest and prompted only a few Italian researchers to investigate Neurath's work. In the end what occurred was an examination not only of the development of the epistemological debate within the 'Viennese Circle' but also of the methodological consequences of its particular contribution to the empirical social sciences. Dating back to 1967 we find a rigorous study regarding Neurath and neo-positivism by Gianni Statera entitled *Logica Linguaggio e Sociologia* where the author identified the tasks of scholars of philosophy charged with taking the results and methods of the individual empirical sciences seriously, starting from the epistemological developments provided by the Viennese circle's *Erkenntnisstheorie*. Statera was the first in Italy to address the need to revisit Neurath's original methodological indications, linking them to Marx Weber and to

Dewey's methodology, where they might be re-valued and/or reappraised in the light of the epistemological results of logical Viennese empiricism, which many considered inconclusive as far as issues- problems posed by the methodology of the social sciences.

In actual fact, back then (1964-1966) Gianni Statera revealed an inclination to propagate the topics dealt with in debates surrounding the logical or linguistic methodology of the 'philosophy of science' as well as positions of some prominent philosophers regarding the kind of dialogue required regarding 'the ideal of science'. It is not surprising that his analysis of logical neo-positivism appeared in in "Documenti e ricerche," a journal featuring the work of pragmatist Charles S. Peirce and books by Abbagnano and Franco Ferrarotti concerning sociology, as well as studies by Carlo Augusto Viano, Paolo Rossi, Norberto Bobbio, Aldo Visalberghi and many others. In short, it was also Statera's intention to foster among the sociologists of the time a comprehensive scrutiny of works by Weber, Dewey, and Neurath, of course. It also served to support the notion that it would be very useful to address the issue of the logic of social research at a time when logical positivism was asserting itself thanks to the spread of translations and of several volumes and up until the publication in 1953 of *Il neopositivismo logico* by the theoretical philosopher Francesco Barone who left the legacy of diffusing a more wide-ranging revival of the problem of relations between philosophy and science to pupils of his like Marcello Pera.

One thing seems fairly certain and should be kept in mind. That is that at a certain point, the historians of scientific sociology did not seem to have a clear picture of the tendency of the time to write much of the historiography of science, from a Marxist and of the inclination for anti-metaphysical epistemology and philosophy to assert neo-empiricism, to continuously latch on to classical empiricism, thus denying empirical deduction to the logical-mathematical and 'formal' disciplines. This way, the historiography of science was able to count among its many European havens, all those disciplines which reduced the validity of knowledge to experience. The other perspective, that of the 'formal' disciplines, whose traits, as we know, are found in neo-positivism and in the doctrines of the English mathematician and philosopher Bertrand Russell, was to follow the destinies of British school of the philosophy of science. In Italy, for many there remained only the discussion of a sociology devoid of a true empiricist tradition or, criticism of Marxism, as in the case of Abbagnano in particular who, as mentioned above, practiced 'positive' existentialism as opposed to Martin Heidegger and Karl Jaspers's negative variety which evolved over the years as affirmation of a neo-illuminist humanism akin to Dewey's and aimed at considering human reason in its critical sense and the 'techniques' it was capable of providing to renew and improve the individual and collective existence of human beings. Antonio Aliotta's disciple formulated -not without attracting the interest of the social scientists of the time- his concept of 'existence' as 'structure' to the point of acknowledging the consistency and validity of the choices for human beings are responsible, without annulling their problematic and corrigible dimensions. In two essays on *Esistenzialismo positive,* Abbagnano linked the problematic dimension of scientific human knowledge, and, as a result, of scientific research, to existence. All told, this philosophy was intended, by historians too, as problematic knowledge capable of expressing ways human beings existed as finite entities, with existence as a mode of being, differing both from science and knowledge by excluding from its horizon the logical-contemplative dimension of humans professed by philosophies inspired ideally by both positivism and phenomenology. The horizon acknowledged to philosophy was that of 'possibilities' which humans, who '*are* not but *become* history', are continually called upon to choose between both as far as they themselves, others within their communities and to the 'things' of the world are concerned.

In any case, the main responsibility of existentialism was described (as we have tried to point out here) in a collection of writings published in 1948, as a sort of anti-dogmatic and anti-sceptic elucidation capable of close association with a number of research endeavours and historiographical clarifications. Italian admirers of Dewey of the second half of the century or aspiring social scientists of the time could not fail to notice that those highly-appreciated theories regarding the problematic nature of history corresponded likewise to the problem of existence itself, in all the multifarious possibilities it provided for consolidation and continual renewal. In reality, a broader treatment of those themes was presented in Turin, in 1942 by Abbagnano in an introductory volume on existentialism, used to introduce an exhibition featuring, among others, contributions by Uberto Scarpelli on Marxism, writings by Pietro Chiodi on Heidegger and those by another student of Banfi's, Enzo Paci, as well as books by authors like Armando Vedaldi, Luigi Pareyson, Giovanni Cairola, Guido Pedroli, David Invrea and Patrick Romanelli, the naturalist philosopher, who, at the Turin Conference on methodological studies held in 1952, dealt with Dewey's scientific ethics. Furthermore, in 1947, a volume promoted the year before by Turin's Cultural Union, was published bringing together two 'methodological' essays by Geymonat and physicist Enrico Persico as well as studies by Adriano Buzzati-Traverso, Piero Buzano and Eugenio Frola on the rational bases of science, *Fondamenti logici della scienza,* a volume where Abbagnano himself addressed the philosophical issue of science in relation to the experimental nature of the latter.

The not overly 'mechanical' alignment with this discourse, might have meant for certain kinds of Italian scholars of philosophy and for social scientists, interjection of Drewey's study of procedures and the logic, the veritable backbone of his investigations which, however, in the many variations of instrumentalist philosophy which landed in Europe from the early 1900's until after 1938, gradually became the bearer of new theory of knowledge and concepts, seen more as an attempt to make philosophical knowledge aware of the methodological revolution taking place in science. This is very important. John Dewey was always took care to refer to the historical development of science by taking an interest in logical epistemology and, from the beginning of instrumentalism- around the time when his articles were published in "Mind" (1886-1887)- asserting the practical character of knowledge as a process of 'idealization' of experience and, equally, as a realizational process of intelligence. In addition, around 1903, in one of the four chapters of *Studies in Logical Theory*, Dewey revealed his fondness for logic rather than for formal analyses of thought.

From a strictly naturalistic point of view, logic (not epistemology, therefore) should gradually recognize itself as an authentic social discipline and the social sciences as a branch of the natural sciences, with the term 'natural' used in the broadest sense possible for the time.

By addressing logical-epistemological issues, the young Dewey was establishing his theory of meaning and this marked an important stage in his journey towards instrumentalism and experimentalism, allowing those approaching those studies the possibility of recovering empiricism though as part of an idealistic conception associated with a problematic conception of reality and its rational foundations. It was after 1949 and the publication of the first Italian translation of *Logic, the Theory of inquiry (1938)* edited by Visalberghi, that many authors undertook in-depth investigation of Dewey and philosophers and historians of science began to discuss, more than the issue of 'instrumental reason' epistemological problems in terms of the existence of a scientific method. They began to consider, that is, concepts like procedural means of directing experimental operations. Therefore, conferences on the work of the philosopher Burlington were began to be organised in Italy (like that by Abbagnano with Mario Del Pra and others in 1952), which, on the one hand, attracted the criticism of older Marxists like Lucio Coletti, who complained of the unscrupulous use of naturalistic methods, while, on the other hand, there was the risk of falling into a transcendent critique of the 'instrumentality' of concepts, as in the case, for example, of relations between epistemology and pragmatic knowledge according to the logic of Dewey, discussed in 1951 by Andrea Vasa (see *Ricerca sul razionalismo della prassi).* Meanwhile, Dewey began to admit the existence of conceptual theory as a specific aspect found in the definition of social investigation and seemed to promote consideration of a logical 'point of view' within the historical development of various social disciplines as far as the difficulty of practicing the experimental method was concerned. An open and rational field was made available to sociological studies and the empirical social sciences. An area of interest where one might practice logic was tantamount to examining experimental thinking. Epistemology was expected to assert itself only after a confrontation with matter, with a view to establishing its applicability. This would have meant admitting the possibility of a rational study of its genesis, though only in relation to various ways of treating empirical material. This legacy was for historians of scientific subjects or historians of science.

Chapter 7

SOCIOLOGY AS A SCIENCE OF CONCEPTS

Reception of John Dewey by Italian researchers of the social sciences occurred simultaneously the diffusion of the idea whereby the contemporary conception of science might consider rationality as a process capable of encompassing the analysis of its relationships with various empirical conditions. It was Nicola Abbagnano, who in the 1950's, professed a similar consideration of rational activity while also recognizing, in part, the possibility of using it in sociology, in a context where the results of methodological studies regarding neo-rationalism and the specific character of the Marxist analysis of history had already spread culminating with the publication of Geymonat's *Saggi di filosofia neorazionalista*. A number of authors opposed Dewey's 'instrumental and scientific reason' with 'dialectic reason' destined to influence the 'official' direction of the philosophy of science up to and beyond 1960. In Italy, Dewey's instrumentalism seemed to follow a fate in some respects akin to that of the reception proffered to pragmatism by the historians of scientific subjects, but in opposition to the abstract theories professed by idealism, positivism and logical empiricism.

Obviously, the reception of Dewey by the history of science needs in-depth investigation in relation to the theoretical and conceptual definition of the empirical and social sciences, wherever a certain 'period-bound' interpretation of a conventional nature narrowed the field of research to the point of proposing one particular version. The same may be said of Otto Neurath if we recall that, apart from compiling the *International Encyclopedia of Unified Science,* with the help of Rudolph Carnap and Charles Morris, his anti-foundational vision of knowledge and the presence of some conventional instances, overcame the empiricist perspective within which Neurath was always relegated by historians of scientific subjects. It is the conceptual genesis of social and empirical sciences that actually seems to cause particular problems to the historians of science during second half of twentieth century, most of all in ambits where the theoretical problems regarded philosophical historiography more closely and delegated to represent scientific culture and the interests of scientists. From a certain angle, the issue of the conceptual definition of sociology considered as a science appeared quite relevant in that, *in itinere,* it had been deprived by the historians of science themselves of its essential 'period-bound' connotations that cannot necessarily be represented in deviations of philosophical historiography. Despite this, in the 1950's in Italy, a certain interest was shown in some of the peculiarities of the social sciences and treated in collections of essays which represented a veritable occasion for historians of scientific subjects to investigate matters in greater depth as in the case of a number of volumes published between 1949 and 1953 in *Saggi di critica delle scienze* that issued articles by Abbagnano and Geymonat but also by Norberto Bobbio, Cesare Codegone, Eugenio Frola, Prospero Nuvoli, Bruno De Finetti, Piero Buzano, all members of the Centro Studi Metodologici di Turin (Turin's Centre of Methodological Studies). The decade following the 1950's was when less generic versions of Dewey's productions were formulated, partially more in keeping with the real methodological problems of the social sciences, as Gianni Statera pointed out in 1968 (see *La conoscenza sociologica*) on the basis of closer examinations of the the original works of Aldo Visalberghi, Franco Lombardi and most of all, Carlo Tullio-Altan, regarding non-pedagogic reception of Dewey. Abbagnano had already expressed his views to this regard, followed by Franco Ferrarotti who, as Statera had anticipated, would express himself also concerning Weber. We ought not overlook the fact that Altan in two consecutive essays dedicated to instrumentalism and critical functionalism in cultural anthropology, published on the early numbers of "La critica sociologica" in 1967, addressed the issue of Dewey's epistemological assumptions, the matter Visalberghi defined in *esperienza e valutazione* (1958) as a 'transactional' vision, that is, a sort of dynamic and critical approach to the tasks of science. This needed to take place, according to Dewey, within a 'problematic situation' capable of accounting for some conditions where social action manifests itself, not only as a product of a 'relational' and 'interactional' situation. Altan's essay is important in that it brought the first confrontation with American sociology to fruition, by proposing analyses of value models, starting with their problematic existential basis and revealing the task of science, as organized cultural structures aimed at continuous development of their conceptual apparatus, and therefore, of ability to penetrate the reality. A similar idea (though not referring to the same 'model of knowledge') was taking shape among the few representatives of Italian neo-organicism, born within the field of statistics and related to reception of American sociological thinking. It aroused interest in operational developments of the Paretian functionalism, like those of Michele Marotta (1964) who studied for his degree with Antonio Aliotta and was the author of an interesting essay on Talcott Parsons published in "Rassegna Italiana di Sociologia," ["Italian Review of Sociology"], n.2 April-June 1963, addressing the topic of functional analysis of cultural anthropology (on this subject see "Rivista di Sociologia," year. II, n. 3).

Surprisingly, in Italy, Dewey's instrumentalism raised problems of a conceptual nature concerning the issue of scientific sociology at the very moment when the importance of availing of epistemology began to weaken and the problem of referring to the concrete problematic situations of a theory of human relationships and methodology began to emerge. Statera's contribution in this seems quite relevant. Moreover, it is sufficient to follow the debate surrounding sociological objectivity which, critics of scientific concepts opposed to so-called 'objectivistic gnoseology' developed into a given perspective and a whole series of methodological analyses quite unlike those practiced by neo-positivists, and, consequently, by historians of science. It was after the 1940's and 1950's that interest taken primarily in the study of natural scientific procedures was able to replace the themes-problems of empirical social sciences, within an ambit where a decline in neo-positivism corresponded to an assertion of the philosophy of science as well as to an relinquishment of epistemological problems by researchers dedicated to defining the peculiarities of sociological analysis as indicated by Dewey. The birth of sociology as 'conceptually oriented empirical analysis', as Ferrarotti wrote in 1968 in his *Trattato di Sociologia*, did not register among its devotees interests capable of producing an epistemological 'point of view' in the field of the historiography of social sciences, although a debate about post-Weberian writings led one to think otherwise.

On a different front, research was carried out concerning objectivity implying an approach due to Italy's slow reception of Weber and of his ideas regarding historiography, which, all told, rather than history, concerned culture. Many of these topics were referred to by Ferrarotti in his book *Max Weber e il destino della ragione*, 1965, which dealt with Italy's liquidation of historicism by Felice Balbo in consideration of philosophical knowledge being existential, that is historical material. The book was introduced by an interesting essay by Ferrarotti regarding Weber's 'conceptual system' which appeared in "Quaderni di sociologia' the previous year and republished again in this book.

This in-depth analysis of Weber helped the idea that empirical analysis was always influenced by a 'point of view' to circulate among the few sociologists of the time. This idea which sustained that a point of view evolved conceptually appeared in Weber's well-known essay *Die 'Objektivität' sozialwissenschaftlicher und sozialpolitischer Erkenntnis* [The 'cognitive objectivity' of social science and social politics], 1904. The conceptual dimension, according to Weber, was of heuristic value so that the social-historical sciences (already defined by Heinrich Rickert) should be able to rationalize the empirical world, beyond the normative ambit of science which should succeed in freeing itself from values while bestowing direction on 'value relationships' or 'relationships with values' (Wertbeziehung) useful for the achievement of logical operations. It is important to understand that the heuristic function of given 'operations' was kept at due distance by idealist philosophers like Benedetto Croce, who, from the time of the controversy over *hypothesism* feared their being converted into logic. From the time of *Logica come scienza del concetto puro* a purely theoretical vision of scientific activity was specified which it was sustained should have oblige scholars of the social sciences to pose the problem, also in terms of the reception-processing of epistemological concepts, seriously, something which did not happen, if not in a piecemeal manner and only among a handful of authors. Croce's treatment of conceptual activity as opposed to that of the scientist concealed an intuitionistic methodology which Weber, at the time, considered 'naïve' from an epistemological 'point of view', and not only, if we take the empirical social sciences into account but also if we refer to the impressive results of the discussions between colleagues of Croce's and Gentile's concerning logic. The truth is that Croce's epistemology was non-existent also in the face of German historicism, as the depth of the studies carried out by Pietro Rossi show. In 1956, Rossi traced the events back to Italian neo-idealist times, existentialism, Marxism and neo-Illuminism and, in any case, to a period when sociology seemed to flounder around presumed conceptual transformations of reality where the sciences of culture or social sciences took shape and contributed to taking historical-empirical research as related to methodological theoretical and abstract constructs, into due account.

This was a view which tended to reconcile the world of concepts and that of reality, admitting that for Weber it was the conceptual dimension of science that needed rational ordering, following some choices rather than others, which meant, when all came to all, starting from the researcher. It was along this route that sociology established itself, devising ideal concepts of instrumental and historical knowledge, while remaining in a certain sense bound by transformations of historical methodology to such an extent that, in the end, the Italian debate on philosophical historiography was practically dominated by them. Furthermore, this debate managed to have an impact for several decades upon the education of many of its historians and practitioners. The inclusion of studies of the human as a historical world, and, as a result, the reception of the problem of the establishment of the objective validity of the historical and social sciences, became an important fact for Italian philosophical culture during the second half of the 1960's.

At the time it was possible to demonstrate further the scale of possible outcomes of historical research into the empirical social sciences. It suffices to mention a series of papers published in the 'classical philosophy' series edited by Franco Lombardi, including *Storicismo contemporaneo* by Rossi, meant for schools (1968), and containing a broader-ranging interpretational key to 'period-related' absorption of historicism: from a critique of historical reason (W. Dilthey), to relationships with neo-criticism (W. Windelbandt, H. Rickert, Georg Simmel), to the methodology of the social sciences (M. Weber) and relativism (O. Splengler, E. Troeltsch, F. Meinecke) to historical materialism (G. Lukacs, K. Mannheim), Croce's idealism, pragmatist naturalism (J. Dewey. MR Cohen, JH Randall) and neo-positivism (O. Neurath, C. Hempel, K. Popper, E. Nagel). Moreover, Rossi's text provided the definition of the relationship between considerations of historiography and pragmatic naturalism advanced by Dewey beginning with his

Experience and Nature, 1925, where experience too was to be treated as history, up to *Logic, the Theory of the inquiry* where, however, we face the problem of a selective choice of historiographical principles in relation to logical postulates from which assessments should derive. Rossi himself pointed out that American naturalism had, during that period of the century, begun to consider problems of historical knowledge in a manner very different from that of the European historicist movement and idealism, unwittingly, pointing out that one of the points of contact between the social and the natural sciences was to be sought by scholars in the conceptual determination of their aims, so that sociology now had the task of clarifying its own.

It was regarding this precise point that Ferrarotti in an essay of his called *La sociologia come partecipazione* which appeared in 1960, (published as a book in Turin the following year) renewsing Rossi's *Introduction* to the Italian translation of Weber's *Methodology of the social sciences* (1958), discussed the purposes and methods of processing as well as distinctions between historical knowledge and natural science which according to Weber lay in logic's tendency towards individuality. Ferrarotti's aim, already aired in 1952 in works like *Introduzione alla sociologia come scienza,* was to reveal the usefulness of 'understanding' as a process, and of dealing with the possibility of fostering purpose-related sociological research by recognizing the validity of the relationships existing between the researcher and the object of research and the correspondence between the various concepts developed and participation in the research process itself as a problematic situation. The fact is that during the decades that followed, a significant part of sociological production was directed towards fulfilment of empirical-research tasks established according to given assumptions, some of a 'humanistic' nature, and requiring genuine political, critical, civil, and intellectual commitment, as in the cases of the entire period of the so-called 1977 struggles: from the prodromal texts of Ferrarotti 'alternative sociology' (1972) to the many essays published in "La Critica Sociologica" by 'dedicated people' like Corrado Antiochia and Marcello Lelli, the author of *La sociologis degli altri* (1975) confined-a few years later –to the Department of Economics, Institutions and Society at the University of Sassari.

The critical and humanistic vocation of sociology was in keeping with scholars' growing interest in the constituent aims of the scientific disciplines to the point that that they focused on issues of methodology. The later led authors like Gianni Statera to seek continuity between Ferrarotti's proposals of the 1950's and his instrumentalist-critical approach to the social sciences (see *La conoscenza sociologica*) after having partially re-discussed the validity or otherwise of the existentialist instances and of a new form of 'comprehensive sociology'. In some way, it was necessary to return to Weber and his proposals, as Rossi suggested in his essay *Oggettività scientifica e previsione di valore* that followed a study congress held in 1964 and was published the same year in "Quaderni di Sociologia," [Sociology notebooks], a magazine which for over a decade hosted several articles which made an important contribution to the contemporary scientific and philosophical debate. At a time when rational interest was being shown in the fundaments of the development of historical knowledge as well as in the relativity of some of the criteria affecting the choices historical knowledge might avail of, the supremacy of historical studies seemed to encourage an interest in the empirical social sciences but only after crediting itself with the methodological analysis of the problem of objectivity conducted by Weber.

Moreover, to this regard, and returning to the issue addressed above, historians studying the sociology 'of the period' should not overlook the criticism against authors who subscribed to the process of assimilation between different currents of scientific thought and certain reductive positions of the advocates of idealism. One such critique was that which Rossi levelled against Carlo Antoni's approach in *Dallo storicismo alla sociologia,* 1939, where the latter held that contemporary German historicism was a degeneration of the Romantic conception of history in the direction of sociology, sustained that, in the long run, Croce's philosophy was its authentic offshoot and justified the exponents of German historicism as his precursors. And the examples might continue to the point of the close scrutiny to which some scholars like Franco Lombardi submitted the decline of a number of absolute categories professed by historicism (see *Dopo lo storicismo*, 1955) or sociologists and historians of the discipline, who might have been damaged if topics related to the conceptual history of historical knowledge seeped into the imminent debate concerning philosophical historiography which, in Italy, persisted well into the late 1960's, allowing a number of enthusiasts and lesser researchers of sociology to enter the ranks of the 'historians of ideas'.

In short, it was philosophical historiography and some of its most prestigious representatives, prompted more by *Methodonstreit* than by the embers of idealism, that handled the task of discussing whether the epistemological productions of authors like Weber belonged or not to the historiography of scientific subjects, after having known, with Geymonat, that grand period of development following the discovery of the theoretical autonomy and cognitive value of scientific enterprise and having reached the middle of another period where studies regarding the methodology of historiography, based on pragmatism, were coming to light. Some instances were Paolo Rossi's *"Storia e Filosofia,"* 1969, a number of anthologies dealing with the development of scientific thought from Bacon to Galilei and scientific revolutions from Copernicus to Newton, or works by Antonio Santucci on the thinking of David Hume and William James.

The historians of science too, except from some secular Marxist historians, seem to have entered the debate regarding approaches to sociology and certain topics emerging from its conceptual-historical claims, from a more 'official' philosophical slant, like that represented by academia and its future division into historiographical 'schools'. The fact of not adhering, as sociologists, to this or that historiographical position had a significant impact on a whole generation of researchers; it meant their having to relinquish a historical 'point of view' of the empirical social sciences. Many representatives of Italian scientific and philosophical culture, called upon more and more frequently to express their ideas concerning sociology, were gradually concerned with revealing their subordination to this order of ideas, while demands for greater autonomy of scholarly judgement regarding philosophy and history and/or the history of philosophy were incontrovertible. It is no coincidence that it was precisely from the 1950's on, that the necessity of referring to scientific subjects so as to avoid losing sight of philosophical culture itself was felt as well as the exigency of associating it once more with a historiography free from the monopoly of idealistic dilettantism. It was at the end of the 1950's, in fact, that copious compendia of sociological studies were published in Italy (for example by Camillo Pellizzi in 1956) whose sole merit was that of describing the opposition between defence of sciences and empirical subjects and exclusive acknowledgement of the 'cognitive functions' assigned by Croce and his followers to philosophy and its historiography. Many cultural 'ambitions' of the historians of sociological thought were allowed to develop at the time, as they sought to free themselves from the dominant culture by anchoring their in-depth studies and analyses of the 'new materials' produced not only by the history of the human-social sciences but also by the debate surrounding them at the time. At that precise moment the history of sociology differentiated itself from history, from the philosophy of art and, in general, from all non-scientific forms of cultural development, qualifying itself according to verifiable and reproducible concepts based on research and scientific work, not simply on techniques of research. A historian's scientific work might also take the form of his/her efforts to reflect, at theoretical and methodological level, on the relationships between theoretical and empirical research (as in the case of the first classifications regarding theoretical and empirical research drawn up by Filippo Barbano and published in the "Il Politico" journal, in 1954). In part, the debate on 'infirm science' caused a tendency in some scholars and scientists to produce a new historiographical culture, though in the presence of philosophy.

From a strictly chronological point of view, the Italian sociological experience is linked to a succession of scholars who pursued the specific interests of the discipline of the period. These are authors who are usually included in the annals of traditional and / or academic historiography and who recall given events regarding the development of sociology in Italy, like Samuele Sabino Acquaviva (who died in Padua in 2015), Francesco Alberoni, Luigi Amoroso, Achille Ardigò, Guido Baglioni, Filippo Barbano, Norberto Bobbio (who died in 2004), Marcello Boldrini, Giorgio Braga, Antonio Carbonaro (who died in Florence in 1998), Vittorio Castellano (deceased in 1997), Giuseppe De Meo, Giuseppe De Rita, Franco Ferrarotti, Luciano Gallino, Gino Germani, Corrado Gini, Alberto Izzo Livio Livi (Rome 1981, Florence 1969), Michele Marotta, Angelo Pagani, Camillo Pellizzi, Pietro Rossi, Manlio Rossi Doria, Renato Treves and many others whose sociological experiences date back to the mid-twentieth century. The interests of these scholars regarded the conceptual level with a marked inclination towards general sociological theory and its application in areas like statistics and methodology. Many conceptual interests were generated by analyses of sociological thought. Some historical and sociological derivations were, for example, the urban and environmental sociology guidelines used to examine the city in relation to modernity, as presented in works by scholars like Giandomenico Amendola, Gian Franco Elias, Franco Martinelli, Giuseppe Barbero, Fulvio Beato (who died in 2014) and other specialists in the sociology of territories and states which paved the way towards current studies concerning migration.

One the first initiatives promoted by Italian academic circles with a view to discuss similar relations from different perspectives and for heterogeneous purposes, was the Conference held in Bologna in April 1954, where Catholic, secular and Marxist authors confronted each other. References to the event were published by "Il Mulino" that same year in a volume entitled *Filosofis e sociologia Sociology* while an article regarding it written by Pietro Rossi appeared in the "Quaderni di Sociologia" in the autumn of 1954.

The conference was attended by Felice Battaglia, Abbagnano, Pietro Chiodi, Enzo Paci, Renato Treves, Nicola Matteucci, Luigi Pedrazzi, Antonio Santucci, Cesare Luporini, Renato Zangheri. The fundamental topic consisted in the relationships existing between the sociological investigation and philosophical research, where, surrounding the general issue of links between 'Philosophy and Sociology' different visions of scientific autonomy were presented and interpreted from the 'point of view' of phenomenology and of the 'philosophy of values', to the point of arriving at a number of logical and empirical conceptions of sociology and considering philosophical matters and philosophy in general as a truly scientific methodology. Furthermore, witness had already been borne to relationships between phenomenology and sociology in a new collection of essays published in 1951 by the "Archivio di Filosofia," [Archives of philosophy], the journal of the Institute of Philosophical Studies of the University of Padua, edited by Enrico Castelli. We also need to recall that by way of a somewhat more indirect consequence of the debate a number of in-depth studies which treated the issue of phenomenology appeared in the 1950's and, in the following decade, as testified by the *Fondamenti della logica di Husserl*, dated 1965, a work where the author Franco Voltaggio questioned himself for the first time concerning the conceptual evolution of analyses of the fundamentals of phenomenology, following the possibility of discussing Husserl's epistemology, and, in any case, certain 'attitudes of epistemology to avail of as 'criteria' by which to interpret phenomenology; this, beyond the relationships existing between the latter and Marxism, a topic widespread in Italy beginning with the late 1950's. The investigation, outlined by Voltaggio which admitted historicism as a fundamental perspective capable of embracing all possible and imaginable sciences, including phenomenology, as requested by Italian culture philosophical on the whole and capable of foregrounding its need to determine, for example, an epistemological interpretation of Husserl's 'logic' etc.

However, the task of accounting for certain links and questions was taken up again, on the one hand, by some scholars who had always manifested the intention of dealing with philosophy in order to take the greatest advantage from it to discover trends bound to the methodological debate of the time and, on the other hand, by those who had already undertaken the route of the historiography of science or of epistemological historiography. We have already dealt with the former group of scholars bin our treatment of historiography. With reference to the second we should not forget that it was Paolo Rossi who commented on Pietro Rossi's treatment of contemporary sociology and on some of the positions presented during the December 1953 Conference of Milan (see "Rivista critica di storia della filosofia," IX), where the 'point of view' of the philosophers of science aired the need to determine the specific character of sociology in consideration of some of Weber's 'attitudes' while, the question of criticism of philosophical concepts stemming from rear-guard phases of scientific research was also posed. This meeting followed another held in Turin in June of the same year, accounts of which are to be found in a number of journals like "Aut Aut," "Rivista di Filosofia," "Il pensiero critic." Besides, we need to mention other conferences which assume significance within the ambit of a more modern revaluation of the issues addressed and which provided a somewhat spurious 'decennial' definition of relationships between sociology and philosophy. By way of example we should recall The Gallarate Conference of the 3rd and 4th of November 1964, better attended by sociologists than previous meetings and which witnessed the contributions of Filippo Barbano regarding methodological perspectives, of Giorgio Braga on relationships between the philosophy of science and sociology, of Franco Demarchi concerning relationships between sociology and philosophy and, finally, Ferrarotti, on sociological objectivity. The question which informed that Conference was of discussing the admission or not of a purely scientific sociology and consequent claims of the existence or not of a philosophical kind of sociology besides a scientific sociology.

Meanwhile, the 'period-bound' discussion of sociology's conceptual profile as a science continued to instate novel areas of epistemological reflection within the particular context of the philosophical education of historians of scientific subjects, who redefined the relations between science and philosophy and seemed, in the end, to enter fully into the debate surrounding the origins and the evolution of the empirical social sciences, summing up what has happened over the previous two centuries. This seemed to apply to several exponents of philosophical historiography committed not only to re-interpreting French Illumination. This was the case with Pietro Rossi who carried out a number of studies regarding Jean-Jacques Rousseau, relationships between positivism and the industrial society. Studies focusing on evolutionism from a historical-critical 'point of view' were conducted by Vittorio Somenzi, while Italian Illuminism, idealism and even Hegelism, in the light of its specific developments, were addressed by Mario Rossi to his analyses of Francesco De Sanctis, Salvatore Tommasi and Labriola. The historians of science too were enticed to rewrite their own philosophical historiography despite the fact that the range of interests had changed and deepened considerably, while sociology began to reveal its theoretical-conceptual apparatus as well as its technical know-how. As regards the relationship between historiography and sociology *stricto sensu*, Abbagnano devoted himself to the issue in his *Problemi di Sociologia,* 1959, taken up again later by Franco Ferrarotti to discuss some significant matters. The book examined relationships between sociology, charged with outlining aspects of human life intended as 'relative uniformity' and 'possible repetitions', not singly as in the case of historical knowledge and or historiographical research. For Abbagnano: "the object of historiography is that single event to reconstruct in all its individuality unlike the object of sociology which involves a replicable behaviour to acknowledge thanks to its repeatable nature. It was a question, therefore, "not simply of a dualism of the reality or of cognitive forms, but of two possible objects of research, each one claiming the use of particular instruments and appropriate languages."

One actually notices a different critical-philosophical view of historicism, on the part of philosophers, which came to maturity in Italy giving rise to a rebirth of interest on the part of researchers concerning the matter of the historical value of knowledge, as recommended at the time of Karl Mannheim, influenced by Marxism and by Weber's sociology which, meanwhile, came under the scrutiny of sociologists of religion or that of analysts of the origins of capitalism (in this case, the work of Alessandro Cavalli), etc, for example. These were topics that involved the sociology of knowledge or 'the sociology of knowing' as philosophical historiographers called it initially. In brief, for sociologists and for historians of sociology too, this meant reflecting, to some degree, on the epistemological consequences of historicism in order to arrive, on the one hand, at a structural analysis of epistemology, on the other, at an investigation of social conditionings of epistemological production itself. The epistemological route pursued by Mannheim was followed by a certain kind of analyst and the conceptual-theoretical production of sociology related to its nature and true potential. Franco Farrarotti recalls in Alberto Izzo's ground-breaking text, *Sociologia della conoscenza*, 1966, how considerable interest might stem starting with the crisis due to the discovery of social conditioning of thought as a function of a revaluation of critical thinking.

In 1970, Izzo edited an anthology of writings dedicated to a thorough examination of these themes, published in Turin as a philosophical series directed by Franco Lombardi. The volume brought to light issues concerning approaches and philosophical-criticism, containing, furthermore, an 'Appendix' in a text on 'cultural relativism' by Lombardi which underlined above all the fact that the majority of the 'classics' of the sociology of knowledge were not only aware of the problem of relationships between society and thought, but also of the epistemological consequences of the debate surrounding the assertion of the historicity of knowledge among individuals as well as human and social groups. It is in the work of Karl Mannheim, who died in 1940, that many concepts, subjected to historical becoming, were widely availed of and discussed, especially among early 'Marxist' historians of sociological thought (including some Italians) and many intellectuals who, from the beginning of the 1950's, read his most famous work, *Ideology and Utopia,* which denounced the deformations surrounding those very same concepts. In this specific case, the idea was that any theoretical-conceptual theory whatsoever or thinking itself, might be considered, in turn, in relation to its social-theoretical origins and a certain set of existential bases, recognition of which by a considerable number of scholars, began to distinguish Mannheim's approach from that of Marx, assigning a predominant role to sociological analysis while Dilthey's interpretation of historicism and the formative process of Max Scheler's phenomenological metaphysics were being re-evaluated. That text followed the stages of the assertion of *Wissenssoziologie* and of some of the theoretical prospects practiced initially by phenomenology, idealism and Marxism, with reference to the work of Emile Durkheim, George H. Mead, Robert King Merton regarding the sociology of knowledge and contemporary analyses by Kurt H. Wolff and Kurt Lenk, an exponent of the 'critical theory' and German sociology of the time. The decades in question bestowed concreteness not only on its human facet, on the critical aspect of sociology as a science and on the conceptual contours (also in the light of a European historicity and a Marxist outlook, but also revealed a certain 'openness' towards humanistic perspectives on the part of number of experts like Ferraroti in Italy - who continued to tussle academically with the abstractions of philosophy and philosophical historiography as well as with the formalism of juridical subjects, while others like Renato Treves, Angelo Pagani, Alessandro Pizzorno established themselves within the Higher Institute of Social Sciences in Milan, created in 1962, at Trento thanks to the original intention of Giorgio Braga, who died in Milan in 1983, to promote a statistical and mathematical type of sociology (directed by Mario Volpato) or in ambits surrounding the cultural interest of personalities like Franco Leonardi (*Elementi di Sociologia*, 1961) or the above-mentioned Gianni Statera, until the season of the translations of various 'classics' and the introduction into Italy of North-American sociology like that of Robert King Merton thanks to historians of sociological thinking like Filippo Barbano, professor at Trento, from 1964 to 1969, later, at the University of Turin.

Chapter 8

EPISTEMOLOGY AND SOCIOLOGICAL METHOD

At the end of the 1960's, it was possible to propose *Wissensoziologie* to new devotees of scientific historiographical subjects as a modern form of epistemology though a number of those in favour of the latter in no way reduced their claim to locate sociology upon the ontological plane of thought. This led to a series of discussions. The strategic moment harked back, however, to the results of the debate on historicism, and to those stemming from Marxism as well as to the consequences of some of the issues addressed by Weber and filtered through philosophical historicity and the resolution of which were destined to follow a different pathway from that undertaken by philosophers of science who, in the midst of the dissolution of neo-positivism and in the presence of various currents of ideas, seemed, all told, to launch a whole new range of interests like hermeneutics or other issues inherent to a re-proposition of logic as a philosophical methodology applicable to the social sciences etc. The debate on objectivity and/or 'cognitive' objectivity had left many epistemological questions regarding the empirical social sciences unsolved at a time when the conceptual dimension of sociology began to assert asserting and sustain the importance of uniformity of scientific analysis of the social, that is, of the outcome of practical research in the area and not simply of theoretical investigation. As we have already pointed out, especially with regard to Dewey, the relationship between the social and natural sciences was studied in the light of the decline of certain forms of objectivity that had characterized some of the radical phases of the historical development of scientific thought during the first half of the twentieth century, up to assertion of the relevance of research methods and techniques, as in the case of American operationism –beginning with sociologist George Alexander Lundberg- which, in less 'scientific' terms came to Italy with a number of translations like that by Vittorio Somenzi of Bridgman's *The Logic of Modern Physics* (1927) published in 1952 after promoting the Italian Centre for the Methodology and Analysis of Language with Silvio Ceccato in Milan in 1945 and engaged in the debate concerning operationist perspectives with journals like “Analisi,” and “Sigma” followed by “Methods” (1949 – 1964). In other sectors, but with reference to precise interests, it needs to be said that for the decade following 1944-1945, many scholars, in particular, some philosophers of science, were interested acknowledging the links between the use of methods, some instances of statistical-probabilistic research and scientific practice. Many others, sociologists and historians of sociology especially, continued to deal with historicism and its possible conceptual reformulation in the face of Marxism and dialectics, pursuing, on the one hand, the assertion of a ‘critical theory’ and on the other practicing a rejection of forms of indoctrination in the name of previsions of the course of history. The most renowned example in this sense is Karl Raimund Popper and his book *The Poverty of Historicism,* translated by Ferdinando di Fenizio in 1955 at the behest of Franco Ferrarotti and published in the Italian journal “L’industria.” In all cases, especially that of conceptual developments within the historical-social sciences, the impacted considerably on the destiny of *Wissenssoziologie*, in such a way that some methodological insights and a certain criticism survived or influenced research on epistemological-sociological interests in other fields, associated also with pragmatism and neo-positivism. This did not reduce or impede investigation of Mannheim’s scepticism or Dilthey’s and Weber’s historicism, in the end, by jurists, philosophers of law and future political scientists, who transited from the solid bank of historicism to the quicksand of the political and social sciences (one example is represented from Domenico Corradini, a pupil of Kurt Lenk’s.). As stated previously, from another angle, studies were undertaken to look into the methods used to examine issues of probability and the possibility of establishing a new philosophy of probability as far as experimental results were concerned. The studies in question concerned the origin of the modern objectivist approach and practices of the ‘scientific community’ as related, to methodological choices themselves. We find discussions of these questions in a number of essays published after 1940 by Ludovico Geymonat, who addressed a number of methodological considerations regarding the concept of probability that led him to a ‘period-bound’ confrontation with the statistics of Corrado Gini (first president of the Italian Statistics Office- ISTAT) and the latter’s assertions of its links with the other sciences. These issues had already been aired at the Congress on methodological studies held in Turin between the 17th and the 20th of December 1952. Furthermore, in 1955, Geymonat went more thoroughly into Carnap’s analyses of probability and the inductive method published during the same period, in an attempt to give rise to a discussion in which no active part seems to have been taken by methodologists interested in rationally providing the empirical social sciences with results found in neo-positivist logic, especially where Carnap posited inductive logic as a general theory, or as a study of methods aimed at determining hypotheses through proof. Furthermore, the problem did not merely involve testing the techniques by which to detect them but also existing logical relations between hypotheses and proof to be subjected, analogously, to deductive and inductive reasoning. It is a fact that not even at a time when scholarly attention to methods was spreading, many of the authors involved did not show particular interest in the contributions of the philosophy of science to the debate on the methodology of empirical social sciences. One should not forget, however, that it was as a philosopher of science that Bridgman sustained operationism, the theory whereby scientific terms and the concepts that have not been defined, at least not directly, on the basis of concrete operations performed with instruments, have no meaning. Bridgman's ideas were masterfully illustrated by the philosopher of science, Bruno Cermignani, a student of Somenzi's, in a collection of writings published in *La critica operazionale della scienza,* Turin, 1969. Moreover, it needs to be added that these methodological and gnoseological issues met with a certain hostile resistance on the part of the Italian academic and cultural world, as the Scuola Operative Italiana group set up by the aforementioned Ceccato, Somenzi and Giuseppe Vaccarino, shows. This group extended Bridgman's narrow views of the physical sciences to favour new perspectives like those of Hugo Dingler and Jean Piaget, to the point of developing an evolutionistic epistemology, developed by Somenzi as well as his future interest in cybernetics and neuro- psychology (see *La filosofia degli autonomi* 1965).

Referring again to the same Mantuan author, we need to recall that some of his epistemological studies led to analyses of the development of topics discussed by authors like Norman Robert Campbell, Konrad Lorentz and Popper. Other issues he examined appeared in *La scienza nel suo svilupo storico,* published in 1960. This publication held that research into the history based on the logical development of ideas related to the fact that the history of science itself, ought to provide material for studies regarding the cognitive value of the sciences and therefore for all studies regarding the philosophy of science, methodology and epistemology. The author emphasised that investigations of methodological questions, linked also to those of a philosophical nature following principles of pure logic or of the theory of knowledge, should be practiced, above all, by mathematicians, in physics, biology and in psychology.

This probably aimed at reconstructing relationships between the history of science and philosophy as well as the conceptual reasons underscoring evolutionist epistemology in a moment of particular significance in the development of social, empirical sciences in Italy, alongside a number of 'restricted' issues and views of a methodological nature rather than of historiography. Sociologists had long felt that these, together with other basic themes of the discipline, indicated the need to clarify its shortcomings from within as well as attempts made to indicate the purpose or specific method of the social sciences, in relation to the other scientific disciplines.

The late 1960's were also the years of the crisis of sociology from the 'point of view' of its methodological perspectives which were rethought a decade later during meetings and/or the exchanges of opinions between young authors and scholars of 'first generation' who once they became university chairs intended urging the new generations to reflect on sociology's independence from the other social sciences as well as on the objectivity of its techniques, its social research and the role of the sociologist. It was with these issues in mind that Pietro Rossi and Norberto Bobbio's Centre of Methodological Studies which organized an important Convention on *La crisi del metodo sociologico* [The crisis of the sociological method] in Turin from the 7th to the 9th May 1971. The opening papers were presented by Vittorio Capecchi of University of Bolognia, Guido Martinotti of the University of Milan, Alessandro Cavalli of the University of Pavia while the meeting closed with speeches delivered by other eminent Italian sociologists and social scientists, like Luciano Gallino of the University of Turin (who died in 2016) and Alessandro Pizzorno of the University of Urbino. Among those who presented papers were Francesco Alberoni of the Trent Institute of Social Sciences, the methodologist Laura Balbo Ceccarelli of the University of Milan, Marzio Barbagli of the *Carlo Cattaneo* institute, Bologna, Luciano Cavalli of the University of Florence, the methodologist Antonio Delillo of the Higher Institute, Trent, Massimo Paci of the University of Urbino, Angelo Pagani of the University of Milan (died in Milan in 1972), Antonio Santucci of the University of Bologna, the law philosopher Renato Treves of the University of Milan and the director of Rome's *Luigi Sturzo* Institute, Ignazio Ughi. The Convention featured a debate regarding the speakers' papers. Among those who took part in the discussion we hasten to recall the methodologist Paolo Ammassari, Antonio Carbonaro of the University of Florence, the economist Bruno Contini, the Napolitan epistemologist Antonio Drago, the economic sociologist Alberto Martinelli of the University of Trent, Michele Salvati of the University of Modena, the law philosopher Uberto Scarpelli and the Florentine anthropologist Tullio Seppilli. In Ancona, in 1962, the Italian Association of Social Sciences held another convention in rather a different atmosphere due to the topics dealt with and the intensity of the discussion. The focal theme of the Ancona convention was relationships between sociologists and centres of power in Italy, at a time when sociology was undergoing transformation and finding a place within academia and the institutions, something which aroused long-awaited expectations, without, however lessening the strong need for sociological integration felt among economists, political scientists, jurists, law historians and philosophers etc.

Discussions of the crisis of sociology during the second half of the 1960's helped to lead to the debate on issues like defining the criteria of science or defining empirical research and what conditioned it, also with regard to the call, at the time, for interdisciplinary integration within the social sciences. No less important was the need to find a new and unexplored perspective capable of keeping in mind the relationship between criticism of epistemology and social-research methods at the very moment when the crisis was experiencing a redefinition of 'method' or 'methods', etc. One important problem for that period seems to have been that of bringing the academic debate on objects and methods to generate an outline of methodological studies based on the 'point of view' of a possible renewal of the bases of a scientific sociological discipline related to the other disciplines within the context of the history of social, empirical science in Italy. In this sense, the future contribution of Italian historians of scientific sociology towards the development of an epistemological 'point of view' with regard to the empirical social sciences turned out to be practically non-existent because it lacked developed empirical procedures while it remained dependent –as regards specific characters of the problem and 'internal' to the debate- on the historiography of science which was developing in countries of Anglo-Saxon tradition which trickled into Italy thanks to a handful of authors and the usual philosophers of science. And not only. What happened did not seem to be, as yet, of strategic interest for the definition of a more solid outlook concerning the genesis of conceptual materials in science. This delayed response by the European historians of science was inevitable despite the fact that in the 1970's some sociologists had already been aware of it, like Luciano Gallino, in Turin, who sought to draw attention to particular lacunae caused by the fact that Italian social scientists had failed to take on board methodological studies capable of combining 'dialectics' with formal logic, as in the case of traditional logical studies carried out in countries, like Poland, heirs to Marxist scientific culture. Luciano Gallino proposed examining studies concerning the formalization of 'dialectics' in an attempt at rooting investigations of its processibility and its reference objects in logic and epistemology, that is, in formal logic, so as to not to create a division between the various fields of investigation and contribute, instead, to enlarging the analytical outlooks of sociology's on-going, theoretical and methodological crisis. Contrariwise, authors like Capecchi, especially during the early stages of his mathematic-statistical training and his keen interest in the American 'quantophrenics' and in Paul Felix Lazarsfeld (up until the end of 1967 and his well-known 'second thoughts' regarding techniques, etc.) who might be considered the official representatives of a theory, largely diffused and strongly opposed within the world of academia. For this reason if a veritable crisis of sociology had cropped up, a crisis of formulation would have occurred in this very field. During this period a number of case and procedural studies appeared, like those published in Padua, in a series of methodological volumes containing works regarding sociological research and analysis issued in America during the 1950's and 1960's and translated into Italian in the 1970's. Among others we wish to mention, Paul F. Lazarsfeld's introduction to Herbert Hyman or that of Capecchi to Hubert M. Blalock, texts edited by Guido Martinotti, Laura Balbo and others, containing references to the techniques and various tools availed of in important Italian investigations and empirical research projects. The years when Vittorio Capecchi was editor of the dedicated journal "Quality & Quantity" and wrote articles for the 'Bollettino del Centro per la ricerca operativa' [Bulletin of the operative research centre] were important ones, as Gallino recalls in an article dated 1971. The period witnessed no dearth of publications capable of expressing the keenly-felt need, well before the well-known 1968 protests, to broadcast the theories and research findings of the empirical social sciences, not only starting with collections of early methodological studies concerning the results and the applicative fields of sociology like the *Antologia delle scienze sociali,* published by Bologna's 'Il Mulino' and edited by Angelo Pagani between 1960 and 1963. Many more essays, publications and monographs by methodologists, in Italian and translated from other languages, appeared during this period. They addressed the Italian context and problems regarding a veritable, formal-methodological tradition born locally in the 1950's. We also wish to mention the fact that *Antologia,* edited by Pagani (volume I, 1960 -*Teoria e Ricerca*- to be precise) presented some essays containing definitions like that regarding modelling in the social sciences by pioneers such as Francesco Brambilla, head of 'Statistics' at Milan's *Bocconi* University. Another series of facts to remember regard the articles and essays addressing topics regarding relationships between general sociology and sociology already 'sectorialised' beyond the classes into which so-called 'special sociology' was divided. An example of this is a work by Franco Leonardi called *Theoria etecnica sociometrica,* published in the first number, 1959, of 'Sociologia', whose editor, at the time, was Mario D'Addio.

This occurred during the maximum diffusion of so-called 'quantitative sociology' above all among those who took a one-way view of the American debate regarding empiricism, various techniques and methods, whose analytical-empirical results were accompanied by Italy's comparative historical 'qualitative-sociological' or even by perspectives of a hermeneutic nature that had come to the fore in the 1950'5 and 1960's and destined, in part, to momentary failure. The quantitative approach increased in Italy especially when it appeared clear why other outlooks derived their qualitative approach to sociological research from assumptions of a philosophical nature connectable to research conducted in the field with outcomes of a hermeneutical, biographical, etc. nature, so that the analysis might lead to an epistemological vision, a sort of narration of its development, its way of becoming history and experience, rational development and a reconstruction of memory. Qualitative methods were investigated by various authors in the years between 1997-2008: Luca Ricolfi, Costantino Cipolla, Antonio De Lillo, Roberto Cipriani to name but a few. The issues covered in their publications questioned the qualitative methods in relation to the development of sociology. We can also say that these topics are now present in the manuals of Italian sociology of the new millennium and have replaced the themes once central to the academic quality/quantity debate fashionable in previous years.

In short, where a first-time epistemological version of the social sciences - that is, one favouring a rediscovery of conceptual, qualitative analyses- might have emerged, mathematical models, scales of measurement and research techniques obliging many to vie with the Americans, took over. This began from an examination of those scientific concepts considered synonymous with the corresponding 'operations' etc. To this regard one should not underrate the fact that certain 'period-bound' criticism of the doctrine of so-called 'operative definitions', did not spring from the scholarly studies of the historical development of the empirical social sciences but from methodologists, epistemologists and philosophers of science capable of pinpointing their incompleteness and some of the 'logical' limits within the historiography of the theoretical methods of modern science. In the early 1960's the discussion on 'theory' and/or 'scientific theory' animated the debate between those aware of its relevance as a means by which to define an exclusive conceptual and epistemological historiography of the empirical sciences. Among the sociologists there not many were those who were aware, even less if we think of the historians of sociology of the tine or of the 'historians of sociological thought'. Back then, many took an interest in the use of tools used to measure sociological research and above all, those who appreciated early theoretical confrontations concerning differentiations between qualitative and quantitative analysis (on this topic see some of the studies by Vittorio Capecchi published in "Quaderni di Sociologia," [Sociology notebooks], n. 2 and n. 3, April-June 1962 and 1963, as well as the essay by Luciano Gallino on autobiography and investigative tools in number 1, 1962, of the same journal).

Topics like those concerning the formation of the concepts and theories in the natural or real sciences represented in Italy the interests of those bent on demonstrating how the history of scientific theory might bear witness to the relevance of theme-problems of a methodological order discussed by authors belonging to logical-scientific trends like the methodologist and philosopher of science Carl Gustav Hempel presented in 1961 by Alberto Pasquinelli, who edited a 1952 publication of his, dedicated to the study of the function of concepts in scientific theories and another one dated 1958. In 1961 too, Pasquinelli, who ten years earlier presented the 'Viennese Circle' in an essay published in 'Quaderni di Sociologia', with Domenico Parisi co-edited the Italian translation of the Reichenbach's aforementioned *The Rise of Scientific Philosophy.* As the editor himself of Hempel's essays himself recalls, some of the limitations of the definitions of Bridgman's operationism, might be traced back to articles by Carnap written in 1936-1937, before being attributed to the well-known thesis on 'liberalised' empiricism, that is, to admission that scientific concepts may be defined in terms of observational concepts although in an incomplete fashion. From the 'point of view' of analyses of scientific theories, some of reformulations by Carnap, which Pasquinelli addressed also in consideration of 'symbolic logic' at the time when his *La sociologia del neopositivismo (1953)* was published, appear also in *The Methodological Character of Theoretical Concepts (1956),* and in an article in Italian language, where Carnap investigated relationships between 'observational language' and 'theoretical language', published in 1959 in the "Rivista di Filososfia."

Hempel's book, instead, brought the discussion back to acknowledging - also with regard to the historical development of the empirical sciences- the need to avail more and more widely of general abstract theoretical constructs to the point of considering scientific theories themselves as veritable sets of concepts or of theoretical terms associated with propositions capable of explaining and forecasting phenomena etc. or, eventually, of inferring judgements and suggestions concerning past, even future, events.

His methodology, derived from the maturation of an 'open' attitude in logical-linguistic and general logical-scientific approaches, served to demonstrate how the problem of the formation of concepts and theories in the empirical sciences was linked to analyses of the procedures themselves.

Hempel's reception by social scientists revealed how traditional methodology and its antithesis managed to arouse considerable interests, especially if related contemporaneously to the development of science seen as conceptual matter and, eventually, as continuous renewal of the various notions and processes typical of its very definition. In a different way, we need to view again the interest shown in questions inherent to the development of theoretical constructs related to the experiential world of the philosophers of science who, precisely due to the existence of an entire series of reflections and progress concerning these very issues, have been able to elaborate a kind of *ex-novo* version of the research conducted within the epistemological field by historians of science themselves or, in any case investigations of the history of scientific subjects. This was, in part, the case with Pasquinelli whose *Nuovi principi di epistemologia (1964)* were resumed and developed, towards the end of the 1960's, in his Bolognese lectures on methodology, epistemology and the philosophy of science.

As Gianni Statera noted towards the middle of the 1970's, this text, dated 1964, was extremely useful to historians of the development of scientific sociology as it led to a debate surrounding the concepts of scientific 'explanation' and 'description', following the historiographical interpretation of science. Never had anything of this kind occurred before, not even with some monographic 'reconstruction' texts like *Modern Sociological* by Pitirim Alexandrovitch Sorokin which appeared in two volumes in Italy in 1974 as *Storia delle teorie sociologiche*. Pasquinelli showed how important it was to insist on the importance of the contributions made by some scholars of the philosophy of science to methodological issues and in-depth investigations of the theoretical nature of sociological explanation and its logic. Furthermore, Pasquinelli once again discussed his interest in Hempel and his study of theoretical methods in science in the *Preface* to the Italian version of *Aspects of Scientific Explanation*, published in 1965.

This work of Hempel's too underwent the scrutiny of those who linked their own interest in the philosophy of science to the empirical social sciences and, pursuing, in part, an epistemological version of historiography within context of the same historical explanation, by means of a quest for research laws (general or otherwise).

Alongside the major debate about German historicism and 'beyond' logical positivism (meaning the various rational choices to make in the field of the history of science), Hempel might have been assimilated gradually, especially because he obliged philosophers and social scientists reflect on the link existing between epistemology and history. In reality, this should have happened between the 1940's and 1962-1963, thanks precisely to the original publications regarding the nomological-deductive model, which gave rise to interest among Anglophone and even some Italian authors (like Dario

Antiseri, who spread the ideas of Popper, Pietro Rossi etc.) who included his ideas in their essays of the early 1970's and 1980's. Hempel actually changed his theory of 1942 regarding deductive models 20 years later, continuing to assert the nomological character of historical explanation, linked to a probabilistic variable model, based on inductively derivable statistical generalizations. This occurred in a context of plurality of models of scientific explanation, in the wake of Ernest Nagel's *The Structure of Science,* 1961- translated in Italy in 1968- reviewed in 1962 in 'Rivista di Filosofia" by Nicola Abbagnano as a true summary of the philosophy of science in twentieth century.

An examination of the procedures used by scientists, also when construing their own language, as well as some of the most important tendencies in logic positivism, was assimilated in Italy by some who addressed conceptual revisions introduced into the field and wrote in the presence of a vast panorama of critical-historical studies on sociological methodology. Some of the problems tackled by Nagel in 1961 and proposed as belonging to the logic of historical investigation, or in particular, as an analysis of the rapport between probabilistic and genetic explanations, were availed of and discussed in relation to the sociological method and its 'period-bound' perspectives, used to inform work carried out by scientific historiography and to investigate possible objects and historical accounts of individual or collective actions etc. The methodologists too had to deal with the problem of discussing the nature of historiographical explanation, starting from the conceptual re-proposal of the differences existing between historiography and science including sociological explanation. The problem was also that of choosing one perspective rather than another, to account not only for the rational development of science but also to assert that it implied a 'precise intellectual discipline, that is, logic', Statera claimed in his *La conoscenza sociologica,* where he made every effort to obtain recognition for the general logical structure of a nomological-inferential type to use to define scientific explanation.

Statera's personal contribution seemed, in truth, to sum up events deriving from the historical-critical debate surrounding sociological methodology, implying in the future, a strong need of autonomy on the part of sociological research besides admission of the fact that only by availing of scientific investigation might one produce –at last- a historiography of the human and social sciences as its logical-empirical consequence. Sociologists might have availed of the history of science only as a conceptual referential-developmental framework implying reasoning of that kind or remained outside of its problematic ambit.

So, the historiographical and scientific problem managed to assert itself within the sphere of sociology in terms of analyses of the method and the true nature of scientific investigation, beyond the dichotomies generated within different settings, which appeared 'abstract' if compared with the original setting which claimed uniqueness of logic, as a 'general theory of investigation', according to Dewey.

The problem of epistemology continued to be entrusted to Dewey's logic, discussed above and as Statera pointed out in an anticipatory article which appeared in “La Cultura” seven years before the publication, in 1974, of his book *Epistemologia a il metodo sociologica*. The topics surrounding knowledge and method in the social sciences were discussed in the light of some of the contributions made by American researchers concerning methodological and epistemological questions, so that, finally, Italian researchers and historians of the scientific sociology were provided with investigative materials to permit them to investigate the difficult relationships existing between some of the guidelines and premises of epistemology, sociological research methodology and theory. On that occasion it emerged that the influence of epistemological premises over sociological theory was in no way marginal, especially when governed by a certain kind of epistemological (at times even ontological) ‘duality’ present, according to of Statera, in many of Talcott Parsons’s works and before that in those of Emile Durkheim and the systematic German sociological tradition. This way, Mannheim’s ‘relationism’ came to be seen as a form of relativism and, as mentioned above, European *Wissenssoziologie* accused of ‘theoretical abstraction’ by Robert King Merton in the third part of his *Social Theory and Social Structure* published in Italy in October 1959 and, as a second edition- similar in structure to the second American edition- published in May 1966.

PART 2

Chapter 9

The Crisis of Rationality and Historians of Science

For a certain period the interests of methodologists of the social sciences and the humanities were attracted by the scholarly debate regarding explanation and description, while in Italy manuals of sociology abounded along with a tendency to accept certain American theories which, if on the one hand, permitted students to acquire tools they might avail of in research, on the other hand, while they left many stranded on a pathway that was an end in itself. This was the thesis circulating not only among the most rational of methodologists of social research but also among the so-called 'qualitative sociologists', even after the age of crisis of sociology, in any case after 1968. Meanwhile, the existence of sociology as an autonomous discipline compared to the other social sciences, posed questions that crossed all approaches and found a major impulse in the epistemological debate which in Italy had aroused interest among methodologists rather than encourage historians of sociology to make choices from among the sources and routes in favour of a rational and conceptual development of scientific subjects, while keeping in mind Popper's notion that scholarly descriptions of 'historical facts' prove selective in the long run. This was the precise direction we need to follow to view the series of monographs which attempted, if only marginally, to outline the field of a possible historical and conceptual definition of the human and social sciences, where no attempts were lacking, however, to restrict, the range of action of those who sought to produce a history of the discipline, supplemented at times, by issues of a different nature, as is the case of a number of works published in the 1960's (like Sabine Acquaviva's *Il problema della logica nelle scienze umane,* Marsilio, Venice, 1964).

The issue leads us, actually, to the fact that the choice of 'sources' Italian historians of scientific sociology made then might have occurred as early as the dissolution of neo-positivism, in any case, during the slow assertion of the philosophy of science and the debate on method which led to the theoretical and perspectives contemporaries of the historiography of the 1980's. The latter debate informed by a project for the methodological renovation of the historiography of the human and social sciences, and, in any case, within the framework of epistemological developments regarding all the sciences.

In other words, the dissolution of neo-positivism and the contemporary emergence of the philosophy of science in the early 1960's had the task of providing a comparison between different conceptions of epistemology. This was an area which should have increased the interests of historians of scientific sociology of interest given the new conceptions of the history of the humanities and social sciences that arose. The moment was crucial, if we consider the confrontation taking place at the time between the conceptual tools of philosophy and those of science, where the task of epistemology referred back to the 'quest for the basis' of its methodological structures as proposed by the Bergamasque Evandro Agazzi, pupil of Gustavo Bontadini and editor of *Storia della scienza* (Rome, 1984), a volume featuring an article on the social sciences by Giuliano Di Bernardo (see Part V, Novecento, chapter XXVIII); or, contrariwise, divested of its foundational status as advocated by Aldo Gargani in the late 1970's during science's full-blown 'crisis of rationality.

Epistemology, as Agazzi intended it, meant science as objectively rigorous knowledge in the constructivist sense of the word, so that objectivity would be related to concepts developed by the intellect, as specified in *Temi e problem di filosofia della fisica (1969).* Science he held should be characterised by constant reference to reality, relative to its object and having logical and linguistic value. The development of Agazzi's 'scientific realism' was confirmed by the general philosophy of science and by specific sciences like mathematics and physics, but also by the social sciences, psychology and logic etc. The period that followed, on the contrary, presented conflicting versions of epistemology with a variety of ideas which led, in the 1970's to an increase of interest in the history of science, towards which Italian sociologists were less receptive than others, because the destabilizing atmosphere of the time condemned their rational and conceptual quests. Not to mention the outcome of how the analytical philosophy of European origin and the gradual definition of various conceptual tools used by philosophers and scientists was received.

Il sapere senza fondamenti (1975) and *La crisi della ragione* (1979) by Aldo Gargani contain the results of investigations of the 'crisis of rationality' within the ambit of the development of new horizons. These were constructed view of a new way of conceiving transformations and involved a novel conception of theoretical and epistemological tools, as in the case of the 'rational models' of the final theory of relativity and, ultimately, of the so-called 'crisis of science' which, between the nineteenth and twentieth centuries, led to a profound change in how to consider the nature of scientific theories and their logical structure. Gargani's thesis circulated quite a lot among theorists, sociologists and scholars of historical sociology because the alleged 'crisis of rationality' was not presented as a true model of historiography but rather as an ever-changing repertoire of images of the history of philosophy, the history of science and logic. It was one of the outcomes of the crisis of foundational-type epistemological strategies and their reference to historically defined and/or definable premises which did not appear to perform cognitive functions and might, therefore, refer to conceptual schemata which failed to take logical-linguistic analyses into account. In brief, during the mid-1970's, the epistemological model and its reference to the solidity of theoretical and conceptual apparati appeared to be experiencing a crisis.

On the contrary, this was not the case of reference by a certain kind of scholar to ways of availing of the work of historians of science concerning issues of the scientific formation of ideas and those, sometimes regarded as coincident to logic and the philosophy of science. It was in this direction that the historians of scientific sociology were able to make use of other stimuli and several contributions to transform the structures of the developmental context of ideas including those of a scientific nature. By way of example we find the establishment in Italy of studies of the sociology of science and the renewal of aspects of historiography as far as the methodology of the social sciences and humanities were concerned. In the early 1980's that, referring to these latter aspects, that a *résumé* of historical studies was achieved at a Conference in Turin on *La teoria della storiografia negli ultimi venti anni* [The theory of historiography in the last twenty years], which gave rise to a variety of research trends and laid the basis for a hoped-for methodological renewal at the behest of the social sciences themselves.

On their part, Italian historians of science were able to take advantage of this crucial moment and, in the face of definitive detachment from traditional views, declare themselves in favour of scientific culture or professed belonging to contemporary, modern approaches, for example, Marxism. Authors like the afore-mentioned Paolo Rossi devoted their efforts to the study of relationships between science and philosophy without ever concealing their interest in methodological issues. The early 1970's marked a massive spread of historical studies of science which assumed the task- explicitly and according to the 'editors' of sectors or series like those issued by the Milanese publishing house Feltrinelli and entrusted to the care of Rossi himself- of bridging the gap that still existed between the sciences and the humanities. It was required to take into account that separation between science and society, between science, philosophy and the cultural environment was untenable, while considering the impossibility of practicing a history of society which did not consider how it and/or its developments were influenced by science. Many translations of works by historians of ideas, historians of philosophy, cultural historians and historians of science, from Alexandre Koyré (died in 1963) to some remote and authoritative scholars like the historian of astronomy John Louis Emil Dreyer to the historian of science Alistair Cameron Crombie, etc. appeared in that period, wherever a request to enlarge the horizons of traditional historiography proved attractive, beginning with the roots of scientific thought themselves.

Other personalities contributed with a number of topics in contrast with more philosophical issues regarding the crisis of rationality and which tended to claim, instead, the need to produce rational reconstructions of the history of science in line with what had been stated a decade previously by authors such as Thomas Samuel Kuhn (1969), and Imre Lakatos (1978), who represented -especially for scholars of the social sciences - new 'period-bound' perspectives for the philosophy of science. On the one hand we need to recall the studies and insights regarding dialectical materialism intended as a renewed form of scientific conception, drawn up in the mid-1970's in the wake of Ludovico Geymonat by authors like Enrico Bellone, Giulio Giorello and Silvano Tagliagambe who, among others, went as far as to interpret materialistic physics. The thinking of Ludovico Geymonat (who died in 1991) was pursued by several scholars including Evandro Agazzi, Enrico Bellone, Mario Capanna, Giulio Giorello, Corrado Mangione, Silvano Tagliagambe, Salvatore Veca, Mario Vegetti, Fabio Minazzi Niccolò Guicciardini Corsi Salviati, who fostered a connection between the philosophy of science and other disciplines, including the social sciences. Other scholars devoted themselves to receiving the debate on materialism and others again, in keeping with their strictly scientific training, to studying the relationships to be found between the development of science, history and social contexts, as is the case of Vincenzo Cappelletti (1972) author besides, of the important entry 'Epistemology' drawn up for the Institute of the Italian Encyclopedia (1977) - of which he was director from 1970 to 1992, vice-president from 1993 to 2002 and professor of the History of science at Rome's Sapienza and Roma Tre Universities from 1980 to 2002.

The particular objective of some scholars like Marco Mondadori (1978) Corrado Mangione (1971) was surely that of discussing epistemological programmes and the dynamics of scientific knowledge regarding issues related to images and the rational development of science as well as scientific theories and the history of logic. We also hasten to recall the appearance of the fine reconnaissance work carried out by Pietro Redondo, 1978, worthy representative of a series of 'period-bound' historical journals which dealt with theoretical epistemology and the history of science. Furthermore, several scholarly works met with the favour of a number of publishing houses and Rome witnessed the publication of a series on methodology, the philosophy of language and the historiography of science, directed first by Dario Antiseri, later by Vittorio Somenzi and Francesco Barone. This series published important studies, including some on theoretical-conceptual themes-problems, like those in Larry Laudan's *Il progresso scientifico*, 1979, or in titles translated by Italian authors like Marcello Pera, Franco Voltage, Cowgirl Emanuele, the 'reductionist', Anthony Verdino, Massimo Baldini, author of the book *Teoria e storia della scienza* (1975) and author of the *Introduction* a text by the Popperist, Joseph Agassi, *Epistemology, metaphysics and the history of science* translated in 1978 by Adelino Cattani.

As mentioned already, upon different planes of reflection, methodological studies of the sociology of science also asserted themselves, in particular, with regards to the links between knowledge and social structure and change. These studies emerged within the ambit of the developments stemming from reception by some Italian scholars of the immense work carried out by British and American sociologists in the 1960's and 1970's and promoted by Gianni Statera in the volume called *Sociologia della scienza* which he edited in 1978. It is important to note that that according to Statera, the sociology of science was interconnected with the 'movement for the social relationships of science', with Marxism, scientific humanism and instrumentalism, triggered by Kuhn's historical and sociological perspective and his analysis of shifts in ideas in science. Statera was brought to admit that Kuhn's vision was to be identified with the outlook of historians of science capable of enlarging the horizons of interest of sociologists of science and remind them that Merton obtained his Ph.D. in 1938 with a thesis on the history of science of seventeenth-century England republished in book form in 1970 and translated into the Italian in 1975 by Pasquale De Gaetano, edited by Alberto Izzo and with an *Introduction* by Filippo Barbano.

Reference to Kuhn and to this precise period appears strategic, because it shows that during the analyses carried out by historians highlighted the difficulty of outlining an explanation of the concept of development by accumulation, especially as far as historiographical matters of science were concerned. It emerged, therefore, that science did not proceed in a linear may through an accumulation of individual discoveries and inventions. That was the point. The author intended to explore the historical character of knowledge and therefore the importance of the context that produced it, making an effort useful for an examination of the image of science during the crisis of rationality.

It is in this sense that most of Kuhn's work represented a turning point in the consideration of the image of science and the redefinition of the stage upon which philosophers and sociologists of science might act. The crisis which began in the 1960's certainly underlined the inadequacies of philosophical models of rationality and led to a process of revision by which the achievements of scientific activity might and needed to be explained by historians without recourse to philosophy.

It was the fate of historians of science during the second half of the twentieth century to be criticized by philosophers of science, for their lack of adhesion to a normative methodology, at a time when for over 35 years, American universities like Pittsburgh had been self-sufficient. In that context historians of science might be ignored by social historians for never having conducted specific research into history. On the other hand they might be appreciated in the field of the social sciences and later in that of the so-called sociology of scientific knowledge, which found valid tools in a number of conceptualizations.

The reception in Italy of social studies of science constitutes a significant event from the 'point of view' of a redefinition of relationships between science and its social environment. These studies addressed the dynamics existing between science and technology and the changes that characterized the relationship between science and society. Robert King Merton was the first to take an interest in the delay of the sociology of science compared to other disciplines, and as most of today's historians of scientific sociology are aware, it was not until 1978 that the American Sociological Association assigned a special section to the sociology of science and Merton could support a theory regarding the institutional conception of science, due to an increase in the process of 'verified' knowledge unlike that of Kuhn as Statera pointed out in his pioneering book of the late 1970's. Kuhn took a stand against the cumulative vision of the development of scientific knowledge when he defined 'normal science' and its crisis by introducing the concept of 'paradigm' drawn from examples of scientific practice, including laws, theories and tools capable of providing real 'models of scientific research', as posited in his famous work on 'scientific revolutions', translated into Italian in 1969 by Adriano Carugo (winner, with historian Crombie, of the 'Domus Galileiana' prize, Pisa, awarded for the best monograph on Galileo Galilei and author of the *Introduction* to Preti's *Storia del pensiero scientifico* 1974). In this book, Kuhn recalled Ptolemaic astronomy, Aristotle's dynamics or corpuscular optics. The paradigm was capable of introducing the historian to the 'scientific community' even during periods of cyclically alternating phases which created incommensurability or, on the contrary, as Ludwick Fleck argued, where problems of 'communication' between the members of different 'communities' might be overcome. In any case, it was no longer the scientist capable, to some extent, of accepting a paradigm, but the 'compiler of handbooks' who was now left with the task of justifying the historiography of concepts. It was manuals which concealed 'scientific revolutions' because of being their outcome, or at least, because produced later on. In practice, Kuhn believed that knowledge was transmitted by means of a process that allowed the scientist to adhere to a series of beliefs, categorizations and even 'solutions' whose principal tool was the handbook. Kuhn did not believe, for example, that learning through verbal definitions and rules could be complete. A complete process he believed was capable of constituting a *corpus* of shared or, at least, shareable knowledge. In particular, referring to the social sciences, Kuhn asserted that the phase capable of distinguishing them was of a pre-paradigmatic kind rather than a phase of 'schools' expressing widespread consensus for a vision of the world and/or postulates that may belong to a mature discipline. Despite this, Kuhn received much support from social scientists, who believed it normal that their discipline be characterised by some kind of paradigmatic incompleteness and a sort of limited extension of rigorous procedures. It was possible to address some differences in the light of the normal course of ideas and scientific subjects.

History, according to Kuhn, was full of examples capable of revealing differences between paradigms. The example provided by Kuhn was chemistry. But what seemed to interest the historian most were the changes which took place in science and the so-called changes of perception induced by new paradigms, which might also be traced back to conceptual and epistemological upheavals. In short, traditional epistemology might be subject to change, as well as interpretations of scientific theories during studies of the development of science (in relation to what might or did occur in relation to a paradigm).

It is certainly interesting to follow Statera when he focused at the time on the more or less marked convergences between Merton's objectivist perspective and that of Thomas Samuel Kuhn, for whom, ultimately, the epistemology produced by the historiography of science was based on the sociological, even 'irrational' bases, corresponding, however, to the various phases of tension and crisis inherent in the progress of science itself. But the same interest would have been aroused, in our opinion, by a combination of Merton's formulation of a paradigm for the sociologists of science and the relationship of the latter with a number of social expectations in keeping with the ethos (as a "sociological invariant") of the scientific community that were diffused at the time. It is useful to remember that for Merton there was only a sole normative order at the basis of the autonomy of scientific knowledge capable of safeguarding it from potential social influences. The fact that some authors like him asserted that the natural sciences were based upon knowledge demonstrated by means of universal methods, probably posed a limit, seeing that the task of establishing or otherwise criteria regarding rationality in science and/or in scientific knowledge had always belonged to the realm of philosophy, not of sociology or the sociology of science. For this reason, the crisis of rationality which hit philosophers during the 1960's impacted on the fact that many issues were the result of the attitude towards science shown previously by scholars of philosophy. Obviously, that was sufficiently clear to those who grasped the problematic of questions then being debated and which closely regarded a certain approach to philosophy and scientific historiography. All told, the task of the philosopher seemed to lie within the domain of matters regarding the "justification" of scientific theories, while explanations of why "discoveries" were achieved by certain individuals and not others at a given time, was the responsibility of the historian, psychologist and sociologist. It was in this context of "crisis" that authors like Kuhn and his *The Copernican Revolution* of 1957 translated into Italian in 1972 or Nortwood Russel Hanson with his *Patterns of Discovery (1958*) translated in 1978 containing the results of research on the conceptual fundaments of science and others still, could claim a new historical perspective sensitive towards the social factors of scientific knowledge. In cases opposed to the assertion of method when dealing with historiographical considerations of the work of scientists, we find "the anarchist of knowledge" Paul K. Feyerabend and his *Against Method (1975).* The idea was that of replacing philosophy with representations of the work of scientists and of forming a 'point of view' from which researchers, historians of scientific subjects in particular, would continue to reflect on rationality and its alternatives during the years that followed. There was no turning back. The image of science seemed to have truly changed.

It was Kuhn, in particular, who contested the traditional philosophical vision and therefore, the distinction between 'facts' and theories, uniqueness of method and the unification of science. Unlike the philosophers, the approach he proposed to historians of science was naturalistic-descriptive, not prescriptive and descriptive. In practice, instead of considering the end result, the social process of science was investigated, also referring back to the judgment of particular sub-cultures, their traditions of scientific research. Research might be taken to mean a practice of what is known applied to the solution of problems.

Various stages of scientific change, or rather, of the progress of science were revealed to historians. Normal science was the phase when research followed the specific, accepted, developmental lines like the concept of paradigm, which belongs to this normal phase capable of solving, when all comes to all, the 'dilemma' of the paradigm. When scientists encountered anomalies they did not question the paradigm but set them aside to arrive at a solution by changing the theory. We need to remember that, from the 'point of view' of developments within the historiography of scientific subjects, and, therefore, from that of the new late-twentieth-century historians of science, the concept of paradigm, thanks to a combination of social and cognitive elements, managed to provide an important category to use when exploring scientific communities, their ideas, their research and cultural traditions. It must also be said that having proposed this with a view to exploring science by availing of sociology and psychology, Kuhn was criticised by Popper and by the philosopher of science Feyerabend, as acceptance of their criticism by Italian social scientists too in the early 1980's shows. Popper reproached Kuhn attempting to derive objective descriptions of scientific subjects from disciplines like sociology and psychology, while witnessing a decline of rationality in science from the time of the studies regarding the development of scientific knowledge, (1963) published in *Conjectures and Refutations,* issues in Italy in 1972 translated by Giuliano Pancaldi. Feyerabend held that while Thomas Samuel Kuhn's work sought to rebuild a pathway of science leading to an objective kind of description, in actual fact, it simply outlined some of the principals leading to scientific activity. In any case, disciplines like sociology, acquired some credit compared to epistemology and epistemology based on historiographical grounds. The traditional approach of historians of sociology (of both European and Anglo-American formation) were able to change, considering that from that moment on, many historians learned to take a look at some of the paradigmatic figures and situations of philosophy and the history of science, learning to reflect more on conceptual instruments to use to build their own 'visions of the world'. It was probably at that time that conceptual and epistemological history too, began to reveal the usefulness within a rational perspective of historiographical studies of the development of science. There was no turning back since much had been said and much had also been done.

Chapter 10

POPPER, THE POPPERIANS AND A NEW METHODOLOGICAL THEORY FOR THE SOCIAL SCIENCES

Between the second half of the 1970's and beginning of the 1980's historians of scientific subjects were keenly aware of the discussion between Popper and Kuhn, a series of *debacles* for authors who were prompted to avail of the progress of the debate surrounding the history of epistemology and revolutions in science, especially as what emerged from 'period-bound' works. Without referring to these 'period-bound' works, the task of carrying out rational investigations of problems which interested historians of science grew arduous. It must be said, in fact, that at the beginning of the 1980's the work of scholars regarding the conceptual themes-problems imposed by a continuous re-definition of the relationships between various approaches to/in the philosophy of science became quite relevant.

From the 'point of view' of social scientists, this may have been true in the case of conceptions emerging after the publication of Kuhn's work concerning scientific revolutions, that is, changes regarding both systematic concepts and the history of science. It seems that even in this case the tensions produced by the confrontation between Popper and Kuhn played a key role. The defence and the critical interpretation of Karl Popper during that period of ideological crisis and accounts of yet another condemnation of neo-positivism by the Austrian philosopher, (like that provided by Marcello Pera in his *Popper e la scienza su palafitte,* 1981) are typical of the interest shown by a number of scholars in the question of contemporary relationships between science and philosophy (and referring again to the work of Marcello Pera) as well as the formulation of a theory of science or epistemology based on Popper's acquisitions and aimed at asserting an objective epistemology founded on rational reconstruction.

The exceptional nature of the debate had an impact within several areas including Italy, where attentive authors and translators like Massimiliano Baldini and Franco Voltage took the trouble to draw useful conclusions not only from among scholars of various philosophical and historiographical theories of the sciences but also from among methodologists and from among those who, in Italy, who had gradually developed an interest in philosophical themes-problems restricted to the empirical sciences or the formalist approach. As regards the latter we should not overlook the information provided by Kuhn in his *Postscript* to *The structure of Scientific Revolutions,* 1969, which, in a certain sense, introduced a veritable reform, claiming a correspondence between the conceptual aspects of the historical-genetic theory of science and logical formalization.

Kuhn's assertions embodied one of those fundamental issues which historians of the social sciences operating between the second half of the 1970's and the beginning of the 1980's came across along their way: the problem of how to link the theoretical genisis of the empirical disciplines to shifting developments within historical scientific disciplines. In practice, the problem arose between historians of sociology, historians of anthropology or of psychological and social-psychological issues, regarding how to discuss auto-legitimize epistemology, for example. Relations between historians and the history of science became a key issue of discussion, so that Massimo Baldini, the author of *Epistemologia e la storia della scienza,* Turin, 1974, was able to comment on the quarrel between Popper and Kuhn which had emerged within a particular theoretical context, the international colloquium on the history of science held at the Bedford College, London in 1965 treating on the *Theory and history of science,* the relationship between scientific methodology and the historiographical experience. "The epistemologist today" Baldini held in 1975, "is well aware that if he wants to hone his network theories he cannot help but seek the collaboration of the historians of science." In other words, the historian of science, when all came to all, might borrow a conceptual apparatus which belongs to him in some way, from the epistemologist in order to interpret historical facts on the basis of the acquisition of certain assumptions. The history of science would be entitled to use epistemological theories, transforming its 'prejudice' in its own favour.

In an essay of 1971 - translated into Italian by Silvio Morigi and published in 1983 in *Paradigmi e rivoluzioni nella scienza* where, in addition to the writings by Kuhn there appeared an essay by Joseph D. Sneed, of Stanford University California, and another by Wolfgang Stegmuller, author of *The Structure and Dynamics of Theories* - Kuhn invited historians to reconsider the history of science fully, while providing an overview of developments for the following decade. In actual fact, even formal recognition of the role played by science in the development of the west did not seem to arouse particular interest in science as a historical subject. Kuhn complained that the appearance of the word 'science' alone in the title of a university course or book was capable of causing disinterest or refusal on the part of history students. And not only that. Further considerations may prove useful when undertaking a detailed analysis of the relationship between the tasks of historians and historians of science. First of all, there is the problem of the so-called 'history of ideas' which had indeed surrounded European sociology for many generations and which in Kuhn's opinion should be kept separate from those related to socio-economic history etc. He held that when dealing with historians of ideas it was far easier to find interests corresponding to those of historians of science. One example regarded misunderstandings due to a certain attitude on the part of historians when it came to 'technical' topics, like those presented and discussed at length by the sociologist Robert King Merton in his thesis regarding the development of science in England in the seventeenth century.

All told, the fault that Thomas Samuel Kuhn found at the time with those who were interested in placing science within the history of ideas, was that they ignored technical sources, while the fault with historians of socio-economic matters was due mainly to non-mastery of conceptual ambits in which to grasp the importance of the social aspects of science. In either case, 'techniques' or technology needed to be controlled by a conceptual development of their differentiation within a framework of surrounding change. Almost at the end of his 1971 essay, Kuhn mentioned the almost manifest disdain of scientists towards historical studies in the strict sense, so that their interest was directed more towards specialist field of the history of science. For Kuhn, philosophers, like historians tended to consider their enterprise as being of a cognitive nature, therefore similar to that of science.

These statements were related, in the context of the development-reception to Kuhn's ideas, to a period when an unscientific atmosphere seemed to produce a remarkable flowering of historical studies concerning science, as well as the spread of numerous questions among scholars of the social sciences and of philosophical subjects, including philosophers of science as well as critics and followers of Popper's and of his epistemological version of the historical social sciences.

In Italy, at the beginning of the 1980's, these topics managed to produce a number of monographic studies, essays and papers destined to develop over time, and, in a certain manner, to remain beyond the reaches of institutionalized teaching of the history of science in history departments, a very important issue investigated by Kuhn. In practice, if Popper and his followers aimed above all at focusing their ideas on the problem of the debate regarding the nature of science, Kuhn endeavoured to clarify its 'rational' development and rectify it as a notion when applied to history and to other empirical disciplines of our acquaintance.

Even before the decade previous to the that discussed above, when the emergence of comparisons between Popper's and Kuhn's theses came to a head, in particular 'within' the ambit of specialist discussion which sought to influence, from 'without', contemporary reception of the entire debate over the development of epistemology, the need to discuss the new theory of the scientific method known as 'falsificationism' emerged and was proposed by Popper as a criterion useful to researchers so that they might relate to empirical generalizations. As was fairly well known in Italy thanks to the 1970 Italian translation by Mario Trinchero (author of *Neopositivismo logico*, 1978) of *The logic of scientific discovery,* a second edition issued in London in 1968 and enlarged compared to the first 1959 English-language edition of *Die Logik der Forschung* (published, surprisingly in Vienna in 1934-1935), the complex problem of induction and of the validity of certain laws was no longer an issue. Empirical assertions might be submitted to systematic attempts to detect their errors, as Abbagnano wrote in his *Dictionary of Philosophy*, 1961, under the heading 'falsicability'.

Unlike the inductivist, who from the era of the philosopher Francis Bacon to the end of the nineteenth century, held that science started with observation aimed at (objectively) inferring laws and making 'predictions', Popper believed that it was not possible to carry out observations without a theoretical background and that observations being selective they required a 'point of view'. In this sense, Popper criticised those who tended to justify the principle of induction through experience, demonstrating how it was possible to find oneself a vicious circle or in a condition of infinite regression. Any ulterior attempt at justifying the principle independently of experience, that is a priori, would be more like an act of faith. This was when Popper proposed a non-inductive theory of the scientific method, starting from what philosopher David Hume had pointed out concerning the impossibility of inferring a generalised truth through observation alone: according to the well-known example of scientific literature, regardless of the number of swans we observe we can never infer that all swans are white.

In other words, for the historians of science of the second half of the twentieth century it may have appeared useful to re-examine some aspects of Hume's criticism of inductivism resumed by the Viennese philosopher, where he showed how some of the inferences derived from observation of the laws were not obtained through logical deduction, so that the laws could not–in the long run- be obtained from the data obtained through induction or deductive inference but on the contrary, through induction or inductive inference. Later on, it was Bertrand Russel who affirmed that this method did not guarantee certainty, but rather probability, reducing mathematics to logic: mathematical truths, he held, were knowable independently of experience, only in so far as they concerned truths-by-definition, mere linguistic convictions.

The bases of empiricism were established, as well as the various links between future reflection by historians of scientific matters and criticism of induction by Popper's followers which generated considerable interest by attentive scholars seeking to understand developments within prevailing epistemological guidelines.

In the book *Il Neo-emprismo* edited by Pasquinelli in 1969 for the Italian UTET publishing house, the year when the Italian version of Popper's *Science and Philosophy* was published (and included, besides, writings like *The Aim of Science* which had appeared in journals like "Ratio," in 1957), we already find some ideas tending in this direction, essays on 'scientific criteria' and their relevance, on the topic of induction related to subsequent conjectures/hypothesis of his concerning enunciations etc.

Despite this, Italy did not lack initial reception of issues regarding scientific research and Popper's methodological conception (also associated with the examination of European epistemologies) which led to more or less evident interest on the part of philosophers and human and social scientists. We hasten to recall, for example, the period between 1961 and 1974 when contributions were made by Emanuele Riverso to "Rassegna di scienze filosofiche," by Marcello Pera in *La scienza ad una dimensione*, by Paolo Filiasi Carcano to 'metascience' in the 'Giornale critico della filosofia italiana' and by Dario Antiseri.

The fact is, that both Popper's books of the late 1960's and his future thinking, pointed out that although observations could not guarantee the validity of scientific generalization in that they cannot verify them, they can establish its falsity, that is, confute it or prove its falsity. In practice, the observation 'this swan is black' while it cannot prove the generalization that all swans are white, can prove its falsity.

For Popper, science did not begin, therefore, inductively with observation but rather with conjectures that scientists sought to refute through experimentation and observation. His theory of 'conjectures' and 'refutations', like the text of the same name (translated in Italy in 1972, though some of the chapters had already appeared in reviews like "L'Industria" in 1964) dealt mainly with the problem of induction and led to a distinction between the discovery of scientific hypotheses and their justification. The first part of the theory was not the concern of the philosophy of science but of empirical psychology; there was no such thing as a 'logic of scientific discovery', we could only speak of a logic of scientific scrutiny. It is important to note that Popper expounded his theory of scientific method gradually so that we might perceive it as divided into a number of main theses. First, as stated above, he asserted the impossibility of making inductive inferences with a view to obtaining generalizations and predictions from observational reports. Generalizations and predictions he stated, were conjectures to be subjected to strict control by scientists, regardless of how they were obtained. Secondly, Popper denied the fact that it is possible to calculate the 'probability' of some scientific predictions on the basis of available evidence: conjecture cannot be established as probable according to calculations of probability. Finally, the third thesis held that conjecture could not be justified in a strictly positive manner.

The nullification of induction, reduced practically to the status of myth, carried out by Popper, established another 'point of view' concerning the conception of scientific method among historians of science, so that the history of science itself was to be seen as the history of theories that were changeable and could be replaced by other theories which, if they precedes observation, were verifiable through experimentation. One result of the spread of a new methodological formula to the history of science was admission that there might be many forms of 'verification', as in the case of Newton's method and theory, in the presence of Pierre Duhem's critical induction in 1906 and of Einstein's theory of relativity in 1905. It is no coincidence that in Duhem and Einstein we find issues regarding knowledge of the physical world and the characteristics of the scientific method. Popper always remained particularly attracted by Einstein and especially by the way he submitted his theory to total disproof. There is no doubt that certain aspects of Popper's formulation are associated with corroborations of Einstein's theory without, in truth, overtly influencing some philosophers of science, scholars of physical-mathematical sciences, or, later on, Italians who propagated his epistemological work (for example, the above-mentioned Bruno Cermignani, editor of a popular book on relativity, 1967). From this more or less critical stance the issue is not so very different if we compare Popper's methodological affirmations to the social sciences and, therefore, to the criteria imposed by logical positivism following 'assertions' against metaphysics or those availed of by Marxism to affirm its scientificity and thoroughly disputed by Popper in *The open society and its enemies*, a political work published in Italy thirty years after it was written (volume one December 1973; volume two January 1974) and ignored by the critics belonging to the intelligentsia. The fate of neo-positivist criticism of historicism was different. Diffused as early as the end of the 1950's by researchers like Pietro Rossi in the "Rivista di Filosofia," it was a forerunner as far as contributions during the following decade are concerned, regarding issues related to relations between Marxism and the criticism of historicism we find in Popper.

Apart from the problem of more or less clear 'readings' by various scholars of science of Popper's original approach - so changeable and at times an end in itself- despite the presence of a conspicuous number of professional philosophers operating in these fields, it seems important to emphasize the fact that many of the studies undertaken impacted on the way in which contemporary versions of the epistemological critique of the scientific method were outlined, beginning with periods of considerable transformation and change to arrive at issues which historians of science linked with conceptual redefinitions of empirical science. Again as far as criticism of induction is concerned, a particularly enlightening example resides in the ideas Duhem expressed in his polemic against inductivism which, for him, coincided with the method Newton posited in his *Principia*, when dealing with one of the cornerstones of physics: the acceptance by the scientific community of the three laws of motion. It is worth our while to look at these issues again for a moment and relate them to what has been said about Popper, having first pointed out that the various sections into which *Principia* is divided and in Scolio's considerations of the concepts of space, time and motion, Newton used a number of arguments to indicate how they corresponded to the reality of the planetary system being examined and, ultimately, how to define them as veritable conceptual operations.

It may be useful to recollect the fact that it was Newton himself who declared having adopted the inductive method and affirmed that the aim of natural science was to obtain assertions derivable by means of induction from phenomena. Newton, in practice, declared having derived the law of gravity from Kepler's laws and this was the object of Duhem's controversy. Duhem's criticism was based, in fact, on the inacceptable possibility that Newton had reached his conclusions starting from Kepler's, since, according to Duhem he himself had contradicted them.

According to the laws of Keplero, a planet revolved around the sun tracing a perfect ellipse, while in Newton it followed an elliptical orbit with a few perturbations due to the strength of the gravitational pull of other bodies in the solar system. Newton's theory formally contradicted Kepler's laws. Furthermore it introduced concepts like mass and force not mentioned by Kepler. This would make it difficult, in the end, to derive, either inductively or deductively, a theory involving new concepts of force and mass from a set of laws that fail to avail of the same concepts. Duhem's thinking in *La Théorie Physique* is significant and imposes reflection, also in view of its reaffirmation of the use of the history of science in the study of its conceptual transformations. If Newton's theory is correct then Kepler's laws are necessarily false. This is the point: one possible way of studying the structural changes of the objects of scientific knowledge seems to be enclosed in the conceptual transformations that the history of science aims at tracking down.

It is not difficult to admit, for the umpteenth time, that many of the transformations of concepts concerning the history of scientific subjects always regarded problems of method or, in any case, the affirmation of criteria and concepts dealt with in the place of others etc. This happened to Popper at the very moment he addressed the problem of logic in the social sciences, in a paper dated 1961 which he presented at the German Sociological Congress and which gave rise to serious discussion and to various comments of interest not only to 'dialectical' philosophers like Theodor W. Adorno and upholders of opposite theses but also and in particular to methodologists of social sciences who were able to refer back to that very essay and to debate to take stock of Popper's reception in Italy at the end of the 1960's and the beginning of the 1970's. On this issue, we need to mention a number of original contributions made by historians of sociological thinking, which make comparisons between Popper's ideas on the logic of science and the empirical implications of the latter as well as the criticism it aroused. We refer, in particular, to some writings by Italian authors including Mario Digilio: *Un nuovo Methodenstreit: Popper-Albert contro Adorno-Habermas,* published in "La Critica Sociologica," [Sociological criticism], n. 8, winter 1968-1969.

The essay, part of a vaster range of research, dealt with issues of critical methodology, providing a thorough assessment of the 'analytical' nature of Popper's school while also highlighting Adorno's and Jurgen Habermas's positions of on the dialectical theory of society, contested by Hans Albert. In actual fact, the theses Popper enounced on that occasion may be intended as veritable conceptual reworking seeking to establish themselves in theories and schemata, the nature underlying the concepts themselves and the logical development of some of them, interpreted from the 'point of view' of knowledge and methodology, as is the case of concepts dealt with by Popper in *Logik der Sozialwissenschaften* like 'approximation to truth' or the 'explanatory content' of a theory. These examples may refer to the assertion of the fact that it is only by working on those 'objects' that Popper could come up with a definition of regression or progress (or vice versa) in scientific theories and also find a way to base his reasoning and his statements on a critical resizing of a number of the concepts used in the social sciences.

The theory of scientific objectivity, around which the 'sociology of knowledge' rotated, is one example. This theory, according to Popper, was set up without permitting social scientists to understand that the objectivity of science was not something that might be derived from the exigencies of individual scientists, but from their reciprocal criticism. It was, therefore, not an issue that stemmed from the alleged impartiality of the methods adopted by each, but from the tradition imposed thanks to the division of scientific work and the problems related to it. So, Popper himself examined the concept of 'non evaluability' which, within the social sciences, tied in with that of objectivity, becoming one of its unachievable premises.

He believed that social scientists pursued concepts of objectivity and 'evaluability' achieving results that were somewhat relative compared to those of the natural sciences, although it was important to keep in mind how misunderstandings regarding the methods of the natural sciences mythicized both their inductive nature and their tendency towards objectivity. The same can be said of other concepts around which ideas regarding scientific research spring up like that of 'truth, to be considered as the main scientific value, though not the only one, within the panorama of the different interests and values surrounding criticism of science itself. The concept of 'truth' was very important to Popper because, from the 'point of view' of the logic of social sciences and of the deductive systems (theories) availed of by researchers, he held that the very claim to 'truth' failed to hold up when confronted with its opposite: the main theme, from the methodological 'point of view', was that leading back to our mistakes, which become evident when associated with the idea of 'truth'.

It was modern logic which prompted the advance of these matters and also did much to clarify how the reasons used by scientists might 'explain' phenomena, that is, arrive at solutions to questions by means of a deductive system or theory. For this reason Popper resumed the theory of the Polish mathematician and logician Alfred Tarski, to re-evaluate the concept of 'truth' in his studies, and also to reaffirm it in philosophical terms, tending, therefore, towards an absolute concept of "truth." This way Popper came to reason in such a way as to disavow his own reference to 'explanation' and to the kind of 'consistency' stated by Carl G. Hempel when dealing with the concept of 'truth'. It must be said that among Italian critics and scholars of the time not many pointed out the importance of this contradictory aspect. However, the problems raised by Popper prompted stances among scholars who began to refer to the Popper-Hempel explanation model. Many of these problematic nodes did not emerge from essays by authors who discussed the fate of the model more generally, as was the case with Antiseri, in his article called *Il modello Popper-Hempel* published in 1970 in "La nuova critica." Once again, only a few methodologists proper took an interest in grasping philosophical scientific issues capable of exercising their influence over specific topics of social sciences and the question of the so-called "sociological knowledge" also starting from a re-proposition of Popperian concepts and/or from reformulations of conceptual schemata to be submitted to scholarly criticism.

They were aware of the general model of scientific explanation formulated by the Austrian philosopher in the twelfth paragraph of *The Logic of Scientific Discovery* in the early 1930's and developed by Hempel in his dissertation on *The function of General Laws in History* published in "The Journal of Philosophy" in 1942 and then, systematically in *Aspects of Scientific Explanation (1965).*

Statera writes about these topics in his work on sociological knowledge, emphasising the receipt of 'common' corrosive and/or 'contrasting' elements in empirical social science in the face of some of the positions treated by Popper and those launched by Hempel. His critical contribution to the logic of sociological explanation was aimed, actually, at a general definition of scientific explanation and may be considered illuminating for its time, especially when compared to the plethora of Italian followers of the Austrian philosopher which, in the meanwhile, began to form around a variety of 'minor factions' peoples by philosophers of science, inspired more by the clash between 'critical rationalism' and the principles of logical positivism than by the influence of 'probabilism' over the history of empirical social science. In his book, Statera underlined the fecundity of Hempel's reasoning concerning the 'nomological-deductive' and 'nomological-inferential' (that is 'probabilistic-inductive') models of explication. He also referred to the contribution and formalization made by Hempel and Paul Oppenheim in *Studies in the Logic of Explanation,* 1948, published in "Philosophy of Science."

Hempel's reasoning led, basically, to a distinction between two explicative schemata: the first was the 'nomological-deductive' model, where the *explanandum*, that is the phenomenon to be explained, coincided with the event and certain laws; the second was the 'probabilistic-inductive' model, where, on the contrary, the *explanandum* coincided only with the possibility that a specific event might manifest itself, therefore with the laws of probability. The inductive probabilistic model admitted that the *explanans* might coincide with the empirical conditions in which the event or the phenomenon "X" to be explained appeared as well as determining general propositions as laws.

Following his analysis of Hempel's thinking Gianni Statera reached the conclusion that -in the long run- among social scientists, the idea might take root that 'explaining' and 'predicting' might be considered similar operations, associated with and bound to the formulation of theories. In practice, if social scientists accepted and worked availing of the 'nomological inference' model, then 'prediction' and 'explanation' were likely to appear as aspects of the same process. This kind of approach was to find favour in the scholarly production of the author himself, in the first manuals to systemise logical-methodological discourse within the domain of social research as it established itself in Italy in the early 1980's, at a time when a number of mature philosophers like Abbagnano, launched projects concerning man, revealing a marked lack of trust in the new theories and in the new scientific discoveries and indicating, instead, the limitations of the creation of concepts *ex novo*. Other authors, like the above-mentioned Antiseri, who were trying to unify the scientific method and, therefore, the methodology of science, treasured Popper's teaching, where the debate surrounding 'falsifiability' was supposed to sum up the results of epistemology for contemporaries. Antiseri's motto sounded then very like Popper's: "to avoid sustaining error is a petty ideal." In another sense, and precisely from a conceptual 'point of view', the preoccupations of a theoretical nature stemming from the so-called Popper-Hempel model ended by involving many aspects - and not only those familiar to new-empiricists philosophers like Nagel and others - seemed a quest for insights surpassing the threshold level of reception of the model proposed among philosophers of science and involving the methodological clarification of relations between the physical-natural sciences and the empirical social sciences, according to examples like some of those already discussed here. In some cases, -for example, the philosophical interpretation of science by some authors who dealt with Marx, like the methodologist of social sciences Leszek Nowak, introduced by Giorgio Sandri in a volume dated 1977, called *La scienza come idealizzazione: i fondamenti della metodologia marxista*- an examination of the structure of scientific theories led to Hempel's explanation, separating it from Popper's contribution, so that the conceptual nature of knowledge might appear as a procedure due to the presentation of Marxism as an epistemological tendency.

On another front, Popper's statements on the logic of social sciences, which seemed to solicit further questions concerning the fundamental problems of pure theoretical sociology -as underlined in the concluding theses of 1961, published in his famous *Der Positivismusstreit in der Deutschen Soziologie*, published in Italy in 1972- began circulating within the debate on the theory of knowledge which about to arise among some of the sciences and philosophers of knowledge, as auspicated by Popper himself. In actual fact, these topics managed to arouse interest in a broader discussion concerning the fundamentals of the social sciences, some issues of form or of method (*à la* Popper) and some of a logical-epistemological kind which might directly concern historians of sociology as a science, where the latter dwelt on issues regarding their own ability to keep up to date on theoretical matters - and during a 'short' history of the conceptual transformations of the empirical social sciences - tending towards assertions of advanced positions of research.

In Italy, the early 1980's marked a period of assumption of position by some scholars regarding the fecundity or not of Popper's theories of the conception of scientific knowledge or, in any case, of the identification of a logic of social sciences not due, necessarily, to opposition between dialectics and positivism, as had emerged during the Tubingen Conference of 1961. As regards this last point, it was the sociologists who were urged to take note of the utility or not of pursuing a solution in the wake of some questions instead of foreshadowing other objectives regarding epistemological history, having verified that the sociology's greatest difficulty was establishing an epistemological 'point of view' of its own. All told, the difficulties connected with claiming a theoretical history for sociology would have immediately revealed the impossibility for sociologists to practice an epistemology of sociology. Moreover, the refusal on the part of some authors to narrate the history of scientific sociology starting from the 'mists of time' became evident. By way of testimony we have the contributions of those who, in Italy, felt the need to sum up and report on the state of the debate on sociological knowledge, starting from the results of the previous ten, twenty years. Among these we wish to mention, by way of example, scholars and researchers like Sandro Bernardini, whose volume entitled *La logica della conoscenza scientifica (1980)* and essay *Il secondo Methodenstreit*, appeared in *Sociologio*, a book published in1983 which not only brought together the writings of Ferrarotti, Statera, Franco Leonardi and Luciano Gallino on methodological questions and fundaments, but also outlined a prospectus of sociological analysis in the light of the most recent studies carried out 'after the crisis' of sociological matters which, for some authors like Filippo Barbano, had never ceased to produce their own 'object' and which –having arrived at a certain moment- even at the cost of setting apart traditional theoretical questions on the object of sociology, addressed the establishment of a complex kind of survival, within a context of 'historicity' of science only; never outside of it.

In the long run, this stance of Filippo Barbano's made the issue of social sciences and sociology harken back to the paradigmatic changes undergone by the history of scientific subjects but also to admission of the fact that the 'paradigms' themselves, intended as that set of postulates and general theories commonly accepted and shared by a given community of scholars who referred to a certain type of knowledge and availing of particular tools and conceptual categories, might, finally, be identified with 'historicity'. From here, the different angle from which to view Popper's epistemology as a function of the sociology of knowledge, of the history and sociology of science. From here too, the distance from topics imposed by many on the issue of the rehabilitation of the metaphysical theory by the Austrian philosopher, in opposition to censorship by the 'Vienna circle', a matter which, to some extent, accompanied 'period-bound' reception by Italian scholars of social sciences of Popper's disciple, the Hungarian 'fallibilist' Imre Lakatos and also the writings of some of his pupils like John Watkins and Joseph Agassi. With reference to Agassi, who, furthermore, at the beginning of the 1980's focused on the sociology of science, we ought to recall the book *Le radici metafische delle teorie scientifiche,* 1983 (edited by Riverso), containing an essay written by the same author as far back as 1964, one year after the publication of *Towards an Historiography of Science* and of *Conjectures and refutations* by Popper. We should also remember the diffusion in 1970 of essays by Watkins against 'normal science', published in 1976 in the important Italian *summa*, *Critica e crescita della conoscenza*, a volume edited by Lakatos (who died in 1974) and by that other 'fallibilist' Alan Musgrave, published in Cambridge in 1970. It was with Watkins that many began to believe in the event of a post-Popperian epistemology, as announced by Watkins himself in *Science and Scepticism.*

Chapter 11

FROM AN EXAMINATION OF METHODOLOGICAL INDIVIDUALISM TO THE POST-EMPIRICIST DEBATE

In *The Poverty of Historicism* (translated into Italian by Carlo Monteleone eighteen years after the first English edition appeared in 1957, that is, the same year as his Milanese book *Scienze umane e metodologia, Weber, Popper, Durkheim*) Popper placed greater emphasis on methodological unity within the natural and social sciences, to later change his approach, by admitting the mutual presence of methodological differences, due to the existence of elements of rationality, especially in social situations. In a sense, like Max Weber, Popper was able to consider himself an advocate of methodological individualism, although his conception of human action differed from Weber's teleological instrumental idea, as he made frequent reference to certain non-intentional consequences of action.

En route, he was able to define his reference to a concept, or rather to a 'principle of rationality' within a clearer 'logic of the situation derived from reasoning, as models and economic theories. What appeared to count more was not, in the end, the identification of laws in an explanatory model of the social sciences but rather an 'analysis of the situation' or 'situational analysis' based on the 'principle of rationality' already defined. Again, Popper seemed to submit himself and his thinking regarding the social sciences to conceptual redefinition, where possible, referring, for example, to his position on the deductive-nomological model.

Reception between the late 1970's and the early 1980's of the terms into which Popper had begun assembling, particularly between the late 1940's and the end of the following decade, his conception of methodological individualism, was under discussion. This attracted the attention of those who sought to create, obviously from the 'point of view 'of the social sciences, a post-Weberian theory of methodological individualism. In Italy, this was easier to achieve by relating methodological individualism to the history and philosophy of science and to important logical and methodological issues.

Conceptual analysts, like Franco Leonardi, were then in a position to endeavour to place the results of assimilation of Popper's theories within a framework of a restatement of the problem of knowledge (or the gnoseological problem) with reference to scientific sociology, as a direct derivation from post-Weberian theoretical variants. In short, some sociologists were able to emphasise the importance of having more up-to-date thinking regarding the social sciences depend on the problem of finding a conductor-theme of an epistemological type.

In an article on the foundation of methodological individualism in the social sciences, which appeared in "Quaderni di sociologia" in 1980-1981, Leonardi addressed the problem of discussing, at a certain point, the theoretical-methodological development of individualism in the light of contributions made by authors like Friedrich August von Hayek, Popper and Alfred Schutz, confronted in their renewed 'consistency' with early-twentieth-century Weberian teaching on the social science. This path, the one taken by Leonardi looked like the right one.

Certainly, more than forty years of developments within the social sciences had been able to produce a major debate on the issue of the foundation of the social sciences themselves, capable of summing up the so-called 'history of sociological thought' in a historiography of relationships between logic and the methodological view and, finally, between these and theoretical advancements, surrounding an alleged historiography of the empirical social sciences.

This did not mean the birth in Italy of an epistemological historiography but rather, confirmation of a more pronounced interest on the part of social scientists in a number of methodological paradigms which had, so to speak, generated a sociological tradition, like the one of Weberian origin.

The historical problem seemed to be that of finding a theorisation to act as an anchor from which to start, to which to refer in order to lay claim to the logical and epistemological unity of the social sciences, while relating to all the rest.

It should not be forgotten that in those years the fruits of diagnoses of the relationships between some exponents of neo-positivism and sociology were gathered. This did little more than generate lack of attention towards epistemological issues-problems and, in particular, when the times were changing compared to those when authors like Statera recognized in Dewey alone and in his logic of scientific inquiry, clarification of the terms in which a Neurath might transpose epistemological problems into issues belonging to the 'theory of knowledge', running the risk of appearing truly contradictory. On this issue there remains Statera's essay *Otto Neurath e la sociologia del neopositivismo,* his *Introduction* to a volume of the original writings of the Viennese theorist printed in Rome in 1968. In actual fact, there was a time when it sufficed to follow Dewey's epistemology to convince oneself that epistemology could not simply exist if separated from problematic situations. And this was a purely methodological topic.

The fact is that scholars like Leonardi tried to find a way to bind Italian sociology to the European debate on the nature of logic and the epistemology of the social sciences, by adhering steadfastly to a number of the premises concerning the nature of scientific knowledge. There was no lack, within the context of the above-mentioned essay, of reference to some of the issues upon which Karl Popper had worked a lot during the very early 1970's with papers, essays and conferences until he wrote his *Objective Knowledge*, a theoretical study of 'three worlds' and/or of objective-world new approaches, products of the human mind, etc. ... a 1975 Italian translation of the original Oxford edition of 1972.

According to Leonardi, among the fundamental issues sociologists had to face was the theoretical possibility of discovering not only the existence of a quest for a methodological paradigm but also the fact that the empirical social sciences were practically obliged to adhere to methodological individualism, investigated, besides, even by scholars of Popper like the afore-mentioned Watkins and Agassi, in particular with regard to criticism of the ontological issues used to justify individualism itself and others posited in favour of general criticism of so-called 'holistic' perspectives. One should note that, during this same period, Leonardi, (along the lines of the theoretical research he conducted, while investigating some of the epistemological topics and consequences closest to the admission of the existence of paradigmatic interests in the social sciences) reflected on the opposition between individualism and holism, on the nature of 'constructivism' and on the specific logic of 'structural-functionalist theories'. All this led him to propose analysing the peculiarities of structural logic by carrying out an examination of possible interconnections between Popper and usual methodologists like Raymond Boudon. One should not forget, either, that many of publications issued between the 1970's and the early 1980's were veritable revisions of theoretical approaches informed by North-American structural-functionalist suggestions, as they were introduced to the West as *the* 'Sociological Theory ' *par excellence*.

One of the conclusions proposed by Leonardi in an article *La dissoluzione dello strutturalism olistico* he published in 1981 in the Roman journal, "Sociologia e recerca sociale" founded by Gianni Statera, was one denouncing the logical and methodological weakness of holistic approaches to the social sciences, after conceding scholars of those disciplines a kind of alignment with a quest for paradigms capable of keeping the interests of investigators of social and scientific linguistic issues focused on empirical matters and, in the long run, on the possibility of turning 'meta-empirical lexis into empirical lexis'. The reverse process might turn out to be highly counterproductive.

In this sense, this seemed to lessen the well-known reference by many scholars of social sciences to the fact that -again according to Popper- science seemed to emerge from metaphysics or, at least, to weaken claims that several supporters of Popper sprung up, in line with his idea of 'metaphysical programmes', destined, in large part, with Lakato, to become 'scientific programmes'. This was evident from the perspective of the philosophy of science rather than from that of a quest for aspects of a theoretical-conceptual, paradigmatic philosophy of the social sciences.

The sociologists, on their part, were able to appeal, up to a point, to Popper's discourse regarding metaphysics, leading towards a conceptual definition of the paradigm only in the sense indicated by Kuhn and, as noted before, by Merton too. This meant appealing –in the end- not to a "datum" of a metaphysical or ontological-gnoseological nature but to an earlier development of the 'scientific community', considered more and more as a necessary antecedent to the paradigm itself: except for premises of a methodological order. Epistemology on totally sociological bases could then take place. So, the problem of knowledge could be put to historians of scientific sociology in terms very different from those of the Popperian approach of the time (and of later), and, therefore, in terms truly new to the sociology of science as associated with developments within the fields of the sociology of knowledge and of general sociological theories. A prospect of interests far removed from what Popper had claimed with reference to a continuous quest for 'metaphysical programmes' and distant from his original hostility towards studies informed by 'social conditioning of thought'. Unsurprisingly, the reference harkens back once more to *The Poverty of Historicism.*

Several interests matured, in addition, in relation to ulterior and increasingly possible in-depth investigations, derived, on the one hand, from receipt by social scientists of the 'period-bound' contributions of philosophers of science who had succeeded in tracing a new conceptual itinerary and, on the other hand, from the receipt or refusal of some to accept the epistemological proposal of an 'alternative' pathway towards method. It would not have been difficult to understand that Popper's studies of the nature of concepts and their connection with crucial scientific-methodological issues would have had long-term effects: a new 'point of view' regarding the philosophy of science was establishing itself and with it a new framework for the development of criticism of the theoretical bases of post -Popperian epistemology itself. In Italy, as elsewhere, this meant a momentary and detached involvement of historians of sociology and/or historians of science as far as anarchist methodological issues were concerned, which, more often than not, were delegated to the interest and endeavours of other professional philosophers and epistemologists, due also in part to 'irrationalist' suggestions and 'crises' already examined. Quite often, the question was simply whether method might or not be treated in a different way from that of Popper and Lakatos, starting, for example, from the tradition of Bacon and Descartes. In this sense, some authors, who were more or less intentionally 'apologetic' towards the method, assumed responsibility, more than anything else, for entering into polemics, upon the arrival on the methodological scene of, having noticed the effects of acceptance of Popper's hegemony and of his conception of science. To this regard we recall the Pisan, Marcello Pera, and his studies of Popper carried out in 1981 and 1982, in memory of Paul K. Feyerabend.

It needs to be said that a kind of assimilation of the various problems was experienced by most social scientists within the ambit of academically-provided interpretive schemata. This turned out to be very distant from the theoretical coherency of much of the debate that, at the time, concerned the 'crisis' of the philosophy of science more, while other theoretical-critical 'neo-epistemological' studies were established by Grover Maxwell, Frederick Suppe, Patrick Suppes, Mario Bunge, Hilary Putnam etc. In Italy, reception of the major challenge against method carried out by the new philosophy of science forged in the English-speaking world seemed to circulate among scholars of the empirical social sciences who considered it as yet another aspect of the defence or not of Karl Popper and his theories, as a problem of assuming positions for or against someone or something (the world of the 'rationals' and of 'rationality', for example), when it was clear that, in future, general references to concepts and categorical systems, systematic beliefs, conceptual frameworks etc., would be formulated and constitute unique conditions within which to continue discussing the scientific research. Moreover, historians of science continued to recall the statements presented in *The Logic of Scientific Discovery,* which considered the empirical sciences as nothing more than 'systems of theory' and the theories as nothing more than 'nets cast' to capture 'the world'.

In the end, even the new historians of science were obliged to take note of the fact that a convenient way to continue the discussion was to work within the ambit of the development of 'conceptual frameworks', as the Anglo-Hungarian philosopher and scientist Michael Polanyi began to call them in an era of neo-positivism predominant, or within that of theoretical idealizations' and/or models designed by historians and philosophers of science like Stephen Toulmin Edelston or that of Kuhn's famous 'paradigms'.

That was, practically, what the new vision of the world and how 'to look at it' was all about, to use an expression drawn from Fayerbend's famous essay *Explanation, Reduction and Empiricism*, 1962. From critics of a universal model of the concept of science to 'paradigm seekers', the 1960's and 1970's had been able to bring an entire series of theses on the theory of knowledge to its destination. This way new images of science opened up new horizons of scientific work and epistemology, to the point of reversing or challenging it in works produced in various contexts by the 'sociology' of ethically-committed authors like Polanyi in *Tacit knowledge*, the development and the natural corollary of the famous text *The Personal knowledge,* 1958, translated into Italian by Emanuele Riverso and published the same year as Hanson's *Patterns of Discovery.*

Interest in Toulmin's work *Foresight and Understanding* (1961) written almost a decade after his *The Philosophy of Science,* translated into Italian in 1968, likewise aroused little interest in Italy. In this text the author dedicated himself- during a phase which brought a number of his epistemological theses to fruition- to studying systems of scientific concepts extrapolated from a vision where the concepts themselves related to each other in not too restrictive fashion as scholars of scientific disciplines and 'traditionalist' philosophers had imagined. We refer here to *Human Understanding* published in 1972, which Toulmin completed with his *Collective Use of Concepts.* For him, concepts and theories (later, neither true nor false) were mere 'transitory products', where a degree of hostility against science seemed to occur cyclically to assert both the opportunism of scientists on the one hand and, on the other, the re-emergence of subjectivity (for example, art, literature) and/or individualistic demands. We speak here, in part, of what Stephen Toulmin developed in *The Historical background to the anti-science movement*, this too in 1972.

The opinions expressed regarding this author by Italian scholars of social sciences were never unanimous, because his early interests focused on reason in various ethical systems, in that they were of interest to issues of language, like what Toulmin sustained in a paper issued in 1950 and called *An Examination of the Place of Reason and Ethics*: "There are no ethical words but only ethical functions of words."

His work might be summed up, in this case in particular, as a reconstruction of ethical discourse, corresponding to a rational language derived from analyses introduced by Wittgenstein and presumed to be true analytical 'techniques'. In any case, his contribution was partially accepted by scholars who read him not only from the perspective of the debate surrounding the outcomes of the post-Popperian philosophy of science but also according to a historical approach to the logic of scientific inquiry and the 'context of discovery' which had emerged from some of his observations of historical process by means of which scientific concepts might be transmitted from one generation of scientists to another. The American epistemologist spoke, in particular, of 'conceptual change' to account for the usefulness of the process itself and the changing nature of science and its practitioners, with a view to the development of important concepts and conceptual variants.

In *Human Understanding,* Toulmin, besides, provided historians of science matters with an exemplary critique of the use of the nomological-inferential method in terms of the reduction of the meaning attributed to formal-type arguments applied to define scientific 'explanation'. His intent was, as we shall see, to criticise formal logic and while also undertaking a critique of the epistemological treatment of traditional concepts, like 'cause' or 'causal explanation', for example. His proposal was that of replacing deductive arguments within the historical sciences with 'explanatory procedures' in order to represent them as general categories of historical analysis.

In reality, from the 1970's and up to 1982, Toulmin devoted himself to the proposal of a new theory of knowledge based on an evolutionary-type approach. We should not overlook the fact that his early theories on relationships between the natural sciences and the humanities seem to have given rise to some of the most interesting questions regarding the meta-ethics and metaphysics of science, at a time when the thinking of some Italian scholars and methodologists of the social sciences began to work on possible ways of framing scientific undertakings, with more or less explicit references to certain levels of abstraction, and, ultimately, following a pathway indicated by Toulmin himself in opposition to the hegemony of the formal logical approach. More specifically, terms like 'epistemology' and 'philosophy of science' were questioned and, in some cases, Paul K. Feyerabend's considerations expressed in his *Science in a Free Society*, published in Italy in 1981, concerning the philosophy of science, seen as a 'useless discipline', held for scholars, and, for this reason, when charged with the simple task of 'propagation', far more realistic than Popper's entire production.

We shall discuss these issues later but not before discussing other topics drawn from among the themes of new epistemology which may involve the judgment of social scientists about some of the problematic issues which traverse the life stories of some of the most influential authors until Feyerabend polemically debated issues "contrary to" method.

An Anarchist Parenthesis

One of the issues in question appeared to be precisely the contemporary diffusion-reception of methodological anarchism and relativism, of a problematic vision of science, which, in order to progress needed to know how to break the rules. Another theme discussed was the reception by and the diffusion among social scientists of the correct terms of post-empiricist revolution begun by Kuhn and continued up to Toulmin. We cannot exclude that around these two areas of the debate it may have been possible to develop a hypothesis of interaction between the two areas in order to produce an attitude among scientists towards an increasingly general kind of thinking as far as their own research activity at a given level of abstraction was concerned. But let us begin from the beginning.

Some of the works written by Feyerabend in the 1970's were translated into Italian at the turn of the following decade, though *Problems of empiricism* appeared in Italy in 1971. These works showed analysts that 'rational' categories for the interpretation of science, from empiricism to conventionalism and critical rationalism, might produce a deformation of historical reconstruction of topics related to the course of scientific matters, ending up by favouring simplifications, of use only within certain theoretical and conceptual frameworks strongly influenced by underlying historical conditions. Careful consideration of the matter showed that science lent its side to interpretational refractoriness, in that it might turn out to be, in the end, an undertaking characterized by the ineffable complexity of its construing actors and of its elusive interactions with the structures, prejudices and rules of tradition, or rather, of the 'dominant tradition'.

Feyerabend strongly doubted that the naive and sometimes simplistic rules that methodologists used to guide them were capable of accounting for a similar 'labyrinth of interactions'.

Science, like every other human endeavour, appeared increasingly richer and more complex than any rational reconstruction might explain, so that the very concept of 'scientific community' was destined to encounter a redefinition in a highly rough world where those who sought to speak in order to representing common awareness, might simply witness the continuous extension of the semantic area of 'science' to disciplines and individuals who encountered imponderables like luck, ambition and any other accidental which might, occasionally, condition a research activity and lead it along unforeseen lines.

Undoubtedly, one of Feyerabend's main tasks was to reflect on the possibility of building communicational ambits and creating common awareness regarding positive judgements of the heterogeneity of the 'scientific community', as well as the fact that different characters at play were bound to different, distant, separate worlds. In science, like in every other kind of human activity, there exists a tension between creative power and the need of order, the outcome of which can never be foreseen, considering the provisional nature of scientific work, on the whole. If Feyerabend showed that he was aware that this dialectical relationship might yield positive results, he was determined, nevertheless, to the weapon of awareness for human beings, so that many of those who studied him during the period when the debate between Popper, Lakatos and Kuhn was being received, were able to compare his ideas to an amply democratic conception of a variety if forms of knowledge, admitting that besides these tensions others existed that were simultaneously epistemological, ethical and political, arrayed around the axis of trust in human capacities.

One constant is that it was possible to be convinced too of the need Feyerabend felt to invest in human awareness (and therefore in the awareness of the scientist) as a creative power and creator. This situation made it necessary to recur to 'anarchist' criticism, whose paradoxes were to be bound to the need of awakening the consciousness of philosophers of science and of scientists themselves from a deeply dogmatic sleep. This way it was possible to underline a descriptive as well as a communicative plan, where a rich, prolific and creative humanity existed alongside powers which, due to specific obscurantist worries were tied to calls for order and the maintenance of the *status quo*, while regimenting matters through rules of exclusion. For the moment, obviously, this situation was analysed from the point of view of the world of science. It could not be excluded, however, that the same situation might have been represented within a broader social universe, so that social scientists, in particular, might understand the message and apply it to their own operational mode. Furthermore, the way they divided concepts regarding observation may have been undermined by the particular nature of human cognitive activity. As Feyerabend wrote, human enterprise showed how the same concepts, like perceptions, might remain ambiguous and depend on context, so that scientists would never be able to grasp completely the meaning of a concept because of its being used only in part in the attempt to find its components. To this regard it is important to recall Polanyi's teaching which sought the 'implicitness of knowledge' which revealed its indeterminacy. The attitude of the philosopher also meant making an effort to become a meta-communicator addressing an audience seeking keys to the analysis of given texts.

That is why Feyerabend loved to call himself a "Dadaist" rather than an 'anarchist' in that the former while not following rules, may also give himself some. The Dadaist assumed no position but changed his mind according to circumstance, convenience, drawing from his resources to build and demolish his construct. The Dadaist was, most of all, a meta-Dadaist: he professed, changed, denying his principles both in theory and in practice: 'to be real Dadaist, we have to be anti-Dadaists." Dadaism remained present as mimesis, a sort of protection to face, with greater conceptual freedom, arguments related to science and its history. Mimesis entered the scene as a challenge to critical intelligence and as a public declaration of its own 'influent-metaphysics' that adopted as its extreme defines the hermeneutical motto: "it is well not to take anything very seriously, especially what appears to be evident." Not even scientists were to be taken too seriously: their world was to present itself as changed both to Feyerabend and to his meta-readers. Conventionally, within the world of scientists, theoreticians were considered as ranking at the top of a scale privileged by tradition, followed by phenomenologists and, moving down the steps of the hierarchy, by so-called 'facts finders' and 'empirics, who worked in close contact with 'data' and measurements and whose tasks, in general, did not aim at developing grand epistemological theories. For Feyerabend, instead, the so-called 'theoreticians' who maintained naive conceptions of the truth and were inclined to hide behind the cloak of 'metaphysics', which ought not to have been beyond or outside of scientific thinking, but included as an integral part of the cognitive operation itself: a dull or chauvinist idea may lead to results just the same as the application of specific methodological rules may. At this point, all scientific theories might be treated as 'influent metaphysics', as Lakatos too revealed in the many points he shared with Feyerabend (and, vice versa). He demonstrated, for example, that Popper's methodology might be summarised in the rule "plan conjectures with greater empirical content than previous ones," the metaphysical assumption of which rule is: "an objective progress of knowledge exists"; usual anti-conventionalism postulating the elimination of exceptions embraces the theory: "nature admits no exceptions" and so on.

This version of 'influent metaphysics' led to the well-known and immensely widespread anarchist attitude in epistemology, where 'anarchism' for 'translators' (including some Italians) was taken to mean a true conception of the world founded on certain explanations and having a method all its own. Anyhow, premises related to that kind of position regarding the heuristics of historical appear in Feyerabend's *Against*

Method, where he provided a detailed analysis of the scientific activity of a famous character like Galileo, underling how the Italian scientist revolutionized the Aristotelian conception of physics by acting in an 'anarchist' way compared the consolidated rules of the time, exploiting opportunism, propaganda, creative intelligence and even ignorance, which Popper endowed with 'endless' potential with regard to science.

Galileo Galilei followed truly 'anarchist' precepts by applying new theories, having first invented and generalized them as veritable alternatives to previous conceptions, considered more or less valid. This way, by examining a number of 'historical situations' like Galileo's, this theory was based on the idea that it was possible to investigate the way science advanced, in the belief that epistemological anarchism might prove to be a modern consequence of the history of a critical philosophy highlighting the compelling strength of the 'ideology' concealed in any methodology whatsoever, wherever scientists invested in the absolute freedom of thought and action of humans acting responsibly in the field of science, and, for whom, any theoretical or methodological expedient might seem to provide an opportunity.

Epistemological anarchism asserted that the co-presence of a different theory of the kind was auspicable both for science and the society of citizens. In the 1980's this concept became quite widespread, conveying the 'deep conviction' that all methodologies, even the most evident, might have limitations and professing a critique of science restored thanks to investigation of an entire range of conceptual ambiguities capable of characterising it. The man-scientist seemed to have become at last the hub of epistemological interest, where scientific rules capable of posing concrete obstacles deserved, therefore, to be transgressed and rebuilt. The role of anarchic scientists was of the utmost importance because they contributed to the struggle from within against a kind of behaviour which had infinite repercussions on social life, making even the most surreptitious urges of society conform to its values. Scientists had, therefore, to bear witness to a new kind of emancipating behaviour: they might ignore laws in order to produce their opposites. One of the examples provided by Feyerabend was *counter-induction*. The famous *anything goes* formula was destined to encompass many of the conceptual intentions of 'period-bound' cognitive anarchists of the time while also determining a whole series of dissimilarities in the perception of issues by historians of science. In actual fact, epistemological anarchism ended by contrasting scientific theories with discourse regarding them, like someone seeking to oppose modern astrophysical theories with Aristotle's metaphysics etc. In these terms, some of the new issues of epistemological anarchism, as mere representations of old metaphysical matters, managed to make its way among contemporaries as an umpteenth, exclusive 'discourse regarding science'. Addressing the second of the two topics chosen here, we may go on to discuss some aspects of the post-empirical revolution achieved by the impact of Feyerabend's desecrating books, in the period between Kuhn's and Toulmin's writings. This also because, even in the face of epistemological anarchism, many social scientists, in the 1980's, in Italy too, chose to follow the route traced by Kuhn or Toulmin. For many 'period-bound' authors including Statera, Kuhn's work had triggered a veritable post-empiricist revolution capable of fostering reconciliation between the exact and the social sciences, between philosophy and the sociology of the scientific knowledge, causing a shift in the methodology of social sciences towards variously ' inclusive' and 'qualitative' methodologies. Many of these authors were led to believe that choices between one 'paradigm' and another might prove partially arbitrary because of being linked, as they were, to the decisions of a revered 'scientific community', to ideological environments and/or political pressure. This meant that the presumed 'rationality' of science a might become its weak point. In short, many authors and methodologists ceased to believe in a universally valid logic by means of which science might determine its progress, so that science itself lost that self-corrective character of which Popper had spoken for half a century. One of the problems seemed to be that if a choice were represented by a historical decision made from among alternatives not justifiable by an external, objective criterion, attention needed to be directed towards 'internal' criteria of selection, that is, towards *historical* as well as *sociological* and *psychological* criteria. In this sense, we need to take note of the fact that for Kuhn until consolidated knowledge in these fields was acquired, it was difficult to speak about progress. This view was held, in part, in the case of the new tendencies of the historiography of science, like those advanced by Lakatos. Moreover, these topics appeared to be very close to those both of relativism and of epistemological anarchism: some of Kuhn's conclusions regarded a number of issues which appeared later.

In Toulmin we find a number of characteristic akin to Kuhn's perspective. Some of these insist on the 'historicist' timbre of scientific work characterized by relativist accents present in conceptions of the continuity-discontinuity of scientific progress itself. While for Thomas Samuel Kuhn the progress of knowledge owed its origins to revolutionary phases, for Toulmin the course of science was characterised by a sort of coexistence between 'old and new." The same concept was developed in part by Feyerabend. The confrontation seems to have occurred between issues of methodological anarchism and some of those posited by apologists of science and its logic, as in the case of Paul K. Feyerabend and Imre Lakatos, for example. A theme shared by the two, despite their many divergences, seems to have been their sociological view of the progress of science. Another point of contact may be found in the research carried out by Lakatos concerning the alleged existence of concepts that might not be expressed in terms of metaphysical assumptions, for example, as mentioned above, Popper's methodology.

Is important to note that as far as the matter of development was concerned, the affinity existing between the work of Popper and that of Lakatos, also in the light of how they were interpreted by critics both of the time and after, who considered the latter's work a kind of transposition of the principal ideas of Kuhn into an 'objectivist' framework informed by Popper. For Feyerabend, part of Lakatos's work might be considered as coincident with anarchism and, rather than with Kuhn's theories, despite what appeared at first glance, Lakatos's position seemed more similar to Kuhn's when it came to the theme-problem of alternation between the famous internally and reciprocally integrated 'research programmes'. One should not forget that Lakatos needs to be placed among the *sophisticated falsificationists* who believed in the possibility that science might be able to ensure progress in any case by following fairly clear and well-known criteria so that a 'programme' might not be abandoned until a new 'research programme' was in a position to replace it. But the main dissimilarity between Lakatos's 'research programmes' and 'Kuhn's paradigms' lay in the emphasis the former placed on methodological awareness. This meant the existence of a metaphysical nucleus which due to a methodological 'choice' remained unquestionable, determining the 'nature' of the scientific programme itself. The Copernican theory, for example, was characterised by the basic assumption which remained un-falsifiable by methodological law and coincided with the idea that the Sun was at the centre of the universe. Another difference consists in the fact that 'the programmes of scientific research' prepare the terrain for a change seen not as a breach but as continuity with the preceding period. This was because, unlike the 'paradigm', the 'research programme' permitted a plurality of approaches to diverse problems to proliferate within it. It was in this sense that Lakatos provided for a distinction between 'negative' and a 'positive' heuristics, in defence of his theoretical and conceptual apparatus, which won the consensus of Feyerabend who found an anarchist position in his initial theories as these were characterized by the 'influent metaphysics' we spoke about above. Recourse to formula of 'metaphysics' sounded like 'considering the fact that there is no metaphysical criterion at the basis of science, we allow any kind of idea to spread, protect it and see where it leads us'.

It is a fact that between Kuhn and Toulmin we find the greatest theoretical contributions by social sciences to epistemology (and vice versa), where the 'anarchist parenthesis' in the history of twentieth-century scientific knowledge impacted considerably, especially in relation to a quest for cognitive fundamentals within the various sciences. Many scholars, including sociologists, remained 'seekers of paradigms; others were destined to redirect their quest for scientific knowledge towards other fields of study. The methodological framework appeared to change also with regard to reception of the conceptual problems and argumentations of the new historiography of science, affected by the more specifically 'internal' debate taking place and by issues deriving from cultural anthropology associated, from the beginning, with relativism (see the relationship between anthropology and relativism in the genesis of Feyerabend's thinking), or the psycho-socio cognitive issues dealt with by Polanyi -esteemed by Feyerabend- who introduced new concepts to describe aspects latent in every act of knowledge.

So, concepts like Polanyi's 'tacit knowledge' or that of 'exemplars' introduced by Kuhn in the *Proscript* to his 1969 *Struttura*, might be linked to other concepts produced within the ambit of epistemological, cultural and linguistic relativism, where Feyerabend himself preferred to refer to the world of science as a 'tradition', using 'incommensurability' to redefine the changes and conceptual alterations between its different 'systems'. The philosopher of science was to learn from the anthropologist, because the latter was more aware than other scientists, (or other social scientists) of having to deal with different cosmologies and different visions of the world, conditioned as he was by linguistic translations operating around the birth of new concepts, which were obliged to create their own specific language, as already held by scholars like Edward Sapir and others.

Rebuilding conceptual apparati as well as linking the definition of epistemological and anthropological relativism to the issue of 'incommensurability', was confirmed by some convictions expressed by Feyerabend in the 1970's, whereby the most promising theories of others were able to free themselves of ballast deriving from 'conceptual continuity'. The human and social sciences began to accept the idea of rather ambiguous indefinable knowledge, bound to previous experiences and perceptions, organized according to 'perceptions' of situations linked to the experienced reality as asserted in his studies by the American Norwood R. Hanson, one of the most influent scholars in the field of the empirical *theory laden* data concept which ended by damaging new empiricism in philosophy.

SCHOLARS OF META-SCIENCE

There is no doubt that the mid-twentieth-century construction of an alternative route to 'logic' and 'rationality' in science based on themes like anarchism and, in any case, on a broader and newer epistemological 'point of view previously developed, was able to affect the scientific and philosophical-scientific production of all those scholars who were dealt with in a more or less polemical manner by Popper, Wittgenstein, Joseph Agassi, Arthur Pap, Thomas Samuel Kuhn, Watkins and Lakatos and many others. Within every debate, of an epistemological nature too, historians of the empirical social sciences of the second half of the twentieth century had a unique occasion to achieve cognition in the presence of veritable revolution which impacted on scientific subjects as well as on the old history of scientific ideas. Unlike what had been asserted in 'official' academic circles for many decades of the last century, the role of the philosophy of science did not appeared so negligible, while the development of the discipline had not turned out to be an 'empty sector', at least not for those who sought to comprehend some continuity within scientific progress, no longer meant as an accumulation, sometimes discontinuous, sometimes not, of the 'endeavours' of scientists. This is also true for those who made a great effort to question themselves -more or less rationally- concerning the progress of scientific subjects, regardless of the fact that re-propositions regarding their genesis stemming from epistemological 'anarchism' made a *tabula rasa* of a simply 'rational' pathway, bringing science and its scholars face to face with the dilemma of imposing history upon it, to favour of a democratic conception of society, but for no specific purpose. Very often the political 'anarchism' of scientists themselves seemed to be their only goal. And to think that a number of contemporaries agree that Paul K. Feyerabend was one of the greatest authorities on issues of the history of science.

Reflecting, instead, on methodological 'anarchism' in the 1980's some Italian researchers (for example Antonio Fasanella in his *Anarachismo metodologico e scienze sociali*) were informed by an idea contained in *Science in a Free Society* according to which a servile pursuit of 'rationality' or of a 'theory of rationality' by a branch of science like sociology, might yield concrete results by having an impact on history considered 'sterile' because derived from 'pure methodology'. Here Feyerabend could not avoid expressing lack of trust in the epistemological history of a discipline which referred, instead, to a meta-reflection upon it, more useful, perhaps, when trying to establish perspectives seeking new outlets for post-empirical epistemology. Moreover, according to some Italian researchers, the 'anarchist' perspective in methodology ended by favouring the adherence *tout court* to so-called 'qualitative sociology' in a context of the development of the history of scientific matters which continued to pay dutiful homage to metaphysics, where epistemology tended to find itself partially assimilated by the philosophy of science which, in the meanwhile, had assumed the physiognomy of an 'autonomous discipline'.

The convergence of these two moments, on the one hand of the assimilation of epistemology by the philosophy of science, and, on the other, the irruption of methodological 'anarchism' onto the scene of the debate among philosophers who had already begun to express doubts concerning epistemology's 'foundationalist' programme, is relevant because there issues were to determine a rapprochement between European and Anglo-American philosophy. The event deserves, even if briefly, our attention, also because the opposition between philosophy intended as epistemology and that meant as hermeneutics, is an example of a strategic passage, mentioned, at the end of the 1970's in some books by authors who pointed out how even criticism of traditional ways of studying science, might have been deduced from ideas expressed by Kuhn and Feyerabend in their previous writings. For these authors (the Princeton 'philosopher' Richard Rorty, for example) the conception of epistemology could not escape the criticism from the notion itself of 'analytical philosophy', so that, when all came to all, the 'philosopher' in this new version might oppose epistemology intended as 'ideal of the foundation' by preferring hermeneutics to the latter: the idea whereby 'being rational' meant, precisely, abstaining from epistemology. Topics of this kind were addressed by authors such as Willard Orman Quine, called upon to express himself regarding epistemology and the results of studies of different "ontological" implications and Ronald N. Giere during the 1980's. Hermeneutics was considered simply as a way of getting around the problem of the foundation of a 'philosophical methodology' (epistemology) lying at the centre of philosophy once more, a veritable escape route for the close of the twentieth century. As was the case with the 'period-bound' authors like Rorty with his *Philosophy and the Mirror of Nature* or Asladair Machintyre with his *Contexts of Interpreation* (1976), it seemed far easier to create a combination between Feyerabend's best-known books and of that other censor of method Hans Georg Gadamer, than start from Kant to figure out what was to be known. Kuhn, too, with his 'incommensurability' seemed to count when determining interest in the hermeneutic method. Also significant was critical literature's examination of Kuhn, which sought to determine 'rational' agreement and disagreement concerning his positions which, in some cases, were (according to the critics) far from the tradition.

As regards this, we need to say that the interests of many, in particular sociologists and social scientists, in hermeneutics, were associated with the fact that hermeneutics might be identified, at the time, with everything arrived at by those who ceased once and for all being epistemologists, and, as a result, sought to look to their own disciplines and outline an epistemological history, knowing it to be derived from changes of 'conceptual schemata' quite unlike each other. Obviously the 'anti-foundational' perspective did not provide the only way to show philosophers and scientists how to avail of the new paradigmatic schemata. For some social scientists, hermeneutics was considered the final destination of the well-known controversy over 'explanation and comprehension', a dispute derived from a traditional debate about the 'philosophy of the social sciences'.

It was during this period that, on the one hand, saw the establishment of criticism against so-called 'qualitative' critical sociology ('hermeneutical', 'ethno-methodological', related to symbolic interactionism, etc.) and, on the other, that of the new discourse regarding 'science' introduced by scholars of history-related methodology. In other words, some Italian scholars began to register a growing disaffection towards methodology accompanied by meta-discursive ideas regarding the philosophy of science; two contingent issues which historians of science would have need to control in order to 'take stock' of conceptual developments regarding the sociology of science and its history. This operation was never actually carried out by scholars under the aegis of an epistemological history valid within the field of sociology and in that of the other empirical social sciences. And, as we have seen, the reasons were numerous indeed.

We shall refer to the development of a particular 'period-bound' debate between Italian and foreign scholars hosted by the Roman magazine "Sociologia e ricerca sociale," founded in 1980 with a view to fostering scholarly communications between sociologists, above and beyond 'academic' conditioning, which, to be honest, some of these authors were able to perceive, as time went by, in terms of their membership of the corporation of sociologists (in 1982 the Italian Association of Sociology was set up) rather than of claims tending towards theoretical and methodological options or those developed within the so-called 'schools' or 'minor schools' which sprung up in the 1980's. Nevertheless, at the beginning of the period in question, inter-disciplinary discussions regarding updates in the fields of the various sciences from biology to the social sciences were lacking, especially in areas like social biology and its research programmes, etc. In actual fact, it seemed likely that outside of some rather narrow areas of debate and discussion, the above-mentioned options could never be applied where the tendency of sociology as a discipline belonging to science was to move into the pre-paradigmatic stage of scientific development indicated by Kuhn, that is, into that phase where the so-called 'schools' would certainly play a role, but not in favour of recognition of the peculiarities of a strictly epistemological and methodological nature. In Italy, the existence of the so-called sociological 'schools' indicated a special event within a situation which already presented itself as anomalous, where reference to certain paradigms might yield the same result as having to refer to the existence of a 'scientific community'. One of the many responsibilities of the Italian historians of sociology as a scientific discipline was not having known how to grasp, at the time of the development of the sociological disciplines, the salient features of a variegated paradigmatic model of sociological science.

In many ways some forms of early-1980's debates compensated for a number of *défaillances* on the part of historians of the science of sociology, where there existed a tendency to initiate prospects of interpretations and investigations of scientific subjects against a background vaster than that outlined by 'epistemology' or the 'philosophy of science' and which, even at terminological level, might be resized or otherwise modified by researchers during use. In practice, research might be a part of the effort to bring scientific work within a new area of semantic space at a 'higher level of abstraction' as in the case of the common use of the term 'meta-science' meaning 'discourse about', as in 'meta-language' and 'meta-concepts'. The intentions of some scholars regarding even limited topics and issues are found in the introduction by Alberto Marradi, professor of the methodology of political science at Bologna University, to a monographic issue (April 1984) of "Sociologis e ricerca" dedicated to scientific knowledge.

The publication contained a number of relevant contributions including an essay by Leonardo Cannavò which took stock of the conceptualization and logic of research into sociological methodology, thus completing an examination carried out in 1983 concerning the origins of the operational programme in epistemology, and operational analysis in sociology, starting from an article presented at a Conference on the *Validity and reliability in the construction of sociological variables* held at Trent the previous year. The 'Atti' (minutes) of the conference were published in Milan in a volume edited by Antonio De Lillo, containing the activities of the Methodological Work Groups and essays on specific topics like trust in surveys, by authors of the calibre of Ermelinda Boccuzzi, Joseph Giampaglia, etc.

The two essays written by Leonardo Cannavò in "Sociologia e ricerca sociale," were, in reality, a first 'showdown' within the ambit of reception of operationism in Italy and were related to the era's 'period-bound' criticism of methodologists and of some results due to reception of Lazarsfled's 'paradigm'. Furthermore, they also opposed formalizations by American authors which had contributed to the spread and/or reinforcement of operationism as a component of the *natural science trend* (Cannavò recalls, among others, Stuart Carter Dodd, author of the *Dimensions of society*, a text where he developed the formalized sociological theory expressed in mathematical terms). Another reason why these articles on operationism made an important contribution to the ambit of 'meta-scientific' interest, was the way the author of the essays opposed Italian sociologists' impedimental cultural attitudes towards 'receiving', and, later on, reasoning in terms of definitions of sociological concepts of scientific thinking, and, in any case, of referring to 'internal' development of scientific subjects themselves. This way, it needs to be said, that a certain kind of reflection should have been carried out by historians of sociology, where operationism belonged undoubtedly to the history of science. The same discourse, precisely in terms of definition, began with an essay by Alberto Marradi, published in the same monographic *Metascienza* volume *of* 1984, which focused on the use of the term 'theory' in the social sciences and on a 'type of meanings', seeking room for 'meta science'. Marradi discussed many existing typologies in vogue among researchers as well as the numerous meanings of the term 'theory' as used both in everyday and scientific language.

From an 'interior' developmental 'perspective' of the history of science, the author examined positivism's treatment of 'law' as a replacement for 'theory' up to the decline of the inductive and objective approach, and, simultaneously of the fiscal epistemology of positivism, capable of indicating a veritable breakthrough in the field of the various research procedures and/or activities of scientists, with respect to a higher degree of theorization applicable to the social sciences. Marradi was led simply to insist upon the scarce contribution made by positivism to epistemology, while reaffirming, on the other hand, the progress made, from an anti-positivist perspective, by the debate on the *theory laden* approach, to the point of providing many semantic expansions of the word 'theory' and advancements of treatment of existing relations between concepts and 'theory'. In practice, investigation into the relevance of terms was important for the debate between devotees of 'meta-science' would have been desirable where thinking on the formation and use of concepts within the empirical social sciences had become important from a historical-epistemological point of view. Moreover, those were the days when systematic analyses of the language of social scientists were carried out, leading to publications like *Social Sciences Concepts (1984)*, a book featuring writings informed by the mythical "Committee on Conceptual and Terminological Analysis" (COCTA). The authors ranged from political scientist Giovanni Sartori who was initially devoted to sociology to Fred Riggs, etc. Theirs were essays where the writers indicated not only the relevance of analyses concerning logical-semantic issues and, as a result, the 'terminological confusion' existing between scholars of political science and sociological subjects, but also the importance of applying the conceptual-analytical method to 'bridge' the epistemological-historical gap existing within the disciplines involved, or at least, to report on the progress being made by scientific sociology's examination of concepts and terms. To this regard, and precisely concerning the results of the process made by some of the founders of the COCTA in the 1970's, Marradi, in April 1987, issued an article in the "Rivista Italian di Scienza Politica" (year XVII, n. 1) where he linked concerns originating from the world of social scientists obliged to tackle 'certainist' epistemology.

Returning to the 1984 essay one notes that it was the author's intention to discuss a given perspective following –as closely as possible- the view of historical developments of scientific matters, where pursing, up to a certain point, 'restrictions' and the 'extensions' of the term 'theory' really meant a new examination of the various developmental stages of sociology and the empirical social sciences in the light of new and/or future perspectives. The same might hold for a redefinition of some concepts and the sociological tools concerning relationships between quantity and quality. We should remember that at that time, in Italy, it was Marradi who reignited the debate among methodologists of social and political sciences on the issue of more or less a-critical reception of conceptualizations produced within the North-American tradition. This debate produced a discussion around psycho-metricalist Stanley S. Stevens and the problem of measurement (on this point, see Marradi's *Misurazioni e scale* 1980-81) to the point of involving scholars like Franco Leonardi and Gianni Statera, called upon to intervene, as well as many young researchers including Luca Ricolfi, Leonardo Cannavò, Maria Concetta Pitrone, Paolo Ammassari and many others. On its part, the use of the term 'theory' had managed to connect existing classifications and meanings to taxonomies produced in the 1970's and derivable, in turn, from the use in scientific literature of relevant distinctions, like those between 'theory' and 'hypothesis', between 'propositions' and 'laws' and so on, as anticipated by Marradi in the 1980 *Concetti e Metodi per la Scienza Politica* manual, reprinted the following year in a revised shorter edition entitled *Concetti e Metodi per la Ricerca Sociale*, and, finally, again with this second title in a third enlarged edition published in 1984. This text, republished several times in anthologies, introduced a number of epistemological issues based on the author's intention to clarify some fundamental concepts in use in the field of research within the empirical social sciences and reasoning also with regard to the strategic transition from 'theoretical constructs' to 'observational issues'.

Approaching reconnaissance work on similar topics, paying particular attention to the overwhelming influence of epistemological arguments regarding scientific literature, and in part, the interests of methodologists, for Marradi and other 'period-bound' devotees of 'meta-science' meant being able to follow a debate of a certain importance regarding the treatment and reception of issues of gradually increasing interest to 'new epistemology'.

Not surprisingly, the 1984 special issue devoted to topics of 'meta-science', opened with an essay by Toulmin translated as *Filosofia e teoria della scenza. Storia e sociologia della scienza*. Here Toulmin demonstrated, more clearly perhaps than usual, how it was possible by then to illustrate a particular view of science subjects, by availing also of contrasts between 'points of view', with due reference, however, to the new course of epistemology, which needed to rely on social scientists if it sought to deal with the history and philosophy of science. Toulmin's argument was very simple: the history and philosophy of science had, he believed, the framework within which both the sociology and the theory of science might operate.

It is useful enough to understand how and why Toulmin arrived at a similar proposition. According to him it was necessary to consider a number of observation points leading to the conclusion whereby due to stimuli by factors, 'external' or social and 'interior' to the knowledge of experts, scientific knowledge might be intended as something intellectually and socially inseparable from all the rest, from the political and social life of men, from their everyday activities. The history of science was to be treated, therefore, as an important practical occupation by those who sought to engage in empirical research aimed at showing the world simply for what it was. From the 'point of view' of the theory of science this meant that a conceptual effort made to link propositions, formulate hypotheses or operate following a certain logic etc. might come up against a tendency to practice a theoretical knowledge of society.

Toulmin's message was addressed, in part, to social scientists and methodologists at a time when many topics met with growing disaffection even as far as methodology was concerned, as underlined in an essay by Statera published in the same monographic issue of the above-mentioned journal. Statera denounced the anti-methodological '*Nouvelle Vague*', not so much in contrast with philosophers who questioned themselves regarding the problematic of the cognitive fundamentals of science, as with initiators of new trends, who posited a sort of demarcation between them and the other kinds of knowledge we have just discussed. This essay brought to the surface the quarrel between 'quantitativists' and 'qualitativists', where the paradigmatic contest between the two approaches in the making meant the rebirth of a debate among 'competitors' within Italian sociology and provided new topics for the historians of scientific sociology at a time of anarchist 'drifting'. Historians of scientific issues were permitted to investigate two different conceptions of the purpose of sociological knowledge, where the emphasis was placed, on the one hand, on the need to come up with results in terms of "regularity trends" concerning above all the macro-social environment and, on the other, the tendency to assert that the purpose of scientists was to reach an 'intentional' comprehension of society and of 'social matters', otherwise unreachable by means of methodologies based on questionnaires, on multi-varied analyses, etc.

The fact is that the questions posed by some insisted on the importance of clarifying the type of inter-current relations existing between the debate regarding the post-Popperian scientific method and the birth of new approaches. Sociology might have considered itself involved, where, for example, proposals by Kuhn, Hanson, Lakatos and even Feyerabend seemed to favour a 'qualitative' approach and/or have a direct influence over the 'empirical' way sociologists thought and acted. To understand this point better, one might read the work by Peter Winch called *The idea of Social Science,* 1958. In actual fact, each of these strands revealed their theoretical matrix clearly, however fleeting and complex, as was the case with Raymond Bourdon's methodological individualism and Harold Garfinkel's ethno-methodology, that is those theoretical perspectives based on the work of a master.

Statera's polemic still reflected some of the apprehension regarding solicitations questioning the solid basis of social sciences' 'standard' methodology, in an era of detachment from those difficult acquisitions of the past which both sociology and sociologists, most of all Europeans, considered as ineligible historical references in view of a possible dialogue between the many 'national', theoretical and methodological positions it was made of. It was precisely concerning relations between the results of a given anti-scientific time and the 'weaknesses' of the theoretical/methodological structure of the so-called social sciences that Franco Leonardi wrote in 1986 when dealing with the epistemological problems of social sciences in *Di che parla il sociologo?*, when urged to voice some perplexities concerning the fact that among all those 'sciences' which one might call 'human sciences', only among sociologists was it possible to find a group of "those who deny the trustworthiness of the knowledge *their* science pursue. Economists, psychologists, anthropologists, historians, linguistics, semeiologists (but also logicians, epistemologists, etc.) -generally speaking- had sufficient trust in their cognitive enterprises and their chances of developing their sciences (…). Nothing of the kind – Leonardi concluded- happens among sociologists."

Besides these fears, Statera in his essay showed how certain mainly theoretical problems introduced by post-Popperian epistemology might be considered, broadly speaking, as being well-founded and metabolized by the scientific and collective conscience. Gianni Statera, more than others, underlined the need to consider the inevitability of a potential fracture which, at some levels, might take place between theoretical reflections on method and the practice of empirical research. It appears probable that epistemological instances which led to critical discussions of method might prove equally useful as a guide to the work of scientists as the rules of investigational logic and the operationalization of the conceptual apparatus defining a problematic area. In actual fact, within the logic of the social sciences, the principle of the pre-eminence of theory remained evident, as well as basing the research derived from it upon *data*, also within the ambit of 'qualitative' research.

As to the methodological discussion proper, what Statera found fault with in the 'new', 'comprehensive' and 'participative' approaches to the sociology of his day, was, most of all, the fact of having been too tardy and too swift to address the mature teachings of Alfred Schulz and Wittgenstein in *Philosophical Research*, where the themes of relativism, later developed by the post-empiricists, seemed to be enclosed. One of the cases which, according to the author, might be considered exemplary and for this very reason needed to be discussed, was that of the American sociologist John A. Hughes who drew inspiration from conception Scultz's concept of 'living worlds' and 'finite provinces of meaning', to contrast the cognitive capacity of a uniforming tool like survey by interview. The equally and erroneous target of sociologists *à la* Hughes, was, according to Statera, the positivistic conception of sociological knowledge, not in its 'Vienna circle' version, but rather in that of Emile Durkheim. The problem seemed to have been that of coming up against (in some cases, not) with the absence of methodological awareness. From here, very often, even the refusal of positivism which did not imply a refusal of sociology. Moreover, even a lack of methodological awareness (as in the case of Popper's delimitation of science and the historicity of the principle of truth in philosophy and of Anthony Giddens's in sociology) led qualitative sociologists to follow the propositions of authors like Feyerabend, Derek Philipps and Edgar Morin, not to criticise (something widely acquired, according to Statera) paleo-positivistic 'methodologism and 'sociologism', but simply to escape methodology and science. From here stemmed the effort confer proper perspective on the ideas of philosophies more critical towards scientific methodology as well as the attempt to show how the most profound sense of 'anarchist' and 'relativist' criticism might be defined as not incompatible with the logic of research.

According to Statera, Methodology (with a capital 'M') to which Feyerabend, for example, referred should be a sort of methodology created *ex-post* (like Popper's *Logik*) and not what scientists actually followed during empirical research. This statement is so important and can reveal that prescriptive logic which Feyerabend considered practically 'unreal' may be more appropriately identified with discovery than with investigation. In reality, Statera held, Feyerabend himself would have had no objection to make with canon of a correct logic of investigation which "departed from problems to arrive at problems." From a historical 'point of view' of science, Statera seemed to agree with Feyerabend concerning the non-existence of a "methodology independent from history, psychology and physics inexorably guiding our steps," because method "is linked to the object of investigation, influenced by socio cultural conditions of different kinds; it is closely connected to the paradigm concomitant with which it develops." Statera, agreed also with Lakatos and 'common scientific wisdom' and, more so, with the Kuhn's addendum to *The structure of scientific revolutions* 1969, when the US philosopher clarified the fact that the debated concept of 'social consensus' as an arbiter of scientific progress, was meant in a precise way also because it referred to that particular group, the scientific community.

On the whole, the confused and multiform *vague* denounced by Statera in the above-mentioned essay, may be related to the Italian intellectual reality in a context of the spread of anti-scientific thinking which ripened after 1968 and noticed, with some delay, by 'period-bound' historians of scientific issues, until the debate on 'meta-science and methodology. Statera was undoubtedly one of the few to know that the Italian epistemological reality was running the risk of having different anti-scientific instances from several contexts (France, England, Germany and the USA) converge. Without forgetting the ideological matrix of some of these tendencies analysed, moreover, by a number of Italian scholars closely related to sociology and, as a result, to 'anti-sociology'. To this regard we need to mention an interesting essay by Paolo Ceri, published in the four-monthly review directed by Luciano Gallino called 'Quaderni di sociologia' in 1985, in number 4-5 of that year, dedicated to Renato Treves (who died in Milan in 1992) and to the development of sociology in Italy and following a monographic issue a comparative study of Marx. In reality, Ceri takes stock of the growth of scientific knowledge while also soliciting an exploration of the critical and historical sociological aspects of sociological premises and their results in the scientific field. In short, having arrived at a certain point, Ceri sought to re-examine topics-issues which seemed to constitute classical views of epistemological thinking regarding the social sciences, in the light of a desirable kind of 'theoretical pluralism' capable of questioning previous conceptions of sociological science while underlining new philosophical and historical thinking about both it and scientific matters in general. The author in question actually ended by carrying out another rather important theoretical operation; he went in search of basic references to treat as sociology's conceptual or reference model. Crossing the two dimensions of *individual* and *collective action* he obtained a four-part configuration of the field of *social action* named *possible*, whose determinations are 'social movement' (maximum collective action and maximum individual action); 'community' (maximum collective action and minimum individual action); 'institution' (minimum collective action minimum individual action); 'market' (minimum collective action and maximum individual action).

Similar basic concepts of sociological theory described four types of social action, but often corresponded to some models of social order which might have constituted a *Gestalt* guideline (that is, particular conceptual models) of the entire set of activities involving addressed by social research. It is sufficiently clear that the author of the essay referred also to the crisis of neo-positivism and the substitution of the concept of *Received view* with *Weltanschauungen Analyses*, intended as an interpretative element, on the one hand, and foundational on the other, by representing the various paradigmatic options dealing with the social reality as a true 'object' of analysis. To each of these paradigms corresponded not only a number of different accentuations and/or 'visions' of phenomena and relations within them, but also the conception of potential explications of other *Gestalts*.

This examination places us on a plane which, at times, seems to portray the despicable behaviour described, at that time, by some 'period-bound' scholars like Feyerabend, who signalled chauvinism within science as well as attitudes of intolerance and exclusivity of knowledge, directed not only against extra-scientific forms of knowledge, but also against those within the discipline, against other currents with which it was possible to establish meta-paradigmatic theoretical and methodological convergences. It was Ceri's concern to point out that during scientific and sociological activity, disproportionate adherence to some of the paradigmatic options indicated, might lead to division, to 'polarisation' and generate negative consequences in terms of research. From an epistemological 'point of view', privileging one thematic side to the detriment of others might cause devaluation of the social relations considered by other perspectives. A *Gestalt* closed within a more open *Gestalt* effectively limited research and homologated the relationships forms within an absolutist scheme, reducing the restriction of the border existing between sociology and ideology, thus transforming social research into its opposite, emptying it of the predisposition to open up towards all possible conclusions even if supported by a reference theory.

This, obviously, from a socio-political stance. A second negative effect of 'polarisation' lay in the *a priori* non-legitimation among 'currents' which came to clash with each other polemically, thus missing the opportunity of comparing notes in terms of results and hypotheses. Further profoundly anti-scientific effects were those caused, in some very particular cases, by the de-promblematisation of the object of investigation and hypotheses as well as by interdisciplinary fragmentation, which, in terms of 'community polarisation', might have facilitated the dialogue among some disciplines, excluding some others.

It needs to be said that the author in his essay strongly advocated reacting against this anti-scientific, and, consequently, anti-sociological phase, closely associated with a sort of ideological radicalisation, by identifying some 'de-polarising' rules and using the weapon of *theoretical pluralism,* just mentioned here, as veritable sustenance of science growth on the whole, that is as theory and hypothesis, methods and models of investigation. Ceri's analysis also signalled a particularly important moment in the public definition of a discipline like sociology in relation to its very object, and, in a certain sense, 'periodised' (an experimented term used in the Department of Social Sciences at Turin University by 'period-bound' historians like Barbano and a few others) even in its anti-sociological tendencies. Ceri discussed four historical phases. The first was the early 1960's when Italian sociology responded to the cognitive needs of *social issues*: the reference model was the community, from a universalist and humanitarian perspective and in the absence of true 'polarisation'. According to the author, in this situation anti-sociologists are those most politically conservative, idealists and/or the idealists of a 'school', all of them Marxists. The second phase ended at the beginning of 1968 and has as its fulcrum analyses the grand processes of modernization and secularization then taking place; it was during this phase the diffusion of some reference models like 'community', 'institution' and 'market' took place enriching research confined, however, within a more 'polarized' framework. Anti-sociology, in this case, was present among groups of apologists of progress and among Marxists most of all. The third phase ended in the mid-1970's and was characterized by spreading scepticism regarding the progress of the paradigm, where increasingly social tensions drew the attention of sociologists towards politics rather than towards the development of the concept of class, capable of replacing the notion of group in research. During this time span, the decline in the distrust of Marxists was matched by an increase in 'polarization'. Finally, the fourth phase continued right into the mid-1980's and was characterized by a significant increase in 'polarization' between a sociology of subjectivity and one of repression; 'community' and 'market' became alternatives and dropped connections between the two matrices.

What Ceri denounced in the 1980's, after many phases in/of 'crisis', looks like a process of breakdown in Italian sociology's theoretical and conceptual space and the emergence of eclecticism and of the habit of following the most diverse theoretical combinations leading to yet another and subsequent failure to develop particular historical epistemological references, during a period of important change in the public image of science. By following the route taken by Ceri it was possible to add further arguments to the discussion of the anti-methodological polemic illustrated by Statera, in 1984, without overlooking the existence of certain epistemological situations which might have solicited stances among sociologists converging in a debate which, in reality, in Italy never lasted a lot, not even among historians of scientific issues, despite Statera's umpteenth invitation (see *Le epistemologie post-empiriste e la logica dell'indagine sociologica* "Sociologia e ricerca sociale," volume n. 33, 1990) to note the continuity existing between the various epistemological conceptions of the twentieth century but only in order to affirm the insubstantiality of the epistemological debate within the ambit of sociological action as 'research'.

In this context, the 'official' historians of sociology lent their flank to misleading criticism, sometimes of a purely idealizing nature, in a climate of intellectual 'fads' and reception of particular 'new' currents, very often confused thanks to the production and reproduction of conceptual mirages of sociological theories of the social and cultural processes of modernization. Authors like Luciano Gallino, in their mid-1980's studies, pointed to the reduced potential of some of the sociological theories drawn up in the face of particular contexts of development, like Italy. Moreover, not always sociological theories were used to stimulate historians of science, and not always sociology managed to produce adequate historiographical research regarding the course of scientific issues. A number of analyses were carried out at the time chiefly with regard to the obscuring process of Marxism and the success or otherwise of positivism from the years of the so-called 'renaissance' of the 1970's, etc. One such analysis Filippo Barbano's concerning 'early' and 'new' sociology in Italy, which adhered to Ceri's lengthy and detailed socio-methodological thinking, and was published in the above-mentioned "Quaderni di Sociologia," 1985. Other analyses went into greater depth on the development of historical and epistemological histories of European science, in particular the course of denouncements of misunderstandings of the neo-positivist position in relation to the exaggerations and exterminations expressed by the so-called post-empirical philosophies. This was discussed by Statera when he attempted to take stock of the contestation -more or less advanced in terms of time- of the idea of the existence of an 'objective' science that might, by means of observational assertions and rigid protocols linked to the 'physical' reality, make definitive acquisition of knowledge a reality. This idea, according to Statera, was already incompatible with the positions expressed repeatedly by Otto Neurath and even by Moritz Schilick, the closest to 'realist' views and therefore representative of a minority position –held to be ontological and metaphysical - within the Circle of Vienna. At the time when Statera tackled the issue, he also took stock of a relativist 'change of direction' in epistemology, indicated, in Italy, by a number of 'collectible' works by philosophers of science, like that on contemporary epistemology by Rosaria Egidi, published in 1988, which followed, to a certain extent, the thirty-year-long reflection on logical European empiricism, which, starting from 1930, had characterized its fate. In some of the works mentioned by Statera it was possible to find reference to transformations of the image of science of use to the development of sociology and the social sciences. These were less useful to the development of the philosophical discourse which referred to the transformations themselves and the various meanings of 'conceptual' or 'cognitive' 'epistemological relativism', as well as the critical positions of upholders of 'conceptual schemata' and scholars like American philosopher Donald Davidson, who, referring to major conceptual changes manifesting a particular vision of 'points of view' derived from his treatment of different outlooks. Furthermore, Statera, refered to epistemological relativism related to objectivism and the stance assumed by other epistemologists like the Swedish Gerard Radnitzky against methodology, etc. In the latter case, it would have been necessary to discuss the functions of the theory and procedures of science, as Radnitzky suggests in an essay contained in Egidi's book on epistemic relativism (in the section dedicated to the theory of incommensurability), where the author also discussed Popper's 'alternatives' to logical empiricism.

THE SOCIOLOGIES OF SCIENCE IN THE AGE OF RELATIVISM

In an essay dated 1990 dedicated to so-called post-empirical epistemologies, Statera indicated a link between analyses of certain misunderstandings and post-Popperians and pointed out how it was possible, from 'the point of view' of a given kind of reception of neo-positivism and also by availing of post-empiricists criticism of the scientific method, to find similarity of intent in scholars like Neurath, Carnap and the logician Carl Hempel, which was reassumed, at various stages, by the post-Popperians themselves. In brief, according to Statera, a demonstration of this would mean revealing degrees of continuity between the various methodological issues of the twentieth century and, above all, isolating certain anti-methodological polemics thanks to a reassertion of the solidity of rules governing the autonomous logic of scientific investigation, to the extent of affirming its healthy problematic nature. All told, Statera was obliged to reflect on the results of the entry into epistemology of historical, psychological and sociological analyses, keeping in mind the 'rules' which, not only for him, guaranteed the scientific nature of method. To this regard, it is important to note that science had to move from a need to provide a set of results (statements, laws, theories) producing *controllable* results.

All told, the problem for Statera seems to have been that of addressing the alleged autonomy of the logic of investigation, at the one hand, and the 'pseudo problems' of epistemology, on the other. His 'apology' of logic and procedures tended urged him warn researchers, once and for all, of the danger of evading initial conditions; the controllability and communication of scientific experiences. The classical scheme foresaw moving from an already theory laden *problem*, explicating it within a theoretical framework to achieve a conceptual clarification expressed as a hypothesis, then 'operationalizing' a theoretical plan, anchoring the different dimensions of the concepts to empirical indicators and finally to summative indices expressing measurement.

One of the greatest dangers contained in the vaunted anti-methodological *nouvelle* and its trends was precisely that of denying the need giving rise to scientific discourse and to that resulting from the method and the rules it contained. On its part, the logic of survey (in its more mature version) was seen as a meta-paradigmatic example for the scientific community; whereas, 'problems', 'concepts', 'indicators' and 'hypotheses' constituted a scheme to which to refer thinking refractory to the theory which, in turn, might influence the very 'feel' of the problem. When all comes to all, method constituted a sort of 'logical form' where the investigational process found a fixed reference point, so that the method itself might act as a meta-paradigm of the investigation, as clarified further in *Il motodo e il suo contrario* (1991), by Enzo Campelli, a student Statera's.

This study on the recovery of the issue of method in sociology - in many of its parts taken directly from John Dewey- considerably clarified stances regarding reception of pragmatism which hypostasized the logic of investigation as a point of steadfastness, essential if one sought to remain to within the domain of science, to be regarded as a 'human enterprise' devoid of transcendent attributes and endowed with rules applied so as to avoid losing the identity and peculiarities of its vision. The problem seemed to re-propose, in terms of temporal growth, the never-ending polemic between the natural and social sciences, with privileged referents among scholars of knowledge and historians of scientific sociological matters during the second half of the twentieth century. This in an era when relativism 'distrusted' knowledge and there was a renewal of the methodological debate concerning the presence or not of 'equivocal elements' in communications between different conceptions of 'anti-methodology', as well as many 'weak' aspects and 'strong programmes' in sociology, all topics treated by Campelli in his volume, and, given the premises, in a number of studies and contributions of the late 1980's.

For some historians of scientific matters and methodologists, the assertion of relativism in scientific knowledge, meant, being influenced, in part at least, by Kuhn's relativism which led to a wide-scale debate on the explicative and epistemological possibilities of the 'paradigm' concept, a theme-problem about which many social researchers wrote, following typologies, like that of Margaret Masterman etc. or raised terminological objections, like Rita Cavallarodid (see *Paradigmi e Sociologia. Un dibattito insensato?,* 1984), etc. The fact is that issues of the relativism referred to the reality of scientific knowledge opened up a broad range of studies and research for sociology and psychology: large areas of action capable of investigating scientific production as a process to place under observation, most of all during moments of change. It was then that social scientists, and, in particular, sociologists of scientific knowledge became 'paradigm seekers' as did many historians of science. Sociologists of knowledge, more precisely, some sociologists of science were also able to claim primacy in the face of philosophy and pursue the pathway of reaction and follow a 'strong programme', that is, a coherent project of construction of an analytical view of science capable of departing from Kuhn's instances to arrive at some relativist epistemological positions. This was the position of the 'Edinburgh school' and of scholars like Barry Barnes and David Bloor who, from the end of the 1970's, sought to place epistemology within the confines of the sociology of scientific knowledge, arousing a certain amount of interest in Italy amongst sociologists like Leonardo Cannavò (author, in 1984 of a monograph on the theme), Statera himself, Maria Carmela Agodi, Vittorio Ancarani, and others for a whole decade, starting from 1984-1986 when a number of studies were dedicated to underlining some of the limits of the enounced 'programmes', which became a commonplace for scientists up until the beginning of the 'constructionist' current of the 1990's (from Bruno Latour on, etc.).

Apart from the contribution of some historians and sociologists of science, not so famous in Italy, like Donald MacKenzie, Andrew Pickering and Stephen Shapin, involved in work similar to that of Barnes and Bloor, it must be said that as regards a conceptual re-proposal of the sociology of scientific knowledge had to deal with relationships between the outcomes of relativism and 'the sociologies of science', in that it interested the behaviour of some researchers as far as values regarding the 'rationality' and 'truth' of scientific practice were concerned. In short, some European sociologists of science who had interiorized the age of relativism, suggested reflecting on the issue of using fewer rational, universal criteria and principals to evaluate scientific theories but to opt for positively evaluating some practices which remained within particular social and cultural contexts.

Robert King Merton himself, from the American aspect of his 'explorative' studies of the sociological science, held that the emergence of a number of interests and disciplines would have taken place most of all when they were presented as social issues. It was by no fluke that some of the problems accompanying the development of the 'sociology of science' during the age of relativism resided in the crisis of people's opinions of science and in the fundamental values of modern society. This issue was interpreted by a number of European sociologists, in particular David Bloor, who analysed - during a period between *Knowledge and Social Imagery* (1976) and lesser-known studies which revisited Emile Durkheim and Marcel Mauss (1982) – the endurance of a particular 'exploration' of scientific matters which availed of Durkheim's distinction between sacred and profane. Durkheim held that science should appear to be sacred, in that it represented a transcendent social object. Moreover, it must be said that some of the basic assumptions of Bloor's *strong programme* rotated around the possibility of accounting for the universe of *beliefs*, that is, of the various sources (be they scientific, religious or even magical) from which knowledge may spring as a reflection of the social background of knowledge, like values that vein so-called concepts of 'common sense'. Bloor questioned himself on this point, that is, on how the sociology of knowledge might represent the content and the nature of scientific knowledge.

Bloor's answer - and that of that minority of European researchers – probably took the form of attempts to find a programmatic solution to the question, that was to win consensus for the assertion of a naturalist and relativist conception of epistemology which, according to many 'period-bound' publications by the above-mentioned Italian authors regarding the debates on the results of post-empiricist epistemology which should have arisen surrounding the new notion of resolving epistemology within the boundaries of the sociology of scientific knowledge. An arduous undertaking indeed. Little or nothing was said in the works of the aforementioned Italian authors concerning the contents and considerations of 'everyday' scientific practice and, in any case, beyond actual assimilation of prevailing sociological 'norms' which in the 1970's had spread a conception of the social world dominated by values: a theoretical system or 'structure' where the protagonists, including scientists, were to act in a certain way, governed by rules. To this regard responsibility for the diffusion of Merton's programme-paradigm of the sociology of science among many researchers is evident enough; many of them also felt it was urgent to deal with the dynamics associated with continuous changes in the relationships between considerations of science and technology or economy, etc. for example or what emerged from increases in the number of publications by social researchers of science like David Edge, Barnes himself and Riki G. A. Dolby etc. Generally speaking, these last two authors were so critical of Merton and his normative system that they undermined the solidity of his well-known notion of the existence of an *ethos* of science. For Merton the rules of science should have consisted in social expectations urging scientists to adapt to them, providing that compliance with the *ethos* was achieved within the scientific community. It is interesting to retrace some of the criticism directed against Merton, at both theoretical and empirical level, according to

Barnes and Dolby. For them Merton's error was that of insisting on the norms which should rule science without investigating the processes which caused normative structures to change due to lack of financial support or the use of technology and scientific knowledge itself, etc. And not only that. Authors like Ian I. Mitroff doubted the existence of a sole set of rules which for Merton acted as a true 'sociological invariable', where it was possible to use different rules in different circumstances to appraise the behaviour of others or, again, where one encountered incompatible values within the scientific community ('sociological ambivalence'). As regards the penultimate point, it is important to recall that a book by Barnes dated 1974, translated into Italian as *Conoscenza scientifica e teoria sociologica*, in 1979, held the view (accepted fairly well and, at the same time, 'anticipatory' of future studies like those by Michael J. Mulkay, translated into Italian in 1981) whereby the rules spelled out by Merton should have been able to establish patterns of observable behaviour, thus harking back to verbal formulations targeted for an external audience.

Contestation to some of the 'arrangements' of scientific knowledge due to different, new trends, were quite frequent in the period being examined here, providing more and more examples found in the studies of Barry Barnes, who examined the impact of Thomas Samule Kuhn on the sociology of science while also discussing some of the results of that developmental process, as did other authors less known to the Italian public, for example, Herminio Martinis, Peter Weingart and Richard D. Whitley, who may be considered the forerunner of more radical positions expressed by supporters of the 'strong programme' and was also the author of the famous *Black- Boxism and the Sociology of Science* (1972) where he expressed opposition to Merton's three decades (1938-1968) and highlighted the limitations of new approaches associated with Kuhn's perspective. Other objections were raised in Great Britain over the formulation of a theory of science, which, both in Merton's and Kuhn's cases concerned a 'basic theory', where the influence of served to justify the validity of the sociological approach to scientific knowledge. In this case, I intend to refer to research carried out by anthropologists like Mary Douglas, J. H. Beattie and Robin Horton (see his Italian book on sociological organization (1977) with an introduction by Aldo Visalberghi) on the issue of 'natural beliefs' and their variations within different cultures.

Most of the contributions made between the end of the 1970's and the 1980's aroused doubts concerning the universality of rationality and the scientific categories developed in western societies, a context into which Kuhn's programme, which included disciplines like history, philosophy and the sociology of science, fitted perfectly. Many of these positions were summarized in an essay by Joseph Ben David and Teresa A. Sullivan, *The Sociology of science,* which appeared in 1975 in the "Annual Review of Sociology." It was in that period precisely that developments of Merton's theories focused in particular on issues of the sociology of knowledge, in particular, on scientific knowledge. The interest aroused led to publications like *The Sociology of Science* (1973) *Social*

Ambivalence (1976) and *The Metric of Science* (1976). Moreover, the increase in social studies of science seemed, however, to coincide with an explosion, at times, of different 'spirits' of sociology and 'sociology' of knowledge bearing witness to the logical complexity of the discipline, already underlined by the older Popper in his polemic with Kuhn regarding the sociology of 'normal' science.

In reality, the exponents of the 'strong programme' were not officially divided on the Popper/Kuhn front, on the contrary, were rejected by philosophers of science and traditional epistemology as if they questioned the very nature of epistemology in the face of the emergence of the new sociology of science. In a certain sense, the problem was whether to rely on a rational-universal programme or not. For some this meant responding or not or not to questions like that posed by Larry Laudan in his 1984 polemics against Bloor: *The Pseudo-Science of Science?*

In practice, even with respect to the topic of rationality in science, it was clear that the turn of sociology and the discussion of epistemological *standards* capable of constituting a veritable object of investigation for sociologists or historians of scientific issues, would come. In reality, Italian researchers too proposed discussing the epistemological premises of the 'strong programme' with critical reference to the sociological diktats of knowledge established by the Kuhnian Barnes and by Bloor. In an article published in the journal of social sciences of the Roman *Luigi Sturzo* Institute, "Sociology," in the January-April number of 1986, Maria Carmela Agodi sought to conjugate different themes regarding the 'weaknesses of the strong programme' from the 'point of view' of an implicit reflection concerning the sociology of knowledge and the problem of the fundamentals of sciences, a difficult enterprise useful, nonetheless, to historians of science attempting to 'take stock' of the importance or not of epistemology to the foundation of the programme of sociology of knowledge. Moreover, the theme being examined was destined to arouse more and more active interest among researchers and sociologists dedicated to 'period-bound' studies of the 'methodological consequences' of the 'strong programme' or to carrying out reviews of 'reflective sociology' and relationship with the sociology of scientific knowledge. To this regard there was no lack of works by authors like Carmelo Lombardo and Agodi whose articles appeared in "Sociologia e ricerca sociale" between 1990 and 1991, while other Italian scholars like Vittorio Cotesta and Alessandro Ferrara, questioned the cultural premises of sociological knowledge in the age of the post-Kuhnian debate.

Agodi discovered that the epistemological premises from which authors of the well-known 'programme' departed, were bound, mostly, to a certain pathway of scientific subjects, that is, to intentions of demystifying scientific knowledge, where protagonists might refer to a relativistic view regarding forms of knowledge themselves. According to the authoress of the above-mentioned essay, the 'programme' rested on a number of theoretical premises, whose proximity to post-empiricist epistemology to a certain affinity with methodological irrationalism, testified to the fact that theoretical references to the proposal could be traced back to contributions by Kuhn, Feyerabend and Mary Hesse, authoress of the essay *Socializzare l'epistemologia*, which appeared in "Rassegna italiana di Sociologia" the year after the successful publication of Agodi's article.

In any case, the discussion surrounding the 'strong programme' gave rise to the debate on the agnostic attitude of many authors towards epistemological history as well as towards a view of scientific knowledge corresponding with 'official' epistemology and with the sociology of science of Mertonian inspiration. By way of simple description, we need to point out that Agodi's writings reviewed most of the works of the time and referred to the few Italian scholars, including Franco Cassano, who, in 1983 expressed the urgency felt by social scientists of the time to outline some more or less reduced versions of Barnes and Bloor's original programme. Above all, she stated that the most consistent works to this regard were those of Harry M. Collins (founder of the 'Bath school'), Trevor J. Pinch and Wiebe Bijker, as well as Latour and Steve W. Woolgar, authors of *Laboratory Life: the social Construction of Scientific Facts* (1979), and those of Karin D. Knorr-Cetina, Michael Mulkay, Jonathan Potter, G. Nigel Gilbert, Steve Yearley, Neil Sal Restivo and advocates of Harold Garfinkel's of ethno-methodology.

Examinations of the imposing British production showed that, in general, at the base of the famous 'programme' there existed, first of all, a sharing of the assertion that science was capable of basing itself on its own tools. The second point regarded its inability to carry out a sort of self-description, considering the historical weakness of its quest for criteria to avail of in order to mark its specificity also as a system of knowledge. Thirdly and lastly, this might be considered the premise whereby at the basis of science there existed a choice of value only, meaning that both it and its methods were similar to other systems of knowledge. This referred to the thesis of incommensurability, also given the absence of a 'neutral' perspective when choosing between rival theories.

It was logical that relativistic positions regarding the 'sacredness' of scientific matters expressed by the Edinburgh circle were described from an epistemological 'point of view' as an admission of the fact that, at the long run, some procedures and theories of science and, in any case, scientific results, might be believable only to a social group inclined to be guided by them. The authoress in her essay pointed out how Barnes and Bloor in their works were able to affirm the fact that both 'validity' and 'reality' might be tacitly acknowledged (see *Relativism, Rationalism and Sociology of Knowledge of 1982).* The purpose was also that of showing how the limitations which risk compromising the so-called 'strong' programmatic vision, did not hide, ultimately, a degree of epistemological inconsistency, which seemed to dog the social version of science presented by the authors in question. Moreover, Barnes and Bloor had accomplished an extension of the conceptual issue in order to show how the sociology of science should change or better still, replace epistemology, due to what was happening over the years. That way, historians of science in particular, would have a certain quantity of material at their disposal to rationally discuss the development of the disciplines once more.

This was able to happen in the above-mentioned climate of contrast between 'rationalists' and relativists, where the contributions of the philosophers seemed to anticipate the era which underlined rather than the reasons of philosophy, the major crisis of 'rationality' modern and contemporary philosophers were experiencing, or as, starting from 1979, Carlo Augusto Viano and Aldo Gargani underlined in their works, a veritable *crisis of reason*, denoting not only an overall crisis of the quest for 'fundamentals' but also a delay on the part of the sociological disciplines of scientific knowledge with regard to a number of themes-problems. On the other hand, it seems evident that the sociology of science of the late 1970's represented and was even able to tackle the evident 'crisis of rationality', providing a valid response, also considering the fragmentation taking place due to the birth of the new approaches just mentioned. For the 'period-bound'' historians of sociological and scientific subjects, the greatest problem seemed that be that of addressing in an atmosphere of intellectual ferment, the tacitly acknowledged recovery of 'rationality' (Weberian, economical, etc.). This, however, without having first given rise to a thorough reflection on the epistemological fundaments of science, also due, it must be said, to a progressive lack on the part of some of our philosophers closer to the social sciences, for which sociology was able to compensate in part. In Italy, the sociology of science developed without considering its contribution an important moment in the evolution of all scientific disciplines and, not only, as a mere facet of the debate surrounding the 'crisis' of conceptions of science, which involved its activities as well as taking into consideration different 'cultural' values and different social and historical contexts. The new sociology of science, claiming to be also 'new epistemology' was able to provide researchers with much more food for thought; a different dynamics of social sciences.

In a certain sense by merely contesting the "strong programme" in favour of 'constructivism' demonstrated that the attempt to neutralize the birth of a new 'internal' view of the sociology of science established itself more quickly than the presuppositions contained in the 'programme' itself and of which Agodi spoke. The issue is not all that easy. The fact is that at the beginning of the 1990's, scholars like Statera, did not fail to slate the weaknesses and radicalism of certain assertions and bring to maturity a certain simultaneity of assessment of the epistemological ingenuity of the so-called 'strong programme' in favour of the inhomogeneity of the 'points of view' of sociologists of science and of British cognitive sociologists.

In the 1980's above all, plurality of approach had triggered competition between existing perspectives and the positions of specialists themselves, which were founded mostly upon the epistemological sciences most frequently practiced by post-empiricist philosophers or upon a kind of 'empirical epistemology', as Leonardo Cannavò called it in *Studi sociali della scienza Science. Aspetti e problem,* edited by him (1989), and referring to a 'programme' previously announced by Richard D. Whitley. This meant paying attention to research techniques as a constituent of 'meaning' and of scientific advancement; substantial refusal of distinctions between the 'social' and the 'scientific'; assertion of a mainly 'internal' model of scientific change; affirmation of the central importance of negotiation (linguistic, and not only) when establishing scientific meaning; widespread trust in qualitative analytical methodologies as well as in the sufficiency and representativeness of case studies; conviction of the methodological legitimacy of scientific inference; awareness of the excess importance attributed to *theory ladenness* in observation and scientific practice and, finally, a preference for theoretical dispute and investigation of scientific knowledge, rather than for an accumulation, even if controlled, of cognitive content.

Cannavò's contribution is extremely useful when 'taking stock of' integration the different approaches in existence, where the latter followed a pattern whereby the various fields interacted with other disciplines. One of the most important topics treated by this author in relation to the topic was the relationship between sociological methodology and the sociology of science (see for example, his article in "Sociologia e ricerca sociale, n. 28, 1989). In November 1988 at the International Conference held in Rome on the Quality and Quantity of Social Research Cannavò pointed out how methodology, unlike epistemology, treated science as a product, was directed towards research as a process, and how therefore, accepted considering the aims of investigation as socially conditioned and the tools of scientific knowledge as the result of historical-cognitive mediation bent on integrating operational analyses and the sociology of science. The latter was to be seen as integrated with the theory of investigation within a critical-operative conception definable as 'empirical epistemology' capable of providing sociological methodology with new ways of establishing itself.

During a different phase of reflection, it was held that the 'sociology of science' included the 'social history of science' and empirical epistemology while to the 'methodology of research' belonged the 'history of scientific thought' and the 'philosophy of science'; 'research policies' embraced the 'economy of research' and 'scientometry'. One notes how Cannavò considerations regarding the diffusion of theories and models of sociology of science, highlighted not only integration between contiguous disciplines but also specializations of/within the scientific professions and the numerous relationships 'in, and of socio-cognitive sub-systems'.

The fact is, returning to Robert King Merton within the ambit of the above-mentioned context and to the results of the several doubts advanced by new sociologists of science, the scientist became increasingly committed to defending his/her intuitions or 'innovations' not so much as to follow the norms of 'disinterest' or refer to the 'community' of knowledge as for reasons of economics or prestige, of institutional, academic or professional policies, etc. This within a context of more evident attention paid by scientists to a gradual definition (conjugating the terms) of the 'academic' profession as far as the values of science and ethics were concerned and which, in Italy, meant a tendency to deal with the consistency of certain topics emerging from the work of a number of cognitive sociologists like Cannavò and Ancarani in the 1980's, as well as from that of other sociologists, from Pier Paolo Giglioli and Achille Ardigò (who died in 2008) to Franco Garelli, from Gallino to Statera, to Maria Stella Agnoli, etc., whose studies and empirical research continued right up to the beginning of the 1990's. Their research provided Italian scholars with a rare occasion to discuss changes in the role of the social scientist within a particular context of the social development of the image itself of the scientific *élites* in relation to the institutionalization of science as underlined by Ancarani in an essay dated 1989 and in two appraisals published in numbers 1 and 2 of "Polis," 1991. The first regarded a volume edited in 1989 by Cannavò called *Professione scienziato. Organizzazione della ricerca pubblica e professionalità*, the second, edited by Maria Stella Agnoli and entitled *Il ricercatore nell'università che cambia*, was published in 1990. Generally speaking, those topics proposed absorption of what was posited during the previous decade by scholars like Joseph Ben-David (1975) concerning science-society relationships or by Statera (1978) regarding the sociology of science.

It was within this framework that epistemology dissolved into a confrontation between *ethos* and the ethics of science as well as into the 'heart' of scientific research, as suggested by the title of another essay by Cannavò published in 1990 in "Sociologia e ricerca sociale" where the author identified the Italian University as a model which followed networks of communications directed by values by no means defined once and for all.

Ads to post-Mertonian sociology of science, Cannavò had the occasion to assert, also, that in Italy, as elsewhere, it lacked those very 'middle-range theories' which the American sociologist had advocated to combat 'grand theorists' and, which, he believed provided a more focused vision of social phenomena. Another similarly important aspect underlined in that vision of the sociology of science during the age of relativism, influenced by some of Kuhn's, Winch's and Feyerabend's ideas, as well as those of others, seemed to be that underlining factors like social negotiation of scientific meaning itself or various forms of linguistic mediation and 'convention' which did not necessarily lead, *ipso facto,* to dealings with irrationality. Rationality might also turn out to be investigable by researchers from several angles, from organizational to communicational, while requiring contextualization.

Rationality and irrationality thus became nothing more than the two aspects to be investigated following semantic/contextual clarification of specific determinations referable to them. Moreover, in the age of relativism, as stated above, it was possible to foreground particular suggestions voiced by 'qualitative sociologists' regarding some tendencies and interpretations of knowledge like those in Feyerabend's 'period-bound' works or those of other authors, like L. Derek Philips (see: *Wittgenstein and scientific knowledge: a sociological perspective,* 1977).

From an epochal perspective, characterized, however, by a multiplicity of theories, we should not fail to recall the contribution several of them, while investigating science in the making and pinpointing some of the inextricable mechanisms of the black box, made towards the theoretical and practical development of professionalization and specialization within the fields of science causing ulterior 'closures' which tended to widen the gap between science and society. This helped mystify even more the results of examinations of the *throughput* mechanisms existing within the *black box*, as Bruno Latour from the Centre de Sociologie de l'innovation dell'Ecole Nationale Supérieure des Mines, Paris, pointed out in publications like *Science in Action,* 1987. In later years, Latour dealt with processes leading to the formation of scientific knowledge, privileging the socio-political nature of science and the 'artificiality' of some of its main 'products', as already stated in the 1950's and 1960's by that great philosopher of science Gaston Bachelard, investigated, however, with regard to the consistency of his historical epistemology, rather than for his 'imaginary' side, by the Italian Victor Ancarani in a book issued in 1981 called *Structtura e mutamenti nelle scienze.*

The leading idea guiding Latour's theoretical proposition concerned the 'products' of science (theories, laws and technological supports) which might be analysed during their formation to reveal, at the long run, the mechanisms capable of reversing common perceptions of their very development.

Roughly during the period of the debated reception of the 'strong programme' by some Italian researchers of social sciences, it was possible to establish links between results of some of Latour's specific propositions concerning the work of sociologists of science, considered as people subjected to social, communicative and political circumstances, very often determined decisions, even of an 'arbitrary' nature, regarding what might be deemed a scientific 'product' or not. Particularly challenging for researchers was the 'Actor-Network Theory' developed by Latour with John Law and Michael Callon, author of an interesting study, published in 1989, *La science et ses reseaux; genese et circulation des faits scientifiques,*. The author's relativistic impression was quite clear in this instance, along with his opting for anthropological 'choices' when paying attention to concepts such as 'bonds' or 'convention' to the point of affirming that in science, as in social science, there existed a tendency to construct 'signifying systems' self-binding on participants. Connections and relationships in these contexts are always be investigable by having the possibility to enter various productive mechanisms in an attempt to examine in greater depth the processes rather than the 'products' stemming from them. So, the notion of examining the transformation of a certain assertion into scientific matter gave birth to the hypothesis whereby the conceptual apparati devised by researchers, including all the elements participating in the creation of scientific and technological knowledge, so that they might refer to the union between science and technology (Latour, will call this, 'techno-science') were determined by the intention to enable their purpose to survive because utilised by other researchers: assertions became facts only in the hands of others.

From the 'point of view' of the new horizons of interest emerging for the sociologists of science, we need to say that scientists and engineers grew increasingly too busy changing the surrounding environment to allow their own scientific assertions or technological artefacts to prevail, where interactions between scientists were not intended as purely random but as their 'products', keeping in mind the 'trans-epistemic relations' mentioned by sociologist Knorr-Cetina in his *Science Observed* in 1983, when investigating new perspectives of social studies of science, highlighting the fact that decisions made within the 'laboratories' should certainly take into account interests, funding agencies, investors, industrialists, politicians, etc. In practice, the destiny of assertion might be extended beyond the confines of the 'laboratory', as Latour pointed out in his more specifically sociological research into 'techno-science', conducted by availing of semiotic language which sought to describe the work of the scientist without investigating the contents or the 'whys' of certain conflictual aspects or without trying to think of the 'quality of the objects of knowledge or, least, of the artefacts themselves. The content of scientific statements or scientific theories did not belong to Latour's sphere of interest: the fate of scientific fact now seemed to rest in the hands of its users.

Chapter 12

FROM CONSTRUCTIVISM TO THE META-THEORY

As we have seen, the European debate concerning the sociology of scientific knowledge and its noteworthy spread among researchers in Italy was to become an occasion for historians of scientific matters to discuss the epistemological consistency of the social sciences within a framework of further developments of relativism. We should not forget that some critics of the 'strong programme' like Ernan McMullin sought to establish links between sociology and the epistemic issues governing scientific rationality (see *Scientific Rationality; The Social Turn* edited by James Robert Brown in 1984). This operation meant making a special effort to merger different perspectives and 'approaches' starting from the epistemological debate on relevant issues, proposed by historians of science like Duhem-Quine, Kuhn and Feyerabend, achieving more or less 'weak' results based on the 'strong' programme. It must be said that in the context of the development of certain issues, scholars like Filippo Barbano (see "Sociologia" VII, 1973, and 1980) - 'officially' charged to deal, albeit at times in narrative form, with scientific sociology in Italy and recollective of studies on the 'rebirth' before reception of Mertonian perspectives in the 1950's- managed to outline a history of European sociology destined, in truth, to become more complex and intricate from the epistemological 'point of view' also due to the assimilation which took place between philosophers and philosophers of science according to new propositions regarding the autonomy of sociologists of scientific knowledge. And there is no doubt that developments of these very matters appeared to 'official' historians of science as 'materials' of an epochal turn taking place in the age of relativism, quite some time after neo- positivism, later than entry onto the scene of Merton, Talcott Parsons and Paul Felix Lazarsfeld after completion of the so-called 'post-empiricist' affair. Among the 'period-bound' episodes we hasten to mention some who caused philosophers of science to increase seminars, lectures, meetings and conferences and ban Conferences regarding their 'rebirth' as in the case of Varese in 1985, or international meetings like *The Porté européenne des traditions èpistemoogiques Italiennes*, Paris (1987)-see "Fundamenta Scientiae," vol. 1, n. 1 1989 and Naples (1991), with some 'materials' edited by the Italian Institute for Philosophical Studies. It is also true that for a given period of time a few historians of sociology worried about addressing these events only to surrender immediately afterwards to the temptation to deal critically with possible references to 'specialisation' within the sociological subjects (the sociology of science in this particular case) which might account for the existence of an 'epistemic' statute ('historical and cultural', that is, of the scientific subjects according to Barbano) even in the absence of a cognitive statute. Again, in Barbano's opinion, studies and research interests emerging at that time were to be collocated, though only within the Italian context, as follows: *science*, as such, its system, its functions as well as the activities of scientists and researchers, institutions, etc.. The *sociology of science* as a specialized (cognitive and analytical) discipline, its epistemological statute as related to sociology, the sociology of knowledge and history. Finally there was the *experience* of science as a process and as history. Sociologists had to reflect deeply sociologies to validate a number of concrete issues like those concerning the discovery of 'research programmes' and their inclusion in the historiography of science, something which took place in part. With regard to important scholarly works of the time see Barbano's introduction to the Italian translation of Robert King Merton and Jerry Gaston's well-known *The sociology of science in Europe* published in 1980, included in *Scienza e Cultura. Teoria e programme di ricerca sociale* edited by Barbano himself (part of the "Scienza e società" series directed by him for the Franco Angeli publishing house, Milan, from the mid-1970's on) and containing the Minutes of the Italian Sociological Convention on cultural and cognitive processes, held in Rome in October 1981 called *Consenso e conflitto nella società contemporanea* [Consensus and conflict in contemporary society]. As to the issue of 'research programmes' and the formation of special 'units' of scientific researchers in Italy, Barbano referred to contributions made by Maria Stella Agnoli, Leonardo Cannavò, Lucia Ciampi of the Institute of the Sociology of Rome's State University's *Magistero* [Education] Faculty and by Gianni Losito and the lecturers associated with Chair I of 'Methods and techniques of social research' directed by Gianni Statera, Italian 'connoisseur' of the sociology of science (see his articles in "La Critica Sociologica," 1967 and 1974 as well as his books dating 1977 and 1978). He also mentioned scholars of the early sociologists of science and of the 'scientific community', like Stefania Vergati (see her articles in the volume dated 1981), Alberto Gioannini (author of the book *Paradigmi e sociologis della scienza* issued in "La Critica Sociologica" during the winter of 1978-1979), Laura Visentini (an 'analyst' of scientific change) and others, whose theoretical 'period-bound' operations Barbano also spoke. These ranged from the Torinese Dario Rei and Alessandro Casiccia to the Paduan Mario Santuccio and to some critical investigators of the politics of science and scientific research like Paolo Bisogno, Arcangelo Rossi, Antonino Drago, Giovanni Stalio, etc).

All this took place at a time when Italian publishers and some journals promoted initiatives favouring the debate on sociological and historical issues of science begun by "Critica Marxista," 'notebook' n. 6, 1972, and the *Scienze e storia* (1980) series which appeared in important bookshops alongside others like *La Nuova Ragione*, edited by Paolo Rossi, and published by Il Mulino in collaboration with "Scientia," containing contributions by many Italian and foreign contributors. (1981).

The overall result was that, during the 1980's and the early 1990's, a general effort was being made to undertake a solid epistemological history of social sciences was met by very few adhesions due to a tardy assimilation (since the beginning of philosophical studies of science), or to a lack, after an initial 'internal' boom of the debate regarding the discipline which marked a certain degree of continuity in discussions of topics tabled the previous decade. As regards this, the Italian case was emblematic enough, because it marked the absence of a research tradition which investigated the conceptual development of scientific questions, therefore of those concerning the social and empirical sciences, while also denoting a lack of a new 'period-bound' generation of " of sociologists of science interested (according to the diktats of the different disciplines of the time) in the theoretical-practical development of epistemological issues fully rooted in the historiography of scientific knowledge. This did not mean, as we have seen, that the Italian cultural context lacked authors capable of orienting their critical examinations of the various 'pathways of sociology' by availing of modern and contemporary epistemological debates, as it in the case of further studies carried out by Franco Crespi and published in 1985, in a book called *Le vie della sociologia*, reviewed in the "Sociologia e ricerca sociologica" quarterly (1986, n. 19) by Maria Stella Agnoli who attributed it the merit of having sociology begin when it became customary for it to acknowledge itself as a cognitive activity equipped with *specific* conceptual tools as well as criticisable and controllable procedures.

It was someone else, however, who was to have the merit of following the history of Italian sociology through its crucially vital crucial moments and its gradual development of possible links between the scientific standing of its social theory and its ability to plan interventions aimed at satisfying a growing demand for sociological competencies and, therefore, for professionalism, where, in actual fact, academic sociology, and the sociologists of 'academia' (increasingly more numerous as specialized disciplines grew), seemed to address themselves to their own role as a 'privileged issue', while they ought to have increased attempts to sought 'paradigms' above and beyond 'theory' and analyse relationships between sociology, its new approaches, and history for example etc. As regards the 1980-1990 period, see Enzo Campelli in "Sociologia e ricerca sociale," n. 32, 1990.

Paradoxically, at a certain moment, sociologists seemed as 'empty' towards the needs of empirical research as the possibility that their research might be a part of a general change in scientific matters, in the name of a history shared with previous 'epistemics'. It was easier, at the time, to question ways of receiving social theories, models, etc. than focus the debate on themselves within a broader context of interests and connections that might involve contemporary epistemology and its links to philosophy, logic and the sociology of science. The sudden decline of the debate on the *ethos* of science coincided also with attempts to account for opposing polemics between different disciplines with a view to recovering a sociological, methodological and historical perspective, like that of methodologist Luigi Gabriele Frudà - author of a volume on concepts and tools published in 1988- who, in his epistemological text *Methodo, Logica e scienze Sociali,* 1993 tried to make sociological historians aware of the fact that it was their task to outline new methodological pathways, by availing of the integrations between different fields of knowledge (history, sociology, philosophy) achieved by some of the scholars (like Luigi Sturzo) who sought autonomous conceptual status.

Despite a strong tendency towards self-reflection at that time, sociology in Italy was unable to create an epistemological 'point of view' of its own capable of generating within a broader context of social science, its own specific debate regarding the history of science, in other words, a 'point of view' recognised by the scientific subjects, capable of developing a theme internal to research into paradigmatic unity. Of little use was the attempt to consider an 'integration' between epistemology and the sociology of knowledge, even in terms of a hypothetical formulation of a 'programme' of epistemological research in sociology, as proposed by Vittorio Ancarani who reprimanded the sociologists of science of the early 1980's of leaving the issue in the hands of philosophers of science and epistemology, devoting themselves to aspects regarding discovery and never justification. It seemed essential to refer to certain cognitive frameworks and to the development of conceptual structures as well as to contractions of some sociological theories relating to a number of problems of an epistemological nature and to the very tasks of sociologists of knowledge. It should also be said that some aspects of the latter had assimilated the themes drawn from the age-old European 'epistemological revolution' experienced by philosophers of science like Gaston Bachelard in close contact with the 'revolution of the subject' in the so-called 'hard' sciences, which, in the 1980's, led to a certain reinterpretation of knowledge and know-how in terms of studies of what French sociologist Michele Maffesoli called '*connaissance ordinaire*' (1988) even before Gilbert Durand, a disciple of Bachelard's and founder of '*Centre de recherché sur l'imaginaire*'. The issue was reread more recently in anthropological terms related to complexity by Edar Morin (1994) who advanced more or less marked suggestions regarding the development the universe of the imagination reached by the master, availing of the phenomenological method.

In actual fact, the future outcomes of the sudden entry on stage of 'constructivism' induced many historians of contemporary sciences to replace the consideration of 'concepts' and 'theories' with contested scientific techniques, focusing more on the actions and negotiations of scientists, rather than on a quest for the origins of knowledge or for explanations of scientific theories themselves. This increased the relinquishment of reference to historiographical results which actually piloted a part of the interests of those who wished to study scientific activities or those who vaunted -for example- the old theses of the French philosopher Pierre Duhem which sought to found the philosophy of science itself on historiographical subjects.

Almost eighties years after those 'theses', a number of scholars started to make observations considering considerations regarding 'constructivism' as well as on categories used by authors inspired, for example, by the aforementioned Latour and others. Some attention was paid to the contestation of the dichotomies present in the history of western science like that between 'nature' and 'society', assumed by Latour at a certain point to criticise David Bloor. Another criticism Latour launched against Bloor was that he had analysed true-false, rational-irrational dualism in relativist terms, thus denying science a precise epistemological position of its own.

For scholars of the sociological course of science like Latour, the nature-society dualism derived from a kind of agreement between the social actors and not, as suggested, from the traditional views that many scientists, including 'hunters of paradigms', historians and sociologists, had of scientific developments. Besides the spectre of some kind of future metaphysics, 'techno-science' was able to assert itself as a set containing nature, society, people and machines closely connected with and permitting scientific and technological activity.

In the eyes of the 'constructionists', the conception of science intended as a series of progressive actions close to a given truth entered into a crisis, socially conditioning scientific knowledge that is, socially construing it. This way, the degree of certainty available from a scientific was dependent upon the particular context in which it had been built rather than upon 'thinking'. From this descended the entire theory of scientific 'products' and the beginning of the 'newly' revised 'epistemology' of analysis and reflection.

Along this route aspects of scientific knowledge, extended, to some extent over the years, both methodologically and conceptually, towards the domains of the technology and politics. It was an acquisition for scientists that the study of technology might appear as a phenomenon influenced by the existence of certain social groups who, in accordance with other predominant groups, arrived at a new definition of technological 'products', clearly in sight of marginalized groups who, were to force, as Latour claimed in his book dated 1987, unexpected elements out into the open.

In fact, if the arrival of constructivism spread the idea that scientific knowledge might be considered in relation to negotiable 'products' conditioned by social surroundings, in a version remote from that most familiar to Italian historians of sociology and even to sociologists of knowledge, that is, the Manheimian causal approach to the social conditioning of thought, it was also true that the basis on which this sociological thesis was built, rested on considerations how to intend 'constructed' and 'produced' concepts, whose cognitive and 'reflective' unmentioned aspects like 'scientific discourse or 'texts belonging to literature, expanded to become, in the long run, a veritable epistemology of 'knowing' unlike the flaunted hermeneutics of 'knowledge'. All this generated a 'reflective' vision of scientific knowledge which, during the 1980's, gave rise to approaches to 'analyses of scientific discussion' or of new forms of expression leading to the *New Literary Forms* movement.

The spread of these ideas did not meet with the appreciation of European historians of scientific sociology, who had, instead, welcomed the return of Mertonian-informed socio-scientific investigative approaches, in an effort to clarify the theoretical and analytic visions of the discipline in an era of relativism and of epistemological scepticism. In the end, the role of the 'history of science' at that time might have been assumed in a less strategic way if referred precisely to the uselessness of reference to the social sciences. Few were prepared to suggest, at a given moment, analyses of theoretical systems of the social sciences, of sociology, in particular, in a profoundly altered cultural and historical milieu remote from traditional membership of 'schools' of thought or traditions of research constantly at loggerheads with philosophers for example. This accounts for the influence of the American sociology of the 1970's and 1980's as well as that of the emphatic movement of the 'quantophrenics' which impacted on the fate of the empirical social sciences, especially in Italy where the greatest need of sociologists and researchers seemed to be that of to gain possession of tools instead of dealing with the epistemological turbulences of the history of scientific issues.

Sometimes, it was a problem to refer to the existing debate on the social sciences without recurring to the results of hasty and summary research operations, the outcome of improvised 'students' informed by results very often closer than anything else to the dictates of opposing fronts of academic sociology (the 'qualitatives' on the one hand, the 'quantitatives' on the other) which confined the efforts of researchers to mere classifications of the products of either side, thus passing off hackneyed paradigms as methodological 'options'. It was the task of the historian of the sociology of that period to attribute the ones or the others within this or that sphere, leaving open problems of relevant interest to historians of scientific issues, like the 'incommensurability' existing between qualitative and quantitative sociology. This topic managed, nonetheless, to give rise to a discussion, also in the light of assertions regarding the theoretical nature of choices to be made in a more or less subjective and/or individual manner, advanced by scholars like Franco Ferrarotti, Franco Cassano (1989) or other Italian authors like Alessandro Dal Lago- editor of the Italian version (1987) of Howard Schwartz and Jerry Jacobs's *Qualitative Sociology* (1979) as well as of a previous test on ethno-methodology, with Pier Paolo Giglioli, in 1983-, Franco Crespi (1989) etc. directly taken from the polemics and *Against qualitative analysis* from scientist as Franco Leonardi (1991). Examples would be endless. Even Statera in his 1992 essay entitled *Il mito della ricerca qualitativa,* which preceded the position expressed during the international

Convention in Rome, November-December 1993, asked why, in the history of science, including the socio-scientific methodological debate, it was so very difficult to find - referring also to the times- disciplines so capable of producing strong opposition between 'quality' and 'quantity', 'qualitative and quantitative methods', 'qualitative and quantitative analyses', as sociology. This author, who revealed the mythical not the paradigmatically 'immeasurable' features of so-called 'qualitative research' dealt in part with some of the rare elements of this peculiar aspect of sociology as a scientific discipline.

The fact is that these themes, and as a result, 'period-bound' accounts of a presumed scientific historiography, rarely seemed to circulate in such a way as to foreground contributions made by historians of Italian sociology to the endless debate regarding the social sciences and the circular nature of the epistemology of sociology as defined with regard to the cult of some 'qualitative' sociologists, while, for many years to come, disputes were to arise concerning the only 'biographical method' or the validity of *life stories,* starting from different receptions of William Isaac Thomas and Florian Znaniecki by sociological researchers proper. In reality, the discussion focused again on the use of certain 'tools' in sociological research and their cognitive potentials, as a monographic work by Campelli pointed out in 1977 when dealing with the documentarian-biographical method, a topic he took up again in 1982 and, later on when drawing up a balance (1990). On this particular issue many contributions were made by followers of the 'life-history approach' ranging from Daniel Bertaux in *Biography and Society* (1981) to sociological researchers availing of some theoretical-conceptual ideas, like those posited in *Storie senza storia* by Renato Cavallaro (1981), in some texts by Roberto Cipriani (1987) and by Mathias Finger of the University of Geneva (1989) inspired by life-history methodologies, as well as in several by Franco Ferrarotti (1983) regarding claims of pursing an 'epistemological autonomous' by following the biographical method of sociology and social sciences ('qualitative methods').

The consistence of an increasingly more contemporary debate, regarding the existence of a 'systematics of sources' in the scientific historiography of the sociology was replaced -in a long term- by the inconsistency of the umpteenth discussion regarding a quest for 'unconventional' sources or, better still, the use or not of biography by 'quantitative' sociographs and sociologists of this or that position. The real problem did not seem, when all came to all, that of acknowledging one or other method, or recognising as 'sociological' some of the empirical research regarding workers carried out in the 1970's, by Pietro Crespi- for example. The problem in Italy was the lack of a permanent, operative epistemological clarification regarding methods between one generation of historians of sociology or historians of sociological thinking and a future generation of 'historians of science'. The prophecy of Bruno Latour and others seemed to come true. Quality/quantity 'analyses' cloaked the interests of an entire generation of sociologists while science and/or the sciences became a sort of 'exchange' between different schools of scientists 'on the go.

The moment has come, perhaps, to recall, in brief, that a response by Italian sociologists to a number of issues regarding quantitative research, 'social engineering' and 'socio-techniques' at the service of the hegemonic class was forthcoming, after reaction to the debate on the 'crisis' of the method of the 1970's (with reference, in a sense, to the Turin Congress of 1971). Certainly, Italy lacked no examples of studies with a 'philosophical slant', in keeping with the critical spirit which had always animated Italian sociological culture. These concerned the quantitative methods, 'mathematical models' or, better still, mathematic theories of society, developed on the margins of theoretical and applied sociology, in the wake of teachings first provided by Paul Felix Lazarsfeld in the anthology, *Mathematical Thinking in the Social Sciences,* early 1954, later by Hubert M. Blalock and so on. Departing, in particular, from the bases of social theories, that is, from certain concepts, one arrived at the production of the usual tools availed of to verify theoretical research, before claiming make 'social mathematics' a theory. Moreover, one of the first problems of the so-called 'qualitative methods' was that of rendering the concepts used in social, empirical research observable (see Luca Perrone's systematic work, 1977). Besides, the Turin sociologist Francesco Maria Battisti (who died in 2008) in his *Sociologia matematica,* 1982, dealt with many of these aspects. The book brought together the results of the studies carried out abroad well as of his teaching experience during seminars on the 'Methodology and techniques of the social research' held at the University of Rome from 1975 to 1977. In the foreword to the book he underlined the fact that it was necessary, also with reference to these subjects, to carry out a thorough examination of the origins of these concepts and on how to render them operatives. The author of the book referred too, to the history of social statistics and sociological thought, following the more theoretical-systematic tradition he had acquired during a stay at the Columbia University.

It was then than that some scholars felt the need to remedy a series of semantic issues regarding important subjects like historiography, philosophy and the sociology of science, which would surely have aroused an interest in the development of sociology within a vast panorama of objectives to pursue. Some other scholars sought to define the line between logic and methodology, in an attempt to replace assumptions of 'constructionism' with a kind of 'philosophy of procedures'. In actual fact, the latter seems to have raised again, in Italy, the issue of the relevance to scientific sociology of 'theory' after a decade of opposition towards 'constructivism' during the period between realism and relativism.

We need to mention, in particular, some essays and contributions by Marradi (see "Laboratorio di scienza dell'uomo," II, March-June 1987) whose merit it was to have promoted a thematic redefinition which gradually became of interest to theoreticians, epistemologists and logisticians, as well as methodologists within the panorama of 'scientific studies' and that of connoisseurs of 'meta-science' including Alberto Marradi (1984). The task of reordering some of the 'materials' produced was resumed, to some extent, while the historiography of scientific subjects also demanded it having reached a certain point. To this regard it must be said that Marradi, affected by the impact of taxonomic studies by historians, like Ernan McMullin, proposed, in his essay, that Merton's be absorbed by the synoptic frames of the history of sociology due to the 'systematics of theories''. The influence of Kuhn emerged from his mastery of different disciplines (such as historiography and sociology) he availed of to follow an empirical approach to the study of scientific activity.

In this sense, scientific activity was able to disclose the strength of the epistemological nature of its fundamental issues. Again the history of science was in a position to act as guide to a semantic re-systemisation of the terms between 'neighbouring' disciplines, the course of which undoubtedly influenced sociology and its development among contemporaries, to the point of determining the interests of sociology of science, at time confounded with those of the philosophy of science.

What Marradi proposed reordering, first of all, was the portrayal of the reactions aroused historically from logical empiricism on, up until the appearance of the accredited radical and current alternatives accredited most, in any case, to the arrival of new conceptions of scientific activity and the presence of scientists within processes of change and social progress. If the heirs to logical empiricism had the task of discussing the presumed superiority of their 'point of view' and their way of approaching epistemology and the philosophy of science, the new scholars of scientific subjects were charged with reducing the habit of labelling practiced by various approaches, including 'philosophical' ones, taking other unlabelled fields of study into due account, for example sectors regarding relations between science and society, and others.

Marradi, in 1987, proposed reordering the terminological of studies leaving open, in the long run, the issue of what belonged to the sociology of science considered as a field of studies of objects 'typical' of a sociology subjected to continuous redefinition of its 'object' and 'concepts' and of the many ways of viewing science. The problem was also that of establishing which concepts might reside within the terms used by scientists to pursue given types of approach, in this case, an empirical, that is, a mainly sociological approach. To this regard, one recalls a lesser known work on methodological problems of the psychoanalytical theories by Ernest Nagel, professor of philosophy at Columbia University, where the author defined as 'theoretical' concepts unable to describe anything that might become the object of observation but which, all told, was usable or unused by theory. This for Nagel meant that theory could be formulated in terms of 'theoretical concepts' something that made it absolutely legitimate. Besides this, Nagel noticed that theories needed to comply with certain requirements in order to be verified empirically. The first requirement was subjection to logical reasoning that occurs before the empirical data. The second was that the concepts needed to be linked to observable materials.

Nagel's discourse leads us to the heart of the matter, even if this it is not immediately met by examples directly associated with the empirical social sciences. The fact is that a number of other specific issues are contained within the developmental contexts of some scholars' reasoning. Furthermore, the use of the conceptual materials to re-accommodate the semantic field required to produce results capable of giving rise to a series of reflections regarding a new conceptual panorama in the making. In any case, as Alessandro Bruschi, author of an introductory text to the methodology of social sciences called *Conoscenza e method* (1990), made clear, the sociology and the history of science would be extremely useful to explain the dynamics of science, while it was necessary to assuming the task of distinguishing what scientists do from what they ought to do. Bruschi was led to carry out reconnaissance into the semantic areas which specify the methodology, epistemology and technology ('technical rules' or technology of research) of sociology and the social sciences by recourse to the 'disciplinary identity' of the methodology of the social sciences itself. On this specific issue see his essay *Logica e metodologia* published in n. 35 of "Sociologia e ricerca sociale," [Sociology and social research], 1991.

The effort made by this scholar was also intended as a necessary logical clarification of the relationship existing between what many authors intended to call 'epistemology' and the organisation of a set of methods. By 'method' they meant 'a non-void set of procedures' permitting discernment between the diverse 'methods' to use according to the cognitive questions to answer, problems to solve as well as a whole series of intellectual resources available. Maybe, it should have been made clear to Bruschi that many of the problems regarding the terminological genesis of/in the social sciences, might have been treated as epistemological issues including procedures themselves. Furthermore, the author clarified how some rules of an epistemological nature might be used methodologically.

All told, Bruschi was a scholar who, in 1971 with his *La teoria dei modelli nelle scienze sociali* and closely following from authors like Alberto Pasquinelli, began to say that the discovery of a social dimension might be linked to epistemological studies and their birth, where connections between epistemology, methodology and the social sciences became mandatory when seeking to free them from a still para-scientific domain. Bruschi explained, before carrying out the in-depth discussion contained in parts I and II of the book, how the theory of models needed to refer to epistemology, and at more concrete level, to methodology, where the task required was that of clarifying some points concerning socio-scientific methodology in a context which sought to 'prequalify' studies aimed at marking an important, obligatory phase in the development of scientific matters. In this instance too, Bruschi provided some clarifications regarding the characteristic moments of development, that is, those regarding *fundamental* aspects of the development of science. He discussed the more or less 'restricted' conceptions of epistemology, up to a freer conception of it, following contributions by mathematics, physics, and major changes in philosophical approaches, especially those regarding studies by analytical philosophers. Moreover, his research was able to pass through the series of epistemological ingenuities which had created confusion regarding the use of the term 'logic', meant as referring to the abstract sciences and the formal and explicative systems (deductive), including traditional logic and mathematics.

The work of 'cleaning up' then served to examine the complex systems referred by the author to the social sciences (functionalism, structuralism, etc.)

Bruschi's systemisation discussed the concept of 'theory' in relation to the social disciplines, where they referred to theoretical constructs aimed at empirically controlling control of the 'immediate world' of human beings; this way, the theories defined 'taxonomic', 'descriptive' and 'problematic', whether nomothetic or idiographic, were classified according to the model they adhered to.

It is interesting to follow Bruschi's thinking, especially where it seeks the limits of the reasoning used by social scientists in a context of anticipation and the advancement of studies on 'theory' - with its strong 'period-bound' inclination towards paradigmatic reduction- considered as a logical upshot of interconnected propositions at a permanent level of empirical verification of variables. It is a fact that from a 'point of view' opposite to that of the history of science, 'theory' was considered abstract, where 'abstract' meant construing the empirical reality and controlling the laws governing it, but only after having specified, conceptually, what an essential result might be. Contrariwise the reasoning of some scholars, in particular, cognitive scientists, struck by the idea of questioning the usefulness of experience compared to 'theory' proved useful, without considering the fact that it was the cognitivist model, actually, which sought surpass the limits acknowledged by absolutely rational thinking, by presenting itself as an 'alternative paradigm' to holism, already questioned by the explicative model of methodological individualism, regarding which the social scientists and the Italian sociologists were to discuss once more during the years of the birth of and 'strong' opposition to 'constructivism'. Cognitivism as a research programme, built its models starting from mental representations, convictions and/or beliefs of interacting agents and who precede the action and, therefore, instrumental-rational choices. This interaction between beliefs (resumed, as we have seen, during a different period of the development of a 'new epistemology') deriving, it would seem, from arcane experiences acceptable more or less to a number of authors. They managed to determine, theoretically, events or 'latent happenings', underlined by the so-called 'Thomas Theorem' availed of by Merton in his *Social Theory and Social Structure* and by other authors like Paul Watzlawick at the end of the 1980's.

At that time, Italian sociologists developed an interest in the study of models relating to the social sciences, as shown by a number of theoretical writings regarding them presented during encounters promoted by the universities, such as The November Turin Convention concerning which Luca Ricolfi reported in a volume some years later.

Ricolfi, professor at Modena University, later discussed the sociological 'concepts' in semantic and syntactical interpretational terms, based on the clarification and taxonomic 'style' availed of theoretics and in the field of empirical research in the social sciences. Ricolfi also dealt with the representation of 'quantitative characteristics' in research operations (1985 and also in a book published in 1991) and with 'comparative' logic, referred to the construction of models (quantitative, in particular). Other fundamental studies carried out in Italy were those of Luciano Gallino (1984, 1985) and in the wake of the results of a ten-year study of cognitivism (see, by way of example, articles by authors like Giovanni Francesco Lanzara and Giancarlo Provasi in "Rassegna Italiana di Sociologia" as well as by Federico Neresini in "Sociologia e ricerca sociale," [Sociology and social research], -all published in 1995). Gallino devoted his 'period-bound' studies to 'models of the relationships' between the human and natural sciences, thus arousing interest in the formulation of a scientific model mediated by the history of action, paying some attention to cognitive processes as 'special instances' of action, analysable by means of general theoretical references known precisely to sociologists. A quest for 'alternative paradigms', might have been framed, according to this author, within the context of self-reflection by scholars studying the fundaments of their own disciplines, in line with trans-disciplinary events "taking place within the various sciences, stimulated by evolutionist epistemology, cognitive sociology, and post-positivist approaches to philosophy and the history of science" (as stated in *L'incerta Alleanza,* 1992). Gallino, for example, tended to consider the conventionalist and constructivist models as examples of the fact that concepts, models and theories might be considered arbitrary constructs, never a mirror-image of the reality (or the external world).

The study of models prompted scholars to pay attention to topics capable of creating a whole series of topics for analysis among sociologists on the 'strict' determination of ways by which to produce not only the reality as theory, in a more or less experimental or logic-experimental sense, as exemplified by Merton's *middle range theories*. Moreover, functional analysis and its epistemological and social-scientific derivations were related to artificial models, formal languages, complex sets and historical perspectives. Within the scope of these working hypotheses, scientists found it very useful to investigate the formulation of models like that of Parsons's action, meant as a 'testing ground' for the functionality of the theory-research relationship, a theoretical example of modelling. In general, if the models could stem from investigations capable of following the development of the cognitive sciences in the wake of 'constructivism', sociologists would have been able to sustain these 'undertakings' only as self-reflection on knowledge (whoever knows also becomes an object of knowledge). The interest in models among empirical social scientists and other theoretical-practical opposers of 'constructivism', seems to have produced, among researchers at the beginning of the 1990's, fruits which we may define as promotional of future studies and theoretical research into the re-emergence of theory, replacing a 'new epistemology' which, to tell the truth, might have been reconsidered from the 'point of view' of a new perspective of the sociology of science. In any case, among the scientists of the empirical social sciences, the methodologists, the logisticians and the sociologists of science, 'constructivism' was hardly ever read as it had been by more or less pragmatic relativists *à la* Feyerabend who ended up by seeing epistemology as a kind of 'old swindle'. The results of hostility against 'constructivism' urged researchers, more than everything else, to question processes of 'negotiation', where the concept of scientific knowledge, intended originally as a socially negotiable construct, might regard other spheres of interest like that of political decision-making. The result was a kind of de-construction of scientific knowledge: its development in an environment uncongenial to its diffusion and to the transformation of information. Moreover, this intellectual process upon which authors like Vittorio Ancarani were able to reflect during the 1990's by investigating British and American literature and trying to carry out a kind of epistemological 're-localization' of scientific knowledge (1993-1996).

The Theoretical Question

From another 'point of view', the age-old discussion regarding scientific theories and/or rationality in science never betrayed the expectations of authors contemporary to developments of sociological theory, like those who for a while dominated the scene which we shall now examine: methodological holism and individualism. Some of the authors in question used the study of models to propose a theoretical system to use to formulate theories themselves, so as to investigate the construction of theoretical systems as in the case of analyses of Parsons's structuralism, where only meta-theoretical research could reveal the way in which Parsons was able to deal with the logical and constructive character of his system.

To the functionalist model corresponded a meta-theoretical approach or meta-theory regarding functionalism devised by Talcott Parsons. The same may be said of the discussions regarding cognitivist modelling based on Piaget's or Jerome Seymour Bruner's meta-theory of cognitive psychology or the debate surrounding a scientific experimental model or, finally, the consideration of the meta-theory as an explication in neo-positivism.

The debate on relationships between models and 'rationality' followed a different route. Here research investigated the limits existing in the theorization of the 'irrational' *homo sociologicus* model, also taking into consideration the importance of discussing the *rational choice* (utilitarian) *model* in more modern terms and of recovering the theory of social action inaugurated by Max Weber. In this case, we may be reminded of Raymond Boudon and a whole series of his work translated into Italian (see "Rassegna Italiana di Sociologia" n. 2, 1987).

In Italy, the specific interest aroused with regard to meta-theoretical models represented a ripening of those areas of study which, as we have seen, were informed by methodological individualism and its relationship with a systemic-structural conception of individual or subjective action, otherwise known as 'holistic'. This was a favourite topic among the few who, in the 1980's, frequented logical-epistemological themes, like Franco Leonardi who assumed the task of tracing the developments of methodological individualism itself as a 'meta-theory'. In actual fact, as we shall see, the question was posed precisely by Leonardi, at a time when Italian sociologists were engaged mainly in debating the construction of concepts and their use in methodological terms.

From our perspective and, in a preliminary way, we need to say that the 'meta-theory' might be traced back to mathematical logic, that is, to the theory which has a formal system or theory as the object of its investigation. The meta-theory might be characterized by a universe (the object theory and its task might consist not only in providing overall results applicable to entire classes of propositions belonging to the object theory and the entire formal system, but also to the identification of relations between the theory being considered and other theories. In this sense the sociologists might have used the term 'meta-theory'. The same is true for the interest aroused among scientists of scientific subjects by the use of the term 'meta-sociology', meaning that branch of sociological theory destined to deal with methods and the logic of research, placing on different planes of importance, the study of propositions, principles and generalizations, etc, as suggested in the 1960's by Robert Bierstedt who opposed 'meta-sociological theory' or methodology known as 'meta-sociology' with the 'substantive theory': directly referable to the social empirical world. In actual fact, the diffusion of both expressions was assimilable within our frame of reference.

In a different manner, we need to point out that the problem seemed to be that of catching sight of a complex panorama of studies where an increase of interest in the theoretical and meta-theoretical questions of social sciences was simply a reflection of the problem of defining a theory, as an advanced vision of a possible redefinition. In this sense, we need to say that, in Italy, there had been antecedents, especially during the years of full reception of elaborations on the Parsonsian theory of the 'social system' which, in the end, was strongly recollective of the concepts and/or to the construction of 'schemes of concepts'. As Luciano Gallino explained in his Italian introduction to *The social system*, 1965, Talcott Parsons was able to distinguish himself from his contemporaries by managing to build a theory of the most unitary kind, the most minutely pondered in the history of sociology, regardless of the existing American tradition and the enormous debt owing to European sociological and psychological thinking. We need only recall Parsons's first work, *The structure of Social Action*, 1937 (the Italian translation came out 1962) and three of four scholars: Vilfredo Pareto, Emile Durkheim and Max Weber (the fourth being Alfred Marshall).

The fact that Parsons recurred to conceptual schemata before drawing up his theory, referred, according to Gallino, to the whole history of science, which teaches that the formation of concepts and history develop through an endless process of approximation, from more rudimentary to more sophisticated verifiable concepts that may lead to theories with greater empirical contents and explicative capacity. One of the fundamental issues to be underlined regarded the explanatory scope of Parsons's conceptual scheme, where the enormous task of systematically formulating a general sociological theory as part of a theory of action, set limits also concerning the language and terms to use according to the conceptual approach proposed.

When retracing Parsons's conceptualization of the social system (resolved by means of a highly rigorous scheme of logically classified concepts rather than by an attempt to expound theories of 'concrete' phenomena) Gallino appealed to *theory construction* recalling Berliner Georg Simmel and formalistic sociology admitting that the use of conceptual schemata might have led both authors outside of the ambit of a possible explicative structure.

Besides these problems, there existed a series of epistemological implications regarding a social theory contained in a book he published in 1937, where the author analysed systems of action of positivistic-utilitarian origin to reach the conclusion that the relationship between the social/actor terms of the system could not be associated with the criteria of a kind of 'rationality' considered the sole cognitive element capable of directing action in the name of equilibrium and stability so that it might prove to be the only explicative canon applicable to individual human action.

Parsons returned to this issue almost half a century later after writing this book -- in 1978, one year before his death- in response to an article by R. Stephen Warner who examined this theme of human reaction in relation to cognitive elements. Anyhow, it was the elements contained in his early work that a number of Italian 'translators' had assumed a few years earlier. Among the scholars in question we find Luciano Cavalli in *Il monument sociale*, 1970, a book that distinguished itself from those of other examinations of sociological theories conducted by authors like Gallino (from a 'systematic' as well as a social-theory angle) or Gianfranco Poggi between the mid 1960's and the following decade who had most of the works of sociology produced in Harvard translated and published in Italy.

It was also possible to apply to the study of social systems the general theory of systems which, in the opinion of some, might accentuate the opportunity to creating conceptualizations which, as we shall see later, had an important impact on the development of meta-theoretical interests, especially among Italian researchers. The general theory of systems, in turn, managed to take stock of the aspects inherent to relationships between the theory of society and technology and/or the technocratic society, as pointed out by Luciano Gallino (1978) who resumed works by Alessandro Pizzorno, Franco Rositi (1973) and some foreign scholars.

The theory of systems provided the stimulus for many to discuss the conceptual nature of sociology as a science contemporary to the development of these 'social systems' which in the 1980's might have been seen as an 'existing' subject, at paradigmatic breakthrough level, by proponents of structural functionalism, better known as the followers of the German social jurist and philosopher Niklas Luhmann, who used 'reflexivity' and 'self-reference' to specify his concepts better. In short, if society is 'self-referential' this means it has goals. And not only that. Luhmann was also able to conceive knowledge as a sophisticated form of social action, treating social action itself as 'adaptive,' that is, conditioned by the environment which, in turn, could be represented by the world surrounding the actor. In this process of the definition of automatisms according to which the system was placed in a position to function, Luhmann emphasized the cognitive components, in that, for him, adapted action (or *autopoieis*) could prove effective due to the fact that actors are guided by knowledge and subjected to processes of the learning and acquisition of habits. In this case, it is possible to find different definitions of concepts used to proclaim theories of society, emerged from Luhman's production between 1990 and 1995 and his articles regarding the topic (see for example, "Teoria sociologica," 2, 1993) and accepted in Italy at the time.

In particular, it was precisely the systemic theory that scientists availed of to understand which of the 'concepts' used by sociologists might provide a necessary and 'essential' series of formulations. Furthermore, it was useful when considering the theory of systems as a 'super-theory', especially when confronted with the fundamentals of a 'general theory' of the social and human sciences.

In a certain sense, the study of models and the meta-theoretical approach stemmed from all these elements. As time went by, the treatment of sociological research methods and techniques came to include, above all, among the terms relevant to the disciplines, a noteworthy interest in methodological holism and individualism, as well as in terms such as meta-theory, paradigm, theory and model. Even the most advanced research into existing relations between socio-human and natural models, led analysts back to themes implying the passage of theoretical issues like those regarding methodological holism and individualism, to the theory of systems. One should not exclude a close connection between the quests of models of relations between sciences, the formulation of social theories, inherent in the canonical system and appendices derivable from analyses of language, considered a particular type of action, as sustained by John R. Searle and John L.Austin and taken up by Gallino (1992), even earlier, by Paolo De Nardis (1988) in the consideration of the links existing between 'the theory of linguistic acts', models and prospective views of functionalism (including the study of subjects, institutions and social structure).

In his study *L'equivoco sistema* De Nardis goes more thoroughly into previous elaborations of his on social theory and socio-institutional analysis (1976) – a thesis treated later by authors like William M. Evan who, not only created a theoretical bridge between the sociology of rights and the sociology of organisation, but also claimed an new model of a structural-functional type - devoting more space to the analytical theory of language as related to the possibility of redefining the 'system' by taking a propositional, constructive kind of logic into account and, thus, availing of certain models. The objective was that of continuing to focus on social theory, starting with functionalism and, therefore, on a discussion of methodological holism and individualism as an essential 'theoretical issue' aimed at the use of meta-language, that is, a language useful to a definition of other languages, attributing importance to a certain kind of syntactic structure, tending, in some cases, towards formalization. A preliminary examination of Parsons's writings foresaw previous analysis of the right associated with the issue of 'social order', assimilated by the scientific literature of the time as a focal theme of sociological theory, passing from a theoretical analysis of the structure of social action to a structural-functional analysis of social systems, as a system of action. This, in short was the purpose of the book published in 1937 where Parsons questioned himself on the connection between the empirical world and research.

Besides the results of research into Evan and the abundant material provided by *Organization theory,* De Nardis, after a summary of some critical versions of the structural-functionalist theory produced by Italian authors too, dealt with the functionalist aspects best suited to investigations of the semantics of the theory of systems and its epistemological limitations, admitting, as did the author in question, that, in reality, many American advocates of structural functionalism, including Evan, made no effort to tackle an epistemological and methodological discussion regarding the approach itself.

On the contrary, this did not hold for Parsons, because his efforts aimed mostly at 'systemize' sociological theories and seek to provide it with an epistemological basis. This was the direction we can say his studies of social sciences intended taking: an epistemological conception that took shape thanks also to surveys carried out to examine behaviourism in psychology and useful to the construction of a theory of action investigated at human-science level.

His well-known contestation of radical empiricism passed availed of the concepts and theories needed to obtain scientific knowledge. In practice, the conceptual structure permitted adherence to reality (for example, when thinking up a 'system') in such a way as to select some elements in an 'analytical' way. Parsons's 'analytical realism' (see *La struttura dell'azione sociale* XIX) resolved itself in this accentuation of elements with respect to others, accomplished within the framework of a general theory capable of turning sociology into a full-blown science besides arousing the opposition of a whole series of *theories of the middle range* used by Robert King Merton to demonstrate some minor claims within the field of the development of sociological theory, and, resolved for example, in the theories of small groups, of the social organizations and in a theory of communications.

It actual fact, as De Nardis clarified in his book published in 1988 and, to tell the truth, there would always be very few capable of grasping the importance of re-proposing themes starting from certain concepts which had remained 'trapped', within the theory of systems, without taking into account the fact that certain attempts at resizing the general scope of some Parsonsian theses, were carried out by some of Parson's own students within the contemporary framework of American neo-functionalism up until the late 1990's. The most prestigious example was that of Jeffrey C. Alexander who was particularly well received by Italian scientists of Parsons and some researchers with a limited degree of diffusion of some aspects of his theoretical work. Authors like De Nardis, for example, do not yet mention him when *The New Theoretical Movement* appears (see Neil J. Smelser's, *Handbook of Sociology*, 1988). The opposite may be said in the case of Franco Leonardi when referring to a few years later on or to other who became known in Italy from 1982 on, like Bernhard Giesen or Michael Schmid.

In his first, major work, divided into four volumes, not translated into Italian as yet, *Theoretical Logic in Sociology (*1982-1983) Alexander made an attempt at reconstructing the tradition of the sociology of culture availing of the thinking of Emile Durkheim and Max Weber, to develop a discursive model of the social sciences in contrast with scientific awareness which Talcott Parsons had derive from his own theory. In particular, the fourth volume of the work was totally dedicated to Parsonsian thinking and to criticism of his major works. It concluded ending with the invitation to identify new lines of development starting from the revision of positions expressed by the master. This was an attempt at discussing scientific knowledge of society. This author's view was 'individualistic' and connected to the voluntarist theory of social action taken from Parsons. His work ended by dealing with some epistemological questions and an approach which Randall Collins, in 1992, defined as 'multidimensional' (Collins refers directly to 'multidimensional' logic which specifies the construction of a part of social theory; the other 'logics' are those of 'objectivity' and 'subjectivity', depending on whether human action may be defined by given conditions or normative, 'ideal' elements characteristic of the subject). The epistemological issue was that whereby interest shown by sociologists in the possibility or not of getting to know the social reality. Alexander added the theoretical issue, which, together with the former, permitted sociologists to question the nature of the social reality itself, so that the solution to the epistemological problem might lead directly to the solution of the second. In other words, for the author, to pose questions of an epistemological nature did not mean affirming the autonomy of these theories, but the contrary. Moreover, Alexander supported post-positivistic sociology which he believed would lead to the actual achievement of what Parsons proposed only. His proposal, on the other hand, was to pursue neo-functionalism, as revisionism or as an 'intellectual tendency' inclined to involve sociologists who deemed the elements of the Parsonsian analytical approach useful while seeking to place the notion of 'functional requisites' into proper perspective.

With reference to Italy, it must be said that Alexander was invited by the "Quaderni di Sociologia" in 1987 to write in that year's number 9 issue about the contemporary definition of his 'theoretical logic', in the light of the contribution made by Parsons's *Structure.* In the same number of the magazine, Francois Chazel of the University of Bordeaux dealt with epistemological tendencies emerging from that same work. It was a time when functionalist in sociology and I important critical responses to Parson were declining to the advantage of a degree of reception of schemata to refer to action. Referring to his American contemporaries, Alexander observed the emergence of the problems regarding the conceptually restricted theoretical criticism existing in sociology and regarding differences between the empirical, ideological, and theoretical dimensions of sociological work in general. The author tried to define sociological theory by connecting the emergence of new topics regarding the philosophy of science introduced by authors like Michael Polany and Thomas Samuel Kuhn to claims by sociologists that they might do without non-empirical and/or cognitive elements of analysis, etc. In short, Alexander designed a new scheme for the analysis of theories in sociology, referring to investigations of the classical authors carried out by Parsons 50 years earlier. The result was that the analysis was followed a 'multidimensional' method where certain elements prevailed while others were protected by the theoretical approach.

Returning to the 1990's, we need to recall that some of the 'theoretical syntheses' that stemmed from functionalism or the 'materials' based on the systematic theory, etc., regarded the debate surrounding methodological individualism and holism which assumed the appearance of a confrontation between micro-macro alternatives in sociology. De Nardis himself resumed his theoretical examination of functionalism in terms of a logical-semantic quest for systems in an article published by "Sociologia e ricerche sociale" (n. 43, 1994) proceeded in part, by papers presented at a seminar on *Individualismo metodologico e ricerca sociale* held in Rome in October 1993 during a series of lectures for Ph.D candidates specialising in Methodology of the Social and Political Sciences at Rome's "La Sapienza" University and at the University of Florence.

More than anything else, this general sociologist's 'point of view' tended to confirm the need to accomplish some kind of differentiation between a meta-theoretical approach and the *theory* itself, holding firmly to the 'point of view' of the logic of scientific-social research and methodology, considering also systems analyses in not-realistic terms, that is, without following the later Luhmann's 'gnoseological-realist' line in order to propose investigations of methodological individualism simply to assert the 'constructivism' of systems.

It was De Nardis who pointed out, in one of his talks during another seminar on regarding *Systems and social research* held in Rome in October 1991, that sociology was destined to experience a grave crisis of theoretical production which, in turn, would have allowed the 'theory of systems' to assert itself. The effort at mathematizing it, according to this scholar, would have helped to overcome of the age-old dichotomy between the science of the spirit and the science of nature, just as an evocation of Husserlian and Habermasian *Lebenswelt* would have meant restoring, at a certain moment, more or less historicisable ideal-typical categories. We might cite many more examples. Anyway, the scientific use of the terminology inherent to the system would have re-introduced among scholars dualisms of an epistemological order, which, to tell the truth, seem to have remained unsolved even after the introduction of new categories of analysis. Referring precisely to this issue, De Nardis was urged to explain and critically assess Luhmann, for whom a system should not have been posited either as a tool or as a theoretical hypothesis, but rather as a working 'reality'. A system might refer to the stability due to the invariance of the system with respect to the environment. According to De Nardis, Luhmann's system might easily lead 'connoisseurs' to a prospective view of theoretical and meta-linguistic discourse as well as to evident hostility towards an epistemological discussion, as practiced by 'gnoseological objectivism', that is, that 'realism' which, in Luhmann's own programmatic intentions, that the existence of systems began to take on, admitting a non-verifiable identity between concepts and reality. Furthermore, Luhmann's notion of systems led to evident opposition to the operationalism of the empirical social sciences, even where analyses produced by his critics went in the direction opposite, in any case different, from that practiced when seeking a legitimate use of some categories, like that of the 'system' applied to the logic of scientific research.

For De Nardis the 'theoretical issue' was the heart of a constant meta-theoretical quest for constructs, meanings and not only conceptual opposition which, as Franco Leonardi claimed in the 1980's, as confirmed by his article in number 43 of the above-mentioned magazine, conceals epistemological premises leading to the practical impossibility of establishing "either on the meta-theoretical or theoretical plane [whether] we are in a condition to conceptually distinguish the macro from the micro, the collective from the individual, the whole from its parts," etc.

The conclusion reached by Leonardi seems to have stimulated a methodological discussion which some scholars were obliged to address both with regards to methodological individualism and its 'destiny' as well as to problems associated with holism itself as a 'theoretical model'. They were also obliged to examine some of the epistemological premises and a whole series of logical and semantic inconsistencies found after carrying out detailed analyses. Authors like Gianni Statera (who died in 1999) devoted their efforts to those versions of methodological individualism produced up until the early 1990's and around which assembled problematic epistemologically considerations aimed principally at demonstrating, alongside the other kinds of research carried out by the author, the scares relevance of the individualistic postulate to scientific-social research. Statera's research started from a logical-methodological refusal of holism adopted in the name of a concrete kind of research practice tending to affirmation of an opposite notion, dominated by theoretical postulates, in turn, in contrast and incongruent with affirmations of a general nature. Furthermore, the contributions made by the author regarding these issues reaffirmed the need to refer to scientific theory and elaborations on it, in view of a new and more permanent contestation of ontological-methodological postulates like those of some authors and/or those used by some in hermeneutics (as in the case of Antony Giddens's 'double level'), or before by phenomenological versions of methodological individualism *à la* Schülz, all versions to which some sought to attribute a methodological paternity. On this question, the role of Statera in the essay published in the 43rd number of 'Sociologia e ricerca sociale' was clear and polemical towards methodological individualism, defined as a theme of an 'essentially philosophical' nature even where it was proposed as a unique meta-theoretical model, to be read in different social-scientific keys like that provided by Rosaria Egidi in an article published in the same number of the above-cited magazine: that is, an interpretation of the reasons of epistemological anti-individualism.

Problems of Conceptualization in the Social Sciences

The matters regarding meta-theories which have been summed up here, were linked to conceptualization within the social sciences, which intended to include within the domain of 'theme-problems', vaster implications of both a more theoretical and a more empirical 'point of view'. An entire issue of the magazine cited in the previous paragraph was devoted to two different aspects of possible determinations of the sociological knowledge also within areas where developments and/or 'epistemological questions' drew the attention of methodologists like Alberto Marradi, especially if considered as 'questions relevant' as 'indicators'. We shall return on this later.

It is clear enough that from the 'point of view' of more modern historians of science, the fashionable habit of referring of the 'conceptual' issue, as intended by the Latin term *conceptus*, meaning relative to the concept, might also exclude arguments widely treated by the history of philosophy and, in any case, surpassing ancient philosophical opposition, like that deriving, for example, from the tradition of Socratic inspiration as well as that seeking the ontological *status* of concepts found in Plato and Aristotle, up to William of Ockham's 'dispute on universals' and 'nominalism' in opposition to 'realism' etc. The same fate attended the older Hegel and followers of the *Wissenschaft der Logik*. Historians of scientific subjects appealed undoubtedly to classical empiricism, where by concept was meant an intellectual instrument for the organization of reality as introduced into the human mind by things themselves availing of the senses. It is well worth while to remember that from Hume and on, one of the main problems of empiristic philosophy was that of explaining the formation of concepts starting with sensations. Furthermore, this position developed with the linguistic evolution of empiricism. For many strains of thought contemporary to certain rational guidelines (paradigms, behaviourism, neo-empiricism), concepts were considered as a 'rule' or set of rules governing the correct use of a linguistic sign (for example, a word). This kind of position suspended, to some extent, reasoning regarding the formation of concepts, by focusing on how they worked within the ambit of a language or human behaviour. The function, in general, was supposed to be that of working on other signs and/or objects of experience. This was the heritage received and assimilated by the social scientist once and for all, thanks to a transposition different from that provided by historians of epistemology and connoisseurs of the history of scientific sociology, granted that the latter had followed a given kind of discussion regarding the development of theories within science.

For the moment, we need to realise that the theme-problem of conceptualization in relation to 'theory' within the social sciences might have been even more modern and up-to-date than it was. In any case it acted as 'a testing ground' for a goodly number of scientists and researchers who had grown accustomed to practicing theoretical combinations after 'constructivism' and which tended towards the affirmation of epistemological positions that were less rigid than before. It is with this in mind, that we should interpret some 'period-bound' works regarding conceptualization and referring to links between the latter and sociological explication, a topic to which Antonio Fasanella devoted a book in 1993 and which contained a selection of his own essays published a few years earlier in 'Sociologia e ricerca sociale'. One of these essays concerned the crisis of sociology (1990), another on Max Weber's explanatory sociology and method: *Oltre la sociologia comprendente.* Another essay by the same author regarding the limits of the 'radical individualist' conception appeared in the same issue (n. 43 1994) while the above-mentioned volume on conceptualization (1993) contained the results of his theoretical work on the dynamics of changes in conceptual material and on some of the 'operative' developments which might prove useful to scholars seeking to understand empirical data.

In this same n. 43, 1994, a new series where we find essays that make comparisons between 'macro-sociological' and 'micro-sociological' positions in sociology, a theme which, in Franco Leonardi's words, might have already represented the problem of conceptualization in the social sciences in terms of a re-proposition of methodological individualism as a meta-theory. In this sense, Leonardi referred to 'macro' and 'micro' sociology in terms of analytical research into terms like micro/macro, collective/ individual, total/partial, in a direction which allowed him to frame epistemological perspectives which, no longer reductionist, permitted him, though unsuccessfully, to undertake a quest for conceptual definitions drawn from among the dimensions in question.

As stated above, both from a theoretical and a meta-theoretical 'point of view', it would not have made much sense to distinguish *micro* from *macro* at conceptual level, as these were simply ways of viewing the same phenomenon. This position differed considerably from those regarding interpretation and analysis expressed by authors like Bernhard Giesen who, in 1994, in the well-known issue of 'Sociologia e ricerca sociale', proposed theoretically surpassing micro-macro opposition, thus giving rise to a debate between current positions and stimulating critical responses from Italian authors like Leonardi or Maria Stella Agnoli, authoress of an interesting essay, *Il legame micro-macro nella pratica della ricerca sociale* in the same number of the magazine.

In her article the authoress endeavoured to provide an interpretation of the pragmatics of programmes of 'theoretical synthesis' regarding the existence of micro-macro models and convergences between 'social theory' and claims made, for example, by authors like Jeffrey C. Alexander towards the end of the 1980's and attributable, in part, to attempts at conceptualizations of interrelations existing between micro and macro or, at least, at seeing these models as endowed with a sort of 'ontology'. The same which Giesen, according to Agnoli, derived from two original versions of the various models illustrated: the one corresponding to macro 'ontology' (antagonism and structuralism), the other to micro ontology (individualism and constructivism).

In substance, if Dr. Agnoli tried to prove the existence of a conceptualization of use to research and researchers, Giesen claimed that his micro-macro models provided the possibility of re-accommodating relations between different levels of 'empirical reality' in such a way as to be of interest to meta-theory. Again the meta-theoretical approach proved advantageous with regard also to the debate on 'reductionism' which occupied various levels within the discussion concerning the issue of conceptualization in the social sciences.

Another point was that of recourse to positions contrary to the affirmation of reductionism which and tended to broaden the discussion and render it, eventually, theoretically productive at least as far as some disciplines like sociology were concerned. The case of methodological individualism might act as an eloquent example also with regard to the methodological-philosophical theory and used to compare several problematic situations, as held by Michael Schmid, professor at Angurg University, in his Italian article *I presupposti metodlogici dell'individualismo metodologico* published in the above-mentioned issue of 1994.

Schmid's problem was mostly that of conceptualizing the importance of having to discuss social theory and, from a theory of action angle, to discover the conditions from which they stemmed. This with reference to methodological individualism, valuating it while also taking into account that the heuristic value of options like 'individualism' or 'holism' might depend on 'theories' established on the basis of their own respective doctrines and not on factors from without the theoretical programmes of the social sciences less known to philosophers. This statement of Schmid's was recollective of studies by Karl Dieter Opp from the University of Hamburg, 1979, quoted in the bibliography to his essay, which also referred to many publications by other European and American authors as well as to a series of considerations regarding methodological individualism as a methodological-philosophical theory always compatible with the efforts of social theoreticians.

Schmid noted the indefensibility of the 'ontological' argument when referred to methodological individualism, as Statera and Leonardi both held, although they departed from different bases. He directed his analysis towards the consequences of the discussion as far as the formation of theories and social research were concerned. Schmid also contested all and every kind of 'reductionist' programme also when this type of reference affected methodological holism and its attempts to reduce macro laws. All this meant admitting that, in terms of empirical social research, it was possible to provide permanent control of a methodological type for sociology, as shown by elaborations on 'method' by authors like Lewis A. Coser or Herbert Blumer, without excluding actors and their actions, that is, taking into due consideration the social dynamics within which actors interact.

Schmid's conclusion, already announced, was that opposition between holism and individualism would no longer produce anything useful unless the two theoretical traditions were forced to meet and misunderstandings between scientists and recourse to meta-theory sorted out, especially where delimitation was sought. In his opinion, integration might achieved by pursuing the uniformity of a logical model. He referred to Alexander's neo-functionalist attempts (1988) and to those found both in James S. Coleman's 'theory of rational choice" – expounded in *Foundations of Social Theory* (1992) with several references to Parsons's 1937 publication- and in Bernhard Giesen's 'theory of selection' (1991).

All told, Giesen and Schmid' aimed, however, at championing integration between micro and macro positions, and, therefore they represented a 'point of view' acceptable to philosophical methodology and 'period-bound' scientists interested in investigating into the conceptual organizational aspect of social theory and empirical research. From some observational angles, attempts were made to accomplish more thorough investigations, like that aimed for example, at analysing Luhmann's theory, like that of Francesco Pardi (1994) or those regarding micro-macro relationships in sociological theory carried out by contemporaries, including Nicolò Addario, who studied 'radical constructivism' and problems relating to 'naturalized epistemology' (see 'Teoria sociale' n. 2 1993) while carrying out in-depth examinations of topics regarding the conceptual foundation of theories (besides the 'period' examined in *Autopoiesi della societa*).

For the moment, the debate seemed to provide a kind of convergence on attempts to link, at theoretical level, the two sides of analysis, leaving aside references to relationships between conceptualization and empirical research.

As already anticipated, Marradi wrote on this issue, in a style congenial to his logical-semantic studies, in Part II of the above-mentioned 43 issue of 1994, devoted to the construction and use of concepts. His contribution, derived mostly from research and from endeavours at basing the debate on social, empirical sciences and/or on those 'classifications', 'typologies', taxonomic studies etc. which constituted its preparatory conceptual material along with other analyses of complex constructions and material he prepared produced for the lectures and seminars he held in Florence at the Department of Political Science and Political Sociology (attended by the present author as a Ph. D. candidate specialising in 'Social theory and research', 1992 - 1993 who also heard his lectures on *Realtà pensiero linguaggio* [Reality, thought, language], in January 1994).

In his *Referenti, pensiero e linguaggio: una questione rilevante per gli indicatori* published in 'Sociologia e ricerca sociale" Marradi sought to establish a 'point of view' concerning the nature of indicators as if treating other important methodological issues, by relating his reasoning with three 'spheres' or definable worlds, the "spheres of objects, thought and language" and endeavouring to define relationships between them. His work seemed to be that of retracing the current debate regarding 'indicators' by using methodology to explain the terms of studies into the scientific structure of 'theories' making due reference to some of the epistemological matters introduced by Karl Popper in *Objective Knowledge* (1972) and, with a slant different from that of 'doctrinal' formulation and that of other investigators of the nature of the three 'worlds'.

Marradi pointed out that the nature of concepts might not be that of exclusively having but of 'being' – in terms of the words used- 'a meaning''. He also stated that, one of the most frequent dilemmas researchers had to face might be solved by trying not to confound concepts with terms and, from a less exclusive position, admitting that concepts themselves might enjoy a certain 'flexibility', as claimed by Gordon H Bower from a cognitivist stance, and by scholars of kindred disciplines who posited the impossibility of outlining the perimeter of concepts. Moreover, this 'open' conception led to a refusal on the part of authors like Kuhn, who advocated a quest for other types of 'indeterminateness', more in keeping with the tasks which historians of science proposed carrying out. Marradi to this regard, recalled Kuhn's 'new reflections' on paradigms which appeared in a text published in 1974, care of Frederick Suppe and entitled *The structure of the scientific theories*, published in Italy in *La tensione essenziale,* in 1985. In any case, he held that concepts might and did locate referents and, following the semantic route, combined to create 'statements'. While concepts were neither true nor false, statements might be considered true or false and subjected to empirical control, unlike 'pre-arguments' which included various conceptual constructions like 'classifications', 'typologies' and 'taxonomies' (dealt with above).

The author continued his examination ranging from the sphere of referents (the world of objects) and 'ideas' (the world of thought) to that of signs or the world of language, insisting on the fact that Popper himself provided different, more or less updated, versions of the 'three worlds', with reference to the 1975 Italian translation of the original 1972 work and that of 1983, with an introduction by the translator, Arcangelo Rossi, after the book had been revised in 1981.

Marradi's essay focused in particular on relationships between referents and concepts, and those between referents and language as well as on relationships between the sphere of thought and that of language. The author focused first on mutual conditionings between referents, grasping the differences and distinctions between referents themselves. Secondly, he brought together deep and thorough considerations regarding relations between referents and signs, where the latter had to be interpreted in order to consider how language and the words related to 'things', by means of meanings, concepts etc., but especially with regard to the 'mind' to define the 'identity of structure' posited by Wittgenstein in his 1922 *Tractatus* belonging to "facts and the propositions that describe them." In short, Marradi continued, by examining 'extensional semantics' contributions (referred to language) that dealt with the position of the 'Viennese Circle' and its epistemological consistency with regard to the development of different trends in scientific thought like behaviourism in psychology, logical positivism and operationalism in sociology. The examination was conducted in order to find connections between signs and their more or less direct referents, more or less mediated by thought. The attempt involved, mostly, the history of epistemology. Thirdly, the author of this essay addressed the study of the conceptual developments stemming from the sphere of signs and, in any case, the initial distinctions between concepts and terms proposed by scientific literature with regard to connections between the sphere of thought and that of language.

Marradi recalled how many scholars used the words 'concept' and 'terms' in the same sentence as with the same meaning. Later on, he analysed relations between thought and language exactly the light of tendencies to define concepts and language with this kind of relationship in mind. He also provided precise indications of the framework of terminology to use for concept-term relations referred to scientific language and all associated 'semantic-dispersive' problems examined from a different angle. He was also obliged to deal with the twentieth-century 'linguistic turning' and how it contributed to conceptual development where language and its scholars found that 'central' role claimed by historians of twentieth-century science in the handbooks. The same cannot be said for the issue of thought although it did regard relations between thought and language.

Marradi concluded that the nature of the 'indicators' derived from what was examined. This did not prevent him from re-examining the term 'indicator', keeping in mind the differences between the meaning of the term in everyday language and its technical use by the social sciences. He dwelt on the use made of it by operationalist Stuart Carter Dodd and Paul Felix Lazarsfeld starting from *Evidence and Inference in Social Research* with the essay on '*Deadalus*, 1958, (translated in Italy in 1969) where the Austro-American methodologist proposed making a distinction between 'indicator' and 'index', the latter used to measure a concept or as a combination of indicators. The interest shown by methodologists in this these topics which became so paradigmatic also within the remaining 'schools' that it appeared in a broad 'period-bound' range of writings where many Italian authors debated the use of indicators in the sociological research. Their positions were expressed in publications, at conventions and teaching seminars for example those held two years before the publication of an essay by Alberto Marradi in Rome, care of Leonardo Cannavò in *Modelli e Indicatori Sociali,* for the doctoral school of research on the Methodology of the Social and Political Sciences. In a certain sense, some of these meetings, including other 'period-bound' initiatives organised by other Italian national universities, sanctioned the connection between the foregoing topics and in-depth examination of themes inherent to the conceptual and technical-operative development of scientific sociology. In brief, the future, also in the light of a lack of development of a perspective of epistemological historiography, failed to sanction in its favour an absorption of issues, not only of a methodological kind, concerning the perennial and pre-paradigmatic state of sociology.

Furthermore, a year after the publication of 43 number of "Sociologia e ricerca sociale" the magazine hosted an issue dedicated completely to Paul Felix Lazarsfeld considered, according to Marie Jahoda, a '*marginal classic'* and to be appreciated for his contributions to sociological methodology as well as to issues regarding problems of conceptualization. We shall deal with this later especially with regard to the second point.

Meanwhile, Marradi also dealt with terms regarding the operative definition of indicators, where these referred to the possibility of conceptualizing properties. Moreover, in the essay he exemplified how the properties of concepts coincided with the 'validity' of an indicator and how elements from the sphere of signs might work in relation to the 'validity' of an indicator, upon examination of the relationships between one concept and another considered its indicator. In any case, the problem of the 'choice of indicators' was considered relevant, especially when related to 'theory' and how this interfered with choice. What is more, in this field, Marradi drew up a classification of terms which cropped up most frequently in the language of scientists, something he had already pointed out in the essay in 1984 regarding types of meaning of 'theory', a topic he willing resumed. In other words, the debate on indicators seemed to be able to count on the contribution of 'theory' only if related to the progress made within the field of scientific-language studies, where, Morrafi stated, there was 'dispersion' as far as the use of the term 'theory' was concerned. This is how he directed his efforts also when following the route of 'falsificationism' and, ultimately, recriminating the sociologists and the researchers of his time, especially when considered dependent on 'theories' and/or 'the theory'.

In his opinion, 'theory' might contribute to the selection of indicators especially if related to given meanings: first of all as 'tactical knowledge' of certain processes derivable from phenomenology and a quest for 'linguistic agreements' carried out to define social behaviour; secondly, as a 'conceptual frame or network' relatable to measurement (for example in texts); thirdly, as a series of non-empirical-type 'abstracting reflective activities,' explained –for example- by the use of the term 'thought' by historians of sociology; and, finally, with regard to the meaning considered proper of 'theory' referred to 'statements' or systems of statements 'conceived in order to be submitted to empirical control'.

Returning to Lazarsfeld and arriving at the year 1999, it needs to be said that the considerable interest shown by some scientists of methodology (like Vittorio Capecchi, Enzo Campelli and Alberto Marradi himself and many other Italian researchers) in the monographic number of 'Sociologia e ricerca sociale' following a Convention held in Rome, focused on all those aspects to which reception of a sort of epistemological vision might have given rise as an 'alternative' to the dominant logical-empiricist position and to opposition of the positivistic methodology propagated by the handbooks. This especially when considering the period when Lazarsfeld carried out his research and launched his concept of empirical social sciences. To tell the truth, even in the period which followed, the diffusion of handbooks featuring American research in particular, conditioned the younger sociologists so that they accepted an almost 'official' version of methodology and, to a certain extent, a more or less discounted attitude towards epistemological history. Statera in the introduction to his volume on the logic the strategies and the techniques of social research, *La ricerca sociale (1997),* held that this diffusion occurred Italy with the development of 'special sociologies' beginning with the old manuals by William J Goode, Paul K Hatt and Bernard S Philips in use in the 1970's to the time of the diffusion of Kenneth D Bailey'sand the flourishing of handbooks by Luca Perrone, Paolo Guidicini, Rossanna Memoli and Angelo Saporiti, Costantiono Cipolla. In recent times these were written by authors like Piergiorgio Corbetta who sought, instead, to define, above all, the paradigmatic turn of social research (1999). It is symptomatic, however, that Corbetta brought his introductory discussion back to the times when Lazarsfeld and Morris Rosengerg' edited the text *The language of the Social Research*, that is the mid-1950's.

This discussion might concern connoisseurs and Italian historians of sociology of belonging to two decades, one before the other after, when Lazarsfeld was considered a 'quantophrenic' when it came to conversing with philosophers of science, as Carmelo Lombardo noted in his essay published in a monographic magazine cited in *La formazione e la definizione dei concetti sociologici,* despite his links with authors like Rudolph Carnap, Carl Hempel and Abraham Kaplan quoting Statera in his more advanced books, like *Logica dell'indagine scientific-sociale,* 1997. But there is more. Lombardo's reasoning also regarding Kaplan who defined the well-known analysis of 'latent structure' (see mathematical models in sociology) as a conceptualization technique, is useful when defining the so-called 'indicisation processes' in strict relation to detailed processes of conceptual formation which led Lararsfeld to outline, *despite himself* as the author notes, an epistemological model for the social sciences, inspired more by the concrete and provisional work of scientists than by formal logic itself. This work was quite unlike that interpreted by American sociologists of the so-called 'strong' sciences or by those who had, in any case, caused a recovery of interest in Lazarsfeld within a framework of necessary developments of methodology or theoretical-conceptual and technical-operative apparati to the full advantage of the social sciences, pressing analysis beyond the probabilistic nature of sociology. In this instance and, due to certain solicitations, we might refer to Costantino Cipolla with his *Dopo Popper, published* in 1990.

What the abovementioned essay on sociological concepts led the discussion back to an approach to conceptualization which took into account the multidimensional nature of Lazarsfeld's thinking as well as his propensity, as Enzo Campelli noted in his presentation, to examine, in depth, the relationships existing between the theory of research, without omitting, however, those between methodology and epistemology, within whose sphere the social scientist might have been placed, that is, on the 'boundaries' between numerous different approaches. This also concerned 'a style of the research', as Capecchi sought to demonstrate when discussing the contribution made by Lazarsfeld to the methodology of sociology. Of interest also Capecchi's comment on the theoretical-methodological relationships posited by Lazarsfeld, that is the close weave existing between 'theory' and methodology. Capecchi re-introduced some of the guidelines drawn from empirical research to explain the practice of the Austro-American's methodologist. Lazarsfeld actually operated in such a way as to find such concepts existing in the empirical theory of action or, in other terms, 'the concepts that better explain' it. He also drew up a model of interrelations between concepts, otherwise known as the model of relations between 'intervening variables', considering the 'effects' of contrasting relations, of the time dimension and the structure of general theory. Only the conceptual interweave provided, according to Capecchi, proof of alternation between qualitative and quantitative methods capable of highlighting the predominance in research of the web of concepts to be measured.

Referring again to theoretical-methodological development, we might refer to some of the observations concerning Lazarsfeld made in his short essay by Marradi, who tended to implement a gnoseological-terminological discussion regarding the Austrian bringing to light some 'categorical errors' relating to a number of concepts and therefore to the terminology used by the contemporary community of scientists. For Marradi, the importance of Lazarsfeld's contribution to the invention of the figure of the 'methodologist' was comparable to that of the epistemologist institutionalized by the exponents of 'Circle of Vienna'. However, in his essay, Marradi also tried to define better concrete/variable and the concrete/indicator relationships, where the latter was intended as a 'probabilistic' relationship, while explaining that 'possibility' might be defined by means of multiple relations between events and classes of events, etc.

The fact that Lazarsfeld himself dealt with scientific knowledge in relation to the formation of concepts, to the explicative functions of indicators and processes of formalization may be included among the 'innovative perspectives' of the history of science. It was this interpretation which seemed to stem truly from developments of Italian researchers' thinking regarding the work of the methodologist regarding the issue of scientific subjects. Unlike what was held by other European scholars like Raymond

Boudon who no longer dwelt on the *livre imaginaire* or the representation of the philosophy of social science, as he had done in 1970, but on the language of social science (1993), 'cognitive rationality' (1995) or the empirical analysis of action (1998).

On the whole, Lazarsfel's accepted anti-reductionism, considered a 'classic', by issue n.58/59 of "Sociologia e ricerca sociale," [Sociology and social research], might have led to claims that only questions of a conceptual and technical-operative nature be associated with the history of scientific subjects in such a way as to highlight their development. It might even be admitted that sociology as a discipline seemed bent on becoming an integral part of the modern history of science. It is evident enough that the flexible model of the operationalism of concepts (launched in the United States by Lazarsfeld in the wake of the sociological naturalism of scholars like Stuart C Dodd, Read Bain, F. Stuart Chapin and George A. Lundberg and of Percy W. Bridgman's operationism occurred in an intellectual climate not overly hostile to behaviourism, including Lazarsfeld's work), was already a part of a development of scientific subjects quite unlike the 'history of sociological thinking' or the 'history of the scientific theory' etc. To understand this, it suffices to follow the activities of the American Sociological Society at a time, when the writings of some historians of science of the 1950's informed by what scholars like the already mentioned Read Bain wrote concerning sociometry or by articles published in the "American Journal of Sociology" by F. Stuart Chapin of the University of Minnesota (for example *Measurement in Sociology*, 1935) or by George A. Lundberg of the University of Washington (for example, *Quantitative methods in Sociology: 1920-1960*). The interest shown by Lazarsfeld himself in the psychology and social psychology of the German Kurt Lewin revealed his tendency to study empirical themes underlying a certain type of historiography of concepts, of the development of scientific concepts as well as of theoretical experimentation. The same holds for the transmission of his well-known 'operative definition' model created to pass from indicators of concept to variables ("the vocabulary of the social sciences") on the basis of operations capable of being accomplished through data detection. With time this model grew even more detailed until becoming known as the Popper-Dewey-Lazarsfeld model, dealt with by Italian authors like Leonardo Cannavo and Gianni Statera in the 1990's.

Furthermore, it is possible to interpret in a similar way all the efforts made by those who today claim encamping a 'history of methods', meant as a history of social empirical research. An essay in this sense was that by Mauro Antonio Fabiano which appeared in the above mentioned magazine and which moved from this precise hypothesis.

On the eve of the 1960's, Lazarsfeld and other authors, for example, Nathan Glaser, gave rise to a sort of analysis of the origins and the development of concepts and tools in areas familiar to historians of science. It was during those years that handbooks were published - from another perspective which we shall call the 'history of sociological thought' - presenting different ways of indicating the 'history of sociology'. With reference to this we might mention two works in Italian by Alberto Izzo. One dealt with

Don Martindale's history of sociological theory, the other with the historical reconstruction of the development of John Madge's empirical research theories. But they did not deal with the history of the social sciences, that is, they did not provide a conceptual history of sociology linked to a qualified vision of the development (historiographical too) of scientific subjects.

One 'epistemological' limit (if one may speak of limits in the case of Lazarsfeld) might be that of having reasoned about problems regarding conceptualization within the social sciences without some critique of the concepts that were formed, as someone pointed out. A sort of venial sin committed by the master and his followers.

Chapter 13

Concepts and Revolutions of Knowledge and Know-How

A goodly number of methodology handbooks of the following generation, produced by Italians sociologists in the 1990's, began to discuss the logic of research and the changes in cognitive structures which gave rise to visions and perspectives of a theoretical nature impacting on the empirical social sciences. Although the techniques, and, in general, sets of procedures availed of by the community of researchers, were taken into some consideration, it appeared that one of the most important issues used to frame the discipline's theoretical and epistemological levels, was the debate surrounding paradigmatic unity, also in the light of developments in the history of scientific sociology after three decades of work by Lazarsfeld and his followers, and, in any case, in the wake of American structural functionalism. In some contexts, headway was made by the idea that the study of paradigmatic change in the social sciences resulting from the philosophy and criticism of science might lead scholars to a more precise historiographical treatment of the development of science in such a way as to trace a pathway for sociology similar of that of other disciplines; a veritable history of epistemology accredited among modern and contemporary historians of science. This meant revealing how, in this direction, philosophers of science might produce 'materials' capable of disseminating results more readily among the new generations of historians of sociological and historical subjects and not only within the sphere of philosophical historiography. And, as we have seen so far, this seems to have happened, without the opposite taking place, without historians of sociology applying the same method of research to their own theoretical, historical, conceptual and epistemological development. In fact, very often it happened that certain tasks were entrusted in full to historians of philosophy and to the acceptance of historiography by philosophers. A striking example is that of the reception, in Italy, of Max Weber and his epistemology (the last complete collection of methodological essays was published in 2002). Moreover, it suffices to recall the late 1950's and the tasks philosophical research assumed regarding the social sciences and the 'awareness' of the new generation (I am referring here, by way of example, to the little-known known volume of March 1957 containing some of the 'material' produced by a Congress on the theme held in Bologna that year).

In developmental contexts like ours a certain kind of discourse does not seem to have taken precedence over the task of historians of scientific subjects, who gradually discovered a strong affinity between the enormous work to be done and changes in 'conceptual structures' (what the historian of physics and critic of science, Thomas Samuel Kuhn, called 'paradigms') such as to bestow credit on a kind of paradigmatic historiographical research. In reality this theme seduced the minds of many of our age, without, however, producing affirmative results within many of the social disciplines. As we have in this essay, to be 'seekers of paradigms' and be it only 'periodically' does not mean and has never meant for the sociologists taking a serious interest in the history of science. A lack of methodological unity has caused all the rest. The only result has been, maybe, that of considering, as Kuhn did, the social sciences and sociology in particular, as if they lived immersed in a state of permanent or pre-paradigmatic 'immaturity', at least from a certain angle. But even this thesis of Kuhn's was came progressively up against new perspectives of the sociology of scientific knowledge, not overly akin to 'official' versions of epistemology. In the end, it appears that this whole series of events had considerable influence on the lack of the development of a 'point of view' accredited among the scientific subjects which had once witnessed developments regarding the issue of a quest for paradigmatic unity. That is the main strand of this book regarding the conceptual intricacies of the history of Italian sociology, a topic to be proposed *in primis* to 'producers' of the history of sociology of the second half of the twentieth and the beginning of the twenty-first century.

With reference to the latter and the last part of the twentieth century, it should be noticed that a kind of permanent paradigmatic reflection regarding their own existence in an age of pluralism of sources of sociological knowledge, as well as of reflection on their function within the scientific disciplines, might have benefitted sociological subjects, making it possible to avail of a tradition of research generated by 'producers'. Finding precedents is not an impossible enterprise. It is far less difficult than finding connections between historians of sociology and historians of science, so hostile to the discipline. That was the way things stood until the death of some of the scholars who opened up new routes to sociologists of science. In a work called *Storia della scienza e sociologia della scienza* (1993), and following the historical perspective, Luciano Gallino, recalled the approaches of some new historians of science like Stephen E. Toulmin, Nortwood R. Hanson, Thomas Samuel Kuhn and Imre Lakatos, of whom we have spoken above. In part, the importance of these authors for the rest of scholars of the social and empirical disciplines may be attributed to the fact of having grafted onto the tradition of historiographical studies, scientific issues, concepts and fundamental frameworks reference (the famous "programmes of research") capable of bestowing direction upon the praxis of scientists, that is, of giving them a meaning, including a progressive definition of the limits within which to elaborate scientific theories. Not only, but adherence of historians of sociology to this new kind of reasoning meant releasing them from the then traditional, "hostile" way of conceiving the development of the discipline while influencing the way historians of sociology would propose themselves in decades to come. The historical vicissitudes of sociological science during the twentieth century may be considered illuminating. This is why Gallino was able to point to degree of complementariness between the history of science and the sociology of science, stating that the former had received from sociology a "significant contribution in terms of conceptualization," a systematics of phenomenal types and of the observation of different phenomena.

Links of this kind entered into the purposes of historians of Italian sociology wherever reflection involved wider areas of theoretical and historical-epistemological interests, influenced by too many events tending to assert profoundly changed new socio-scientific horizons. In reality there was no lack of occasions to compare and discuss historical-sociological matters as well as the chronic delay in establishing social epistemology as a discipline. A seminar on the 'history of sociology' organized in May 1993 by the 'Theories' section of the Italian Association of Sociology, was the occasion for some already-known published authors and researches like Mario Aldo Toscano (University of Pisa), Orlando Lentini (University of Naples), Alberto Izzo (University of Rome) and Alessandro Cavalli (University of Pavia) to discuss the state of the discipline and the work of researchers. The relationships between history and sociology, examined on that occasion in a paper by Toscano, were submitted to a theoretical scientific interpretation, starting from the positivistic sociology and a thorough and in-depth investigation of the endeavour made by contemporaries to receive the 'moderns', where sociological science was meant, at the one hand, as an academic achievement and, on the other, as a veritable demonstration of the 'social' as a phenomenon endowed with incommensurability. Furthermore, Toscano pointed out how, from the 'point of view' of the kind of history of the discipline endorsed by Italian sociologists, the methodological tradition might represent a front for those who, renouncing the glut of transformational processes and changes, intended to dedicate themselves totally to the analysis of the forms of cognitive structures. The emergence of sociology was able to express itself everywhere, thanks to an objective historical exigency, where subjective responses would depend on original objective questions. This, briefly speaking, was to be the task of the history of a 'discipline of the future' and of 'modernity' bound to distinctive contemporary traits inherited (as Toscano stated, a little later on in his book *Spirito Sociologico* (1998) from the period of Auguste Comte. Other Italian authors, including Enzo Rutigliano from the University of Trent, who took an interest from an epistemological angle in classical sociological theories (Karl Marx, Georg Simmel, Max

Weber, Talcott Parsons). Another scholar, Carlo Mongardini from Rome's "La Sapienza" University, insisted on a re-proposition of fundamental concepts in *La conoscenza sociologica,* 2001. Both were able to denote in the progress of the history of sociology veritable strains of 'modernity', a concept, often detected and detectable, according to Mongardini, who found in it an initial approach tending towards the ideology of 'modernity'. He also indicated a second one inclined to express criticism of 'modernity', and, finally, a third one, corresponding to sociological analysis. Generally, for these researchers, the link between the history of sociology or the history of sociological thought and contemporary theories produced by sociologists existed above and beyond any kind of epistemological scepticism, starting from the proposal to read the classics once more.

The seminar of 1993 yielded indications from a number of Italian historians of sociology regarding the study and 'non-rhetorical' ways of *reading* the classics (Alessandro Cavalli) in terms of the 'history of thought' and sociology of knowledge and historiographical interest, while admitting some methodological analogies between different ways of studying texts while paying attention to rigorous philological-hermeneutic analytical techniques. A part of the debate concerned the production of scientific -not only 'humanistic'- handbooks and the various difficulties historical-sociological activities encountered within a context of progressive erosion of the disciplinary barriers noticeable from 1980's, with evident reference to the impossible task of categorising knowledge and 'knowledges'.

On these topics Alberto Izzo provided sociology with distinctive traits to avail of when studying how institutional structures conditioned the individual and re-introduced the question of how to train historians of sociology to outline a problematic and critical ambit rather than one of accepted institutionalization. Regarding this, Alessandro Cavalli noticed how sociology had taken steps towards its legitimization by developing increasingly more formalized theories and empirical research during its institutional acknowledgement as discipline, to some extent, independent from other disciplines. This author recalled many of those quoted here, from Nicola Abbagnano and Pietro Rossi to Renato Treves, to Corrado Ghini and many others without forgetting the vicissitudes of either Adriano Olivetti or the centre-left, the 1970's, etc.

For him the making of a 'history of sociology' meant, precisely, studying the practice of institutionalising disciplines, and, therefore, the apparati of knowledge. According to Orlando Lentini, that series of intellectual and rational operations was charged with the task of creating a history of the other sciences, where the history of sociology was to be considered a part of the history of historiography. Sociology might identify itself with one of many modern ways of making history. In the attempt to bestow a historical dimension on social disciplines, Lentini anticipated the concept of 'social analysis', identifying it with any kind of history whatsoever prior to Auguste Comte and his analysis, as the methods used by the historian of sociology were more or less the same. In this sense, the history of sociology dealt with the history of historical and social sciences, both as antecedent and progress. From the same angle, we should remember that even at that time, in the wake of Immanuel Wallerstein, Lentini sought a transition from sociology to the *historical* social sciences, by means of a quest for a methodology based on concepts of long-lasting historical systems, when the thrust of many interdisciplinary approaches had already dissolved the most modern version capitalist 'world-system', as indicated by non-sociologists like Fernand Braudel and Llya Prigogine. This way, a totally interdisciplinary version might become the context within which historians of science themselves would discover links between the social sciences and the humanities in such a way as to be able to rethink knowledge and ways of knowing.

Lentini himself, in his studies, often returned to the issue of changes in structures of cognition, most of all, when carrying out a new examination of the positions posited by Wallerstein (see, for example, "International Sociology," 1998, 13) where he proposed that the social sciences should be particularly 'open minded' towards an uncertain and problem-filled future, after he had proclaimed their paradigmatic limitations in *Unthinking social science*, a text dating from 1991, published in Italy in 1995 and presented by Wallerstein at Rome State University's Department of Sociology. During the conference presided over by Orlando Lentini, the writer of the present text questioned the author regarding the existence or not of a methodology of scientific historiography available for use by sociologists and/or social scientists in Italy, obtaining a not too exhaustive answer, one referring, at the very least, to the experiences of the "Fernando Braudel Center" of Bringhamton (N.Y.). Moreover, it was at that time, that someone spoke about a generalised impact of sociological subjects on the new way of considering history as well as about the influence of 'Lazarsfeldian historians' like Immanuel Wallerstein and Terence K. Hopkins over the sociologist of science Robert K. Merton. The same emerged as regards new concrete elaborations on a number of concepts (the concept of 'hegemony' for example) suitable for analyses of some of the modern world's major transformational processes. In other ways, a number of historians *à la* Wallerstein, cherished the teaching of the radical sociologist Charles Wright Mills at a time when functionalism held sway in the United States of America. From the beginning they carried out elaborations on procedures, re-valuing, to some degree, theory when social scientists sought to construct historical variables. As Wallerstein himself emphasised in *The Modern World System* in 1974 (translated in Italian in 1978 and in 1980 in successive volumes) "Theoretical activity need not be separate from analyses of empirical data. Analyses can be carried out only in terms of schemata and theoretical propositions. On the other hand, analyses of some processes or events have to have as their starting point a whole series of specific evaluations of a number of variables on the basis of which, we are able to explain how we arrived at certain results."

In any case and starting from some of the theses of other European scholars, the Wallerstein's subsequent analysis and its reception by social scientists and historians of sociology were not only able to direct the debate regarding the gnoseological crisis of contemporary thought, but also theoretical considerations concerning cognitive structures. The result was the introduction of a new methodology of uncertainly, heterogeneity and complexity, practically speaking, new epistemological assumptions regarding the social sciences. To assert this, it would be sufficient to arrive at Wallerstein's recent work *The Uncertainties of Knowledge*, 2004, as yet not translated into Italian.

Anticipating some contemporary historians' reconsiderations of the nature of the various methodologies used in the different disciplines, Wallerstein provided a new theoretical approach to the social sciences while hastening to reset methodology to cater for interdisciplinary, a pathway followed and to be followed during the conduction of the many analyses of knowledge and "knowledges" as well as of possible epistemological worlds, no longer dominated by dichotomies -differences, even if overwhelmed in substance by a *surplus* of conceptual developments. Mario Losito and Antonio Scaglia from the University of Trent and others are an example of a group of Italian researchers focused on themes informed by this 'revolution of knowledge', which culminated in a text by Lentini called *Saperi sociali, ricerca sociale* in 2003.

On several occasions, including that of the above-mentioned 1993 AIS seminar, the author presented a definition of historical productions closely linked to the history of epistemology and invited historians to be epistemologically clear on matters of so-called 'knowledge'. His paper on the methods and perspectives of the history of sociology aimed at defining more thoroughly the links existing between knowledge produced and research methods, that is, between different 'knowledges' and research carried out. The same may be said of his propagation work among Italian and European researchers, during the decade in question, his contribution to the organization of international conferences (Pisa and Paris) on topics dear to Wallerstein and to other authors like Maurice Aymard of the *École des Hautes études*, Paris.

A perspective of 'social analysis' began to replace restricted views, divided at discipline level of the social sciences, where the concept of 'scientific' and 'philosophical' knowledge looked like extensions of forms and the activities of human beings, including those derived from their own conceptual considerations. This led to recourse to a particular kind of epistemological history, more and more independent from the paradigmatic revolutions of 'official' historiography. But we need to make some further distinctions when dealing with certain kinds of historians of knowledge and of the social sciences.

It is necessary to clarify, especially when treating the issue of the modern use by the researchers in question, of the terminology of knowledge in the broadest sense. The term 'knowledge' may refer to any kind of cognizance and culture, and more specifically -at least in the Western tradition- a set of specific, technical, scientific notions may not be considered 'knowledge'. This means that knowledge naturally means philosophical cognizance, not so much that relating to a specific discipline (philosophy) but, to its all-embracing, all-founding nature regarding all possible kinds of science (ethics, technology, the law, social issues etc.). This broader meaning of knowledge was practically dominant in Greek philosophy, from Socrates and Aristotle to Hellenism. In modern and contemporary philosophy, this concept, confined by Kant to the dimension of phenomena alone, was reaffirmed by means of the Hegelian notion of absolute knowledge as thorough self-transparency of the life of the spirit. Little by little, this notion made headway among philosophers and scholars of different disciplines, along with a conviction, more difficult to eradicate, whereby knowing, having a more or less vast grasp of a given thing due to study, teaching, tradition etc. was related -ultimately- to science, even though, the prevailing idea was that of absolute knowledge, like that we find in Dante's lines, *Poi ch'innalzai un poco più le ciglia, vidi/ Il maestro di color che sanno*, Inferno, Canto I, line 130, referring to Aristotle which translates as " Then as I raised my brow a little, I saw the Master of those who know." Many historians of culture and philosophical subjects have narrated the history of learning following the route which led them to the modern sociology of knowledges, in a dilated environment where a multidisciplinary interest in philosophy always stands in the background acting as a guide. This was the state of things until the boundaries of knowledge were limited to the field of scientific interests and renounced claims to cognizance by pursuing knowledge as such, ultimately opposing science with changes and a philosophical quest for its negation. What was once the domain of history, political philosophy and theology, ethical studies, etc., began to take an interest in the field of sociology and social sciences.

As time passed, a view of the sociology of 'social knowledges' was assuming a particular position in contrast to the foundational version of epistemology also within a context of post-Kantian reception of the debate on changes in scientists' ideas and the scientific subjects themselves. From the 'point of view' of some historians of sociology, the 'revolution of knowledge' might have opposed 'scientific revolutions' wherever it witnessed a sort of imminent 'conceptual revolution' among contemporaries. The sociologists of knowledge, for example, in different situations and contexts, produced some considerations when confronted with profound changes to ways of considering 'concepts' in the 'modern age', also when the latter was becoming an object of research by some of the most accredited historians of sociology who, thus, assumed the highly important task of 'taking stock' of the transformations taking place. Other changes affected the field of 'methodological knowledge' which we shall treat following considerations regarding the stances assumed by some scholars on the issue of the so-called 'new rhetoric', a new form of rationality meant to be 'alternative' to that scientific rationality propagated as an 'argument' by Chaim Perelman and resumed more recently in Italy by scientific philosopher Marcello Pera. The contact with these topics also produced an idea of conceptual history quite different from the more or less 'official' version produced by historiographical scientific subjects, close to historical-epistemology. Some Italian historians and academics, like Alberto Izzo, tried, in *I percorsi della reagione (1995)* for example, to outline yet another itinerary for the history of sociological thought to follow and within which to inscribe not only the dilemmas of the sociology of knowledge, but also its rationality and the 'crisis' between its moderns and contemporaries, indicating a not too clearly defined route for the theory of sociology and for the intellectuals of his own era still entrapped in 'invincible perplexities' like at the time of Karl Mannheim. Regarding these themes, and, in different ways, topics regarding philosophical comparison with rationality of a scientific kind had been treated by Italian researchers from different educational backgrounds during the previous decade. One such scholar was Enrico Berti with his *Le vie della ragione* (1987) following more arduous publishing endeavours like those of Aldo Gargani, Carlo Augusto Viano, Remo Bodei, Salvatore Veca and others. On the issue of relationships between rationality and 'modernity', and different from those just mentioned, we find the outcomes of Italy's reception of Habermasian developments of procedural rationality, as a 'third way' between the metaphysics and nihilism of modernists within a (no longer Marxist) project aimed at establishing an *ex novo* critical theory of 'modernity'. To this regard, see Alessandro Ferrara's and Pierpaolo Donati's contrasting articles on *Teoria dell'agire comunicativo* (1986) with an introduction by Gian Enrico Rusconi from the University of Turin, published in "Rassegna italiana di sociologia," n. 2 April-June 1987.

By way of parenthesis, as far as some particular studies are concerned, we need to note that the redefinition of 'modernity' penetrated thanks to studies by some sociologists of knowledge to the study and their reception of processes of culture, political and moral philosophy. Some connoisseurs of theory and sociology, as well as a small number of historians of sociological subjects and socio-humanities (those which we believe can unite literature, philosophy, the visual arts and the knowledge related to communications like linguistics, semiotics, the media) have succeeded in carrying out, from this exclusive angle, a broad comparison between modern themes, parallel both to those concerning conceptions of ethics and the ideas of authors like George Edward Moore, Max Scheler, Richard M.Hare, Jurgen Habermas and Alasdair C. McIntyre and those regarding the so-called 'ethics of responsibility' (Max Weber, Karl Otto Apel, Emmanuel Lévinas) and the 'applicative ethics' of authors like Arne Naess, Hans Jonas and Tristam Hugo Endelhardt. To these we need to add questions regarding inter-subjective practices regarding ethical issues (implying amongst others, concepts of 'justice', 'morality', etc.) like those recently raised by Alex Honneth as a result of 'modern' social philosophy and by other theorists of 'acknowledgement' like Charles Taylor. One result of this broad reinterpretation of 'modernity' has been that of foretelling its imminent future, as researchers begin to use the term 'post-modern', echoing Alain Touraine's 'post-industrialism' of a few years ago. This term may be grasped only of we depart from some of the concepts arising from the philosophical debate within historical-concrete contexts where intellectuals, most of all, sought to publicly acknowledge the utility of resorting to a similar kind of terminology to explain, for example, their own origin.

In this direction, rather than that indicated by historians of scientific subjects, the history of sociology positioned itself alongside theoretical-sociological analyses of new institutional setups, imposed by processes of modernization, globalization and by a transition of society from modern to 'late-modern', as it was renamed by some Italian researchers who dealt with some of the themes most frequently featured in handbooks, associating them with the new sociological scenarios of the end of the 1990's. With regard to this issue, we might refer to *Le nuove frontiere della sociologis* edited by Paolo De Nardis) or the development of sociological thinking (see also *La sociologis e la modernità* edited by historian of sociology Enzo Vittorio Trapanese, 1998). Another analytical perspective belongs to that wealth of studies regarding state policy, institutions, law and the sociology of administration which emerged in Italy from the academic 'point of view' in particular. We should also recall the studies on cultural modernization launched in Italy by sociologist Gino Germani, who died in Rome in 1979. Germani actually founded the sociology department at the University of Buenos Aires and promoted a number of initiatives aimed at the internationalization of sociological studies, which led to the creation of the International Sociological Association, of which he was president.

The time was ripe, besides, for the diffusion of issues concerning the new debate in sociology and the social sciences, especially among those who, being interested in the history of sociological thought, endeavoured to discuss authors like Edmund Husserl, Max Weber and Alfred Schutz, who had conceived 'modernity' and informed a reinterpretation of the theory of action and of phenomenological, conceptual changes including epistemology and the study of its relationships with 'methodological constructivism'. With regard to this precise issue, see Luigi Muzzetto's *Fenomenologia, etnometodologia*, 1997.

The fact is that 'concepts' produced in so many monographs and in numerous theoretical and historical sociological theses on topics concerning 'modernity' and then on its liquidation, were asserting themselves more than the 'sociologists of knowledge' might have imagined, due to the proliferation of studies regarding cultural and communicative processes. This latter managed to impose a series of disciplinary modifications that led sociologists to unexpected destinations. In particular, when speaking of analyses of cultural change among moderns and contemporaries, we need to mention sociologists that followed the pathway of new elaborations on cultural processes, beginning with the founding fathers of social morphology, like the French Gérard Namer interpreted and discussed in Italy by some authors who developed an interest classics of sociological history like Paolo Jedlowski (author of *Il Mondo in questione,* 1998) and Maria Rampazi (see "Rassegna Italiana di Sociologia," n. 3, 2001) co-author with Jedlowski of a volume on the theme published in 1991.

Similarly, the philosophy of language and many of the theoretical findings of analytical philosophers of the last generation (in particular Michael Dummett) tried to change, in part, perspectives regarding the issue of the influence of meaning upon philosophy, for example, and on conceptual clarifications regarding it. Professional logicians and philosophers have added a certain number of more technical analytical instruments to their modern toolbox in order to respond to issues left open by Gottlob Frege, Ludwig Wittgenstein, Willard Van Orman Quine, Ilary Putnam, Donald Davidson and Noam Chomsky. To the 'revolution of knowledge' and, subsequently, to that of 'science', was added an unprecedented 'conceptual revolution', not devoid of discussions concerning the epistemological fundaments caused by the web of problems generated by different readings within the social sciences and between them and other disciplines. For historians of sociology, like historians of science, a veritable quest for distinctions between the social sciences seemed to focus on distinctions between 'points of view' and methodologies which experienced the several 'oxymorons' and conceptual developmental procedures applied to the phenomena in question. Without further reference to some foundational aspects, the history of sociology too, was able to cease its spasmodic quest for 'objective' entities, so that epistemology gradually freed itself from the need to seek a precise limit to transfer from philosophy and the philosophy of science to other fields of research, like those of the social sciences.

The many debates on 'identity' - stimulated by autobiographical reports of 'modern' man as well as by the ambit of 'narrative knowledge' opposable to the maturation reached by some levels of 'scientific knowledge' and/or to the possible benefits of a whole group of new 'reflective sociologists' *à la* Alvin W. Gouldner faced with the dilemmas of 'modernity'- seemed not to have noticed the signs of a transition into a historical period of major conceptual change taking place rapidly within the ambit of sociological subjects and capable of making researchers trespass upon different terrains of action, reflection and, in some cases, of 'negotiation'. And it was easier for this to happen where 'modernity', relativism and problems due to the absence of fundamentals, made new interests produce a more specific redefinition of concepts among new scholars.

In the later 1970's, the French philosopher Jean-François Lyotard, (with a book published in 1979 and translated into Italian in 1985), managed to describe the passage from 'modernity' to a condition called, literally, 'post-modern', a transition from one world to another, by characterising it in terms of a kind of polytheism of cognitive models and emerging new 'narrative knowledge', which he deemed fundamental to scientific and philosophical activities alike. It was precisely in the twentieth century that the humanities began to pay true attention to narrations of individuals immersed in their cultural environments. Important authors like the aforementioned Jerome Seymour Bruner (1986) deepened relationships between 'narration' and 'knowledge' which became real objects of research carried out by social scientists and some new historians of 'modern' sociological subjects. For these, knowledge of a narrative kind was certainly a cognitive mode which permitted the individual to order the symbolic reality surrounding him/her within networks of unforeseeable social relationships. The narrative formula led, actually, to some schemata capable of reproducing the intentions of human action. Furthermore, these schemata organized the knowledge of historians too, implying some kind of conceptual function by which to recognize an object in a given context. According to contemporary scholars, individuals try to bestow new order on events in the form of a 'story' by recurring to coherence, as Paul Ricoeur pointed out in *Time and Narrative*, 1986.

Between the dimension of an 'aesthetics of existence' and that of recovery of the 'subject', autobiographical reports and the discovery of different ranges of meaning, new historians of sociological subjects were able to devote themselves to acknowledgement of a considerable amount of intellectual work but only to create an undoubtedly self-reflexive position, which was, therefore, an end in itself, and which remained distant from and out of touch with developments regarding the history of scientific subjects. On the contrary, it seems that the historians of the sociology of the past were unable to free themselves even from their own history, where sociology failed to liberate itself from the history of other sciences. Meanwhile, the crisis of scientific and the decline of dialectic reason, which occurred among the moderns led, in Italy, to various forms of irrationalism and the so-called 'weak thinking' of Gianni Vattimo, Pier Aldo Rovatti, and, in part, of Aldo Gargani himself, and to consequent reception by many social scientists of Richard Rorty's American neo-pragmatism as well as of Jacques Derrida's and Jean-François Lyotard's 'post-modern' thought, mentioned above. This cycle accomplished in the mid-1980's was followed by considerable resource by researchers to the intricate dimension of 'points of view' from which to observe the action of science and/or of sciences, with a series of investigations into so-called 'complexity' where the contribution of epistemology as well as of particular research by sociologists of knowledge (for example Edgar Morin) proved crucial. So, referring once more to the last decade of the twentieth century, it should be noted that from the beginning of the 1990's, works by scholars like Pierpaolo Donati, revealed the need to confront fundamental, methodological and epistemological issues with issues-problems arising from functionalist, post-Parsonian perspectives (Niklas Luhmann, Jeffrey C. Alexander), from various outcomes of constructivism or, in some cases, from the theoretics of self-poiesis, which, in Italy, met with an uncertain destiny, stemming - without reaching any conclusion- from works which sought to connect themselves with those of important authors like Achille Ardigo (*Per una sociologia oltre il post-moderno* 1988). The effort meant for a number of contemporary sociologists meant seeking to associate a new epistemological frontier with a new 'network paradigm' capable of generating operational relational models in sociology (see Donati, in particular). The mere discussion of the opposition between theory and theories of social systems, made it possible to urge specialists of theory and meta-theory to reflect more thoroughly, and from a different slant, on the quest for the epistemological and cognitive 'limits' of sociology, by taking an 'extremist' view, as it were, of the discipline, not only as a basis for comparison but as an 'expressional' instrument by which to solve, once and for all, the ontological and epistemological issue (see Paolo De Nardis, *La sociologia del limite* Rome, Meltemi, 1999).

It should also be noted that the objects of the crisis and especially the effects of regulatory changes in Italian society became a topic for sociologists who studied the transformations of value and new institutional arrangements. This held for almost two decades afterwards. The sociology of educational studies and training enlarged the range of action of theories imported from abroad used to achieve new potential of / in modern society. In Italy contributions by authors like Marcello Dei, Marzio Barbagli, Roberto Moscati, Luca Ribolzi, Laura Bovone, Luciano Benadusi, Antonio Schizzerotto, Elena Besozzi, Carlo Barone, Antonio Cobalti (2006) and Mauro Magatti stand out in relation to modern processes of globalization. In particular, the line followed by studies of the sociology of education, traditionally supports and deals gradually with the sociology of the family, in particular its institutional capacity, registering a considerable increase in perspectives that trust in a certain theoretical framework (see, to this regard, studies by Chiara Saraceno and Pierpaolo Donati).

On the whole, sociology as a part of the social sciences and, then more than ever before, 'sociologies' devoid of a quest for their 'period-bound' epistemological and identitarian origins, continued to pursue scientific historiography, (something it is important to state here) until after 'deconstruction', while other disciplines (from psychology to physics) began to discuss changes in 'conceptual structures' that had marked paradigmatic developments directed, from within, toward the progress of the various sciences. Taking sides with 'modernity' (or, indeed, with sociology directed towards the epistemological level beyond the *cu de sac* imposed by 'post-modernism', as sustained by Achille Ardigo, 1988) meant that some finally gave up the quest and found one or more variously relativistic loopholes of a hermeneutic nature. For others, however, this marked the beginning of a new era in which to start practicing meta-theorist schemata while suspecting radical choices and options of the past, hard defensive stances or positions once given as definitively certain. The same held, as we have said, for age-old 'extremisms' of an ontological, gnoseological nature (considering also that in the Anglo-American world the 'theory of knowledge' corresponded to 'gnoseology', often to 'epistemology') etc. opposed to other radicalizations at work within the community of scientists and researchers.

Some felt the need to impose a theory of 'modernity' as a consequence, in opposition to the assumptions of sociology and of post-structural and 'post-modern' philosophy. This was a kind of surrogate of the theory of the mass society proposed by the Frankfurt school, through some authors like Antony Giddens managed to develop concepts capable of responding to the life conditions of 'modernity' itself with particular reference to reflexivity and 'identity'. The use of the concept of ' double hermeneutics' marked an attempt at understanding a new kind of world according to the need social scientists felt to grasp, in turn, not only ordinary actors but also the way social theories might be interpreted by them, having made some fundamental distinctions when all came to all- between social and natural scientists, meaning that the theories of the former might give rise to changes in social life which social scientists were charged with observing and interpreting. The use of a double activity of a hermeneutical kind recalls a focal point of Giddens's theory which was concerned with reflexivity, considered as essential to the birth and existence of 'modernity'; continual reflection on social and individual change, by means of which to get to know and control the surrounding environment.

The social sciences appeared very different in many ways and from several angles, so much so, that authors of a certain kind were able to historically verify their indeterminate nature by using epistemics, as has emerged from numerous examinations of post-Kantian science carried out by historians over the last twenty years.

Our trust rests in the fact that only this can lead us all to results regarding scientific and non-scientific matters.

The fact is that in some texts we still find specifications concerning a number of concepts (or the use of 'rules') regarding the social sciences, considering the fact that with the passage of time, the debate on certain topics like 'paradigms' and 'incommensurability' or the even more remote issue of the 'natural sciences' and the 'sciences of spirit', etc., may appear again in terms of methodological re-developments of some thematic areas of epistemology. The irruption of the epistemological history into the spheres of research, philosophy of science and of cognitive studies, seems to have produced a definitive abandonment of some schemes of reference, spread among the most rationally motivated social scientists awareness of the risks due to having to consider science, and therefore their own discipline too, beyond its own theoretical and concrete-historical development. It is no fluke that the much-vaunted philosopher-founders of anarchism and of modern and contemporary relativism were more or less liberal or democratic historians of science, so similar to some of their magnificent predecessors. Neither is it a chance that in Italy this figure is completely missing from academic institutions, the ranks of sociologists that of historians of sociology devoted to knowledge or 'knowledge' in an age undergoing a 'crisis of fundamentals' and experiencing a 'challenge of complexity' as well as a 'crisis of reason' and of that rationality which, when all comes to all, has always characterised the activity of scientists, as remarked by some philosophers at the beginning of the 1980's. So, epistemological history in the social sciences is asserting itself more and more as a result of failure to provide a more recent historiography of science capable of replacing the 'history of knowledge' and the sociology of 'social knowledge', the 'history of thought' or 'history of ideas' and also the 'history of culture' and of intellectuals etc. while it is guaranteed circulation within the various 'scientific communities'. In some cases, the history of science has managed to replace these branches of knowledge, others not yet.

Most of all where concrete specific field research has produced a progressive definition of the operational limits of the discipline the significance of the epistemological contribution seems to fade. The work of the Italian sociologists, if considered as research activity, has for many years now, been considered by some as lying outside of epistemological interests, in that the latter have been generally considered unproductive of logical, procedural contributions to modern scientific work. In fact, the other side of conceptual development is, if we like, paradigmatic of the disciplines in question, seeming as it does, to comprise some 'rules' and rules of scientific methodology, intended as a 'logical form' of reference for scientists and the social scientist too, who practices a kind of methodology where rules -unlike the natural sciences- need always to be chosen on the basis of what is to be investigated. Therefore, 'concepts' and 'rules' which may represent two essential spheres of scientific progress, that extraordinary human enterprise where the tasks, on both hands, seem to reduce gradually. The new historians of the twenty-first century need simply to ask themselves whether or not to address conceptual epistemological history as a history of sciences or a history of scientific subjects. Sociology in Italy has recently lost its greatest historians of sociological thought like Filippo Barbano (died in 2011) and Alberto Izzo (who died in 2014). These scholars have left a legacy that needs to be continued, especially by the younger generation of scholars. From another stance, there are studies that reflect relationships between modernity and the history of sociology, in particular, as regards the sociology of religious studies, as in the case of young scholars like Massimo Rosati, who died in 2014. Moreover, studies of modernity are quite common among Italian sociologists and scholars of culture. Here we wish to make specific reference to the Conference held to celebrate the centenary of the publication of Georg Simmel's *Soziologie* organized by the Sociological Theories of Social Transformations Section of the Italian Association of Sociology organised by Vittorio Cotesta and held in Rome from the 5th to the 7th of June 2008, which gave rise to the publication of a volume published in 2010. Now (that is, in 2016) we can say that there exists a generation of researchers who study the historical aspects of the sociological sciences providing an epistemological point of view. But, they do not, however, represent an established new generation of university-level historians of science who associate their research to the development of scientific epistemological. Their interests are sporadic and jeopardize the future of the discipline, at least from this point of view. So, it seems that a permanent revolution among historians of sociological has not as yet taken place in Italy.

MYTHOLOGICAL KNOWLEDGE AND THE HISTORY OF SCIENCE

In Italy, there is no dearth of scholars who depart from that very indetermination which characterises the social sciences and who, by defending a certain *methodological competence*, as the title of a book by Alessandro Bruschi published in 1996 recites as it addresses once again some of the logical-semantic and conceptual delimitations concerning relations between epistemology/methodology and/or, in general, relationships established between those who seek to build knowledge through the practice of research and, in the long run, following in the footsteps of the indefensible dominion of the methodology of the social sciences. Bruschi's book, which mentions preliminarily the relationship already engaged in by science and 'meta-science', discusses 'methodological knowledge' in the light of the results produced by the process of dissolution of neo-positivism up to the realist turning point of contemporary epistemology discussed by Enzo Campelli in his *Da un luogo comune* (1999), a book which sought once more a route to pursue, historically speaking, too, in order to disclose old and new rational pathways useful to analyses of methodology in the social sciences. Methodology was to be placed at multiple intersections capable of defining or delimiting its 'knowledges' and origin, something even more definable from the 'point of view' of the history of the scientific subjects, than the development of social research. In his text, Campelli claims the primacy of the scientific method's temporal quest in the course of a major conceptual effort aimed at a definition which generated a methodological-type of knowledge within a context of the transformation of associated topics and those which, gradually, intertwined with other topics of epistemological interest. The author seems gradually to acknowledge the importance of various contaminations which took place on the methodological and general epistemological sides alike. In actual fact, even in the absence of these, the course of scientific subjects would have run in any case in the direction of discussions by philosophers of science of the theorem of the self-sufficiency of method. It was by no change that it was to philosophers of science that the task was given to bend historical subjects to suit the exigencies of relationships between new epistemology and the age-old encumbrance of so-called 'methodological knowledge'. In the face of Feyerabend-style relativism and anarchism, the world of methodological rules seemed to shatter. The same cannot be said of the kind of epistemological knowledge which stemmed, precisely, from interrogatives posed by philosophers of science who concerned themselves, also in a very critical manner, with new developments in scientific matters. In particular, the methodology of social sciences seemed to emerge from matters surrounding the procedural aspect of the science which, only in this sense, might be linked to historical scientific issues which attracted to it processes of critical revision discussed within the sciences themselves, from positivism to the dawn of the contemporary debate.

On this basis, Campelli aired his opposition to total preoccupation with the construction of rules concerning methodological knowledge found on the route to modern legitimisation of scientific discourse. From a certain 'point of view' as far as historical issues of science were concerned, this might mean continually rethinking the immense conceptual effort made to produce the methodological debate from the beginning of modern science to the present. From another angle, this might have proven useful when seeking to confirm the fact that the development of scientific subjects followed an itinerary, however tortuous; one shared by researchers, with its traditional-historical choices and/or others regarding the transformation of 'official' historiography into a conventional discussion, most of all as if the references regarded the epistemological debate.

A relevant question arises when it is also admitted that for many social and empirical-scientific disciplines, sociology, in particular, the assumption of new distinctive characters is founded on motivations devoid of an epistemological grounds or on issues which do not derive from a series of in-depth examinations of the subject. All that seems to occur at the present period. This too seems to indicate the limitations of contemporary historians of sociology and social sciences who still do not reason in terms of the history of science, because they are in want of a quest for links with the debate which arose at a given moment concerning the pursuit of paradigmatic unity by branches of knowledge, like the physical or biological sciences. The transformation of the much declaimed 'official' version of scientific historiography into a merely conventional debate among researchers and intellectuals, our contemporaries, within specific disciplines and sectors of investigation might help explain this state of affairs. This might also explain thematic absorption with a quest for methodological rules while seeking a hermeneutics of method, as necessary today as it is disoriented according to Campelli, following, in the long run, the historical pathway of the methodology of social sciences. More than a historiography of social sciences, later shown to be inexistent, at least from the 'point of view' of a given historical-systematic version of the empirical social sciences, the author pursued the route of a rational reconstruction of pathways along which scientific knowledge is obliged to address, thanks to the exigencies of rigor, controllability and also 'objectivity', where basic problems underlined the need to establish a specific sociological methodology within a general framework of the methodological development of modern science. This is the umpteenth proof that the social sciences and, sociology, as specific subjects, belong to crucial moments in the development of scientific subjects as well as to particular ones related to epistemological thinking which, very often, was unable to exclude a number of themes pertaining more to the philosophical disciplines.

After the middle of the twentieth century, a plurality of philosophical trends denoted a panorama marked by a very strong tendency towards scientific specialization involving the social and human sciences (sociology and psychology, etc). Moreover, the approach which in the 1990's appears to prevail frequently in philosophy seems to be that most inclined to be open towards the greatest number of cultural contribution and historical solicitations. Recently, in the case of sociologists in particular, this appears to be the result of greater diversification of accredited research approaches, like those registered starting from Hans Georg Gadamer's and Paul Ricoeur's interest in hermeneutics, Karl

Popper's in scientific method or the Frankfurt school and its propensity for social criticism.

Following the epistemological route in the human and social sciences has meant, however, for some researchers, taking either the philosophical way or that of a quest for foundational themes, for those links and theoretical nodes considered as crucial to the contemporary development of the debate. This was the line followed by Davide Sparti in a volume dated 1995 which addressed the epistemology of social sciences, one of the few Italian contributions to the problem deriving from a re-elaboration on the issue of the theory-meta-theory relationship, in favour of a broader theoretical-philosophical confrontation between topics concerning the practice of a discipline.

Sparti sought to frame a number of perspectives within the ambit of the contrasts belonging to the field of the epistemology of the social sciences, with a view to identifying epistemological *styles* (naturalistic, hermeneutic and systematic) able to operate in a salient context as well as a view of theories of action, which from the actor's angle, might act as an observational unit capable of clarifying the functions of the above-indicated styles.

The social sciences had been delimitated in an early empirical version of sociological knowledge, by a sort of naturalistic paradigm of neo-positivistic origin which succeeded in excluding the individual observed from the discovery of the causes, and therefore, capable of generating a technological model of knowledge. By following the second epistemological style, the social sciences would be limited to a quest for meanings while presenting itself with and imposing on itself critical questions regarding the kind of action capable of excluding research from the risks present, on the contrary, in the models proposed by the natural sciences and/or the naturalist model. The hermeneutic model (and its declinations) would have been followed by a systematic, structuralist and classificatory style, inside of which social sciences would address conceptual issues as well as those defining a number of conceptual analytical categories and finally, the last style, that is, the orientation of theories of action, capable of embracing cognitive approaches and paradigms used to seek intentionality in actors.

If we refer, in general terms, to events of that kind, most of all to specific aspects of epistemological reflection, we might confirm some degree of interest shown by some methodologists in discussing the possible consequences of a definite ban on methodological rules in favour of alternatives practiced in areas of the development of the history of science more inclined to meet the requirements of the *disputatio* than invitations to treat it as a rational activity, unlike any other human intellectual act. In the history of science, besides Bacon's motto *scientia est potentia*/science is power, it seems that researchers had also to assume the concept that Descartes posited in his *Discourse* and which, according to many, founded a strong methodological tradition, difficult to forsake, so that alternatives to method may cause those who address them to plummet into the dark world of non-rationality. In reality, authors like Popper and Lakatos, tended to discuss in their books the immense relevance of this 'official' version to the history of scientific matters, most of all when studying the links existing between scientific theories and method. Other authors, for example, Paul Feyerabend, placed their seal on the end of that tradition, until the moment when the new philosophy of science provided its version. This was a period epoch of profound and different rethinking by researchers regarding processes of scientific knowledge as well as that triggered off by numerous 'indictments' of science, in Italian 'Processi' like the title of a 1978 publication by Vasco Ronchi, an Italian who dealt with Galileo's vision of optics, as Feyerabend was to do.

Enzo Campelli and Alessandro Bruschi criticised -in advanced terms- Marcello Pera's *new rhetoric* which admitted recourse even to a rationality alternative to that of science. This notion, which harkened back to the 1980's, undoubtedly deserves further investigation here, most of all because of implications regarding evident relationships between the history of science and methodological knowledge. From a historical-sociological and philosophical angle, a more thorough study should be carried out, originally into the implications produced by the attempts of some authors to generate similar interest in Italy by resuming some philosophical topics and developing a kind of knowledge no longer founded on incontrovertible propositions of scholars like Giulio Preati who proposed it in his *Retorica e Logica*, 1968 and Enrico Berti who aimed at recovery of rhetoric and the universal community of argumentation (1980's). In actual fact studies of the contraposition between rhetoric and logic should have formed the basis of a rebirth of theoretical and methodological interests capable of providing material for the development of a historiography of science which has never seen the light.

What Pera pointed out from the beginning of his *Scienza e retorica* (1991) and over the years following those sanctioning the 'crisis of method', and in any case, as part of a personal effort to reinterpret the image of science and its progress, might be summed up as awareness of the presumed failure of methodological activity seen as a sort of 'enterprise' not suited specifically to the historical purposes of science. That might also mean that the history of science was –in the end- not so very relevant to methodology itself.

In the direction indicated by Pera -with regard to whom it is well worth while to take a few steps backwards- the findings of the history of science would provide examples of rationality which were too often associated with claims of establishing 'guarantees' or methodological justifications, all connectable with scientific subjects. Examinations of the relationships between method and the history of science were, therefore, crucial. However, as was the norm, many researchers questioned themselves concerning the destination of their own discipline or, in any case, regarding the problem of following given pathways with a view to answering certain questions. Their research might be directed, even today, to tracing within the historical course of scientific issues, the existence of a pluralism of purpose and tools no longer concerned with a universal method and/or a quest for *a* purpose better than all others. Subject to frequent changes, a given historical perspective of science might sound like the echo of the words of Larry Laudan, the inventor of the 'traditions of research' resumed by Pera, for whom, the goal of the methodologist tends to become complicated, wherever a sole comprehensive methodology of science seems not to exist. Laudan himself, as we know, avoided referring to methodological knowledge by means of perspective research traditions, considering methodology as 'open' to critical evaluation and to the construction of specific theories.

All this, as is natural, caused historians of the scientific matters numerous problems With a new kind of perspective, the history of science might no longer be able to justify the existence of certain kinds of methods to the point of imposing a strategy of action of its own compatible with the course of events. This surely meant that having to deal with the methodology might cause different problems for the historians of science, most of all when it was no longer possible to open the way to the hypothesis of a quest for a sole 'strategy' capable of asserting itself among scientists themselves. So, control of methodological hypotheses seemed subject to the *défaillances* of a history of science which sought to assert some methodological rules (the same ones that constituted the subject of topics scientists were unable to renounce) relinking them to the empirical hypothesis. Recourse by authors and historians of science, like Imre Lakatos, to 'research programmes', harkened back to definitions of 'rules' which methodological knowledge should continue to obey.

It was possible, as Pera claimed in his book, that science might have no method and that a number of fundamental assessments of an epistemological character might be justified not by reason alone. Unfettering the scientific rationality of the methodology -in the Cartesian sense- meant admitting that the history of science had succeeded in changing the course of events internally by modifying the objectives it had always entertained. In brief, the attempt would have to be directed towards reducing all its methodological knowledge as well as the importance of 'rules' at transition level from the world of rigorous, objective demonstration to that of *argumentations* interlinked by the intervention of interlocutors using the weapons of persuasion and confutation (unlike those of dialectics). This way the passage between scientific method and rhetoric would be sanctioned wherever techniques capable of asserting debate replaced secure references.

It is interesting to note that Pera based his argument in part on Thomas Samuel Kuhn and to some of his writings, even if, according to Pera himself, Kuhn had not developed further some of his intuitions regarding a kind of 'conversion' concerning techniques of persuasion to use among members of a scientific community. Pera's interest in Kuhn aimed mostly at a new interpretation of considerations by Kuhn regarding a number of controversies concerning the world of nineteenth-century scientists, when the Italian author was preparing his book *Scienza e Retorica* (1984). At that time he was at the Center for Philosophy of Science in the University of Pittsburgh carrying out research.

He holds that an exchange of views with Kuhn regarding the animal electricity controversy between Galvani and Volta, gave new impulse to the intention of discussing the passage between methodological knowledge and rhetoric, and also produced a study in in a volume 1986 published in a new American edition, Princeton, issued until 1991. Pera's statements do not seem so convincing, especially when taking into consideration the intense work and historiographical activity Kuhn was carrying out at the University of Princeton at a time when studies of a certain kind of methodological production were making headway, largely sanctioning revolutionary changes within the domains of both scientific development and the scientific subjects.

The problematic character of changes of a revolutionary kind in science produced veritable discoveries for Kuhn, hard to reconcile with some of the concepts in use before these discoveries were made and which only later and in a particular moment were to influence scientists. If a division took place between scientists, it involved choosing between the cumulative development of scientific knowledge and a development with a different origin. This held also for the methodological knowledge produced during a certain period. If scientific revolutions represent exceptional moments unlike those of 'normal science', that is, the practice of research founded on the science of the past and on one or more results obtained and raised to the status of praxis within a scientific community, the role of some of the actors who claim being able to build the referential-discursive fundamentals should always be made clear.

It was not Pera's intention to fully attribute relevance, which Kuhn acknowledged, to *consensus* among scientists regarding the validity of some of the results reached, which might also assume a paradigmatic role, where paradigms were expected to exert persuasive force (thoroughly evaluated again by Pera) but also attract a stable and sufficiently wide circle of followers (paradigms for Kuhn, were not to be considered as rigid 'rules').

Unlike Popper, Kuhn was convinced that scientists did not normally spend their time training themselves in the severe controls to apply to the theories produced. Nor did they produce consensus where this did not derive from strong anomalies, found in changes of paradigm, that is, within a transformation of the entire conceptual structure used by scientists to relate to the rest of the world. The persuasive power of a paradigm does not seem to surpass the importance of obtaining, step by step, acceptation by the scientific community. This meant that for some at least of the social scientists dealt with at that time by Kuhn, the rhetorical apparatus and, therefore, the debate, techniques, persuasion and confutation, etc, related more to the paradigm as a builder of consensus regarding transitory imposition upon the scientific community. We need to recall also that for Kuhn social scientists and sociology, in particular, were still at an initial phase of their paradigmatic development, whereby they still expressed 'schools', not any truly characterized paradigm of their own and/or the dominant one of normal science'.

The use of the art of rhetoric was, however, considered in this sense and corresponded, therefore, to that pointed out by Kuhn in his *Postscript*, a text where the historian of science extended acceptation of the paradigm to the behaviour of different communities. The behaviour of *scientific communities* was not to be considered in a less prejudicial manner than that of single scientists subjected to their own 'points of view'. As Kuhn affirmed, since paradigms became part of a debate regarding choices of an acquired and consolidate set of knowledge, theories etc. their role became circular and every group was aware of being inclined to use its own tools, categories and knowledge etc. to defend that paradigm. Different scientists with different paradigms work differently in different ways.

Anyway, even 'loser' paradigms needed to undergo the 'value judgments' of the community and its members who chose, eventually, which issues required urgent solution and which did not, until the persuasiveness of argument obtained consensus (or not) in favour of alternative paradigms. It is, therefore, quite clear how Kuhn's vision of paradigm change within the history of science attracted the philosopher of science to consider all the (positive or negative) evaluative aspects opposed to a rigorous view of knowledge and methodology. In other words, the impossibility of practicing Weber's Wertfreiheit diktat, at least from a certain analytical angle, especially as far as sociology and the social sciences were concerned, was thus decreed. Here is one good reason why Kuhn had to defend himself from many accusers in the late 1970's, and not only with the publication of *The Essential Tension* (published in full in Italian in 1985). Ultimately, Pera was willingly inclined to recover from Kuhn the fact that the history of science might not be immune from 'value judgments' which, in most cases, seemed to characterize both the choices made by scientists and their prejudicial or ideological attitudes regarding changes in science. In this sense, it was also clear that methodological knowledge could not provide assurance that irrationality may never be a part of the choices made by scientists who trust in questioning all possible demonstrations.

Resumption of the topic of 'conversion' Kuhn addressed and discussed in Pera's book *Scienza e retorica* also concerned comparison between methodological knowledge and the issue of rationality, wherever it was affirmed that methodology tended to exhaustive its role and thus to impose some 'rules of play' when distrust in method and consequent loss of faith appeared on stage. The appearance of that kind of model, which Pera called the *counter-methodological model*, introduced the creation of a prototype affirming its non-obligatory derivation from rationality, stating instead, that scientific activity could claim to be 'rational' just like any other activity or even as a 'social convention'.

A different way of defining these topics is to refer to the particular conveniences, options and personal tastes of scientists assuming a role within the disputation between different models for which no real debate existed as yet.

Reference to an audience and a community of protagonists introduced, on the other hand, a debate playing a leading role in the introduction, in turn, of a model inside which to impose the idea that all the protagonists agreed on the answers to give to certain questions. It was in this field that, according to Pera, the technique of persuasion might win; only this way, the scientific dialectics that governed the debate could provide the correct answer.

It should be noted that Pera's rhetorical model tended to replace the dominance of methodological rules with a sort of 'free dialectical contest' and replace the 'subjective factors and social counter-methodological model' -as defined above- with a 'dialectical selection of topics'. Rhetoric extended its conquests to the method of natural sciences and not only to the method of practical philosophy (ethics, politics and the law). Also their findings and theories needed to an integral part of 'scientific dialectics' derived, in part, from the original formulations of Aristotle based on ancient rhetoric, meant as a work 'specular' (*antistrophos*) to dialectics, as the reiteration of the same argumentations when addressing an audience.

This way, many problems concerning the history of science and/or recognition of the evaluative factors (even retort and confutation) of protagonists would find themselves suddenly plunged into a sphere of mere discussion among scientists, thus determining a specific dimension of those internal/external factors of science indicated by Kuhn in some of his work. According to Pera only debate might realistically determine choices within the community of scientists, without recourse to the artificial constructions of scientific dialectics like those identified within the sociological scientific subjects (the sociology of science) by the Italian author and which tended -differently, according to him- to randomly, that is, in an overly direct manner, link cultural and social conditions with scientific results. The same seemed to apply (again according to this author) also to the hermeneutics applied to science.

With regard to the latter aspects, it should be noted that the author of the book tended to dwell on examining the validity of certain arguments which depended on factors concerning cultural changes and changes complying with them, tending towards a relativistic assertion of the history of science itself. In reality, more than technological knowledge, a 'scale' of relativism seems to comply with answers in the field of a quest for truth and rationality, it believed represented the development of some epistemological themes-problems, of a hermeneutical nature too.

The most marked forms of relativism to be found, according to Pera, are those emerging from the writings of authors like Richard Rorty who considered truth and rationality 'relative' when confronted with society and culture. The same he believed true of Paul Feyerabend and sociologists of science like Barry Barnes and David Bloor, who believed social contexts to be commensurable to given stages of cultural maturation. Less strong were positions like those of Ian Hacking, Hilary Putnam, Peter Winch and of two scholars of 'linguistic relativism', Benjamin Lee Whorf and Edward Sapir while the least intense degree of relativism was Khun's when he asserted the variation of theories and paradigms within the history of science. The variation of factors within a given configuration of history of scientific issues was meant to testify to the importance of creating continuous reflection regarding the relationship between rhetoric and truth, truth and consensus.

The recomposition of the tesserae of the framework of scientific dialectics, urged the philosopher to examine the matter of rationality facing methodological knowledge as a tendency in contrast with historical issues of science, which may occur outside of questions relating exclusively to the rationality of scientific theories. The question which invaded the ambit of methodology and spread to the domain of historians of science might sound as follows: will it ever be possible to break the rules of method without being accused of irrationalism? This was, in part, the problem Kuhn had to tackle when defending himself from the accusation of professing irrationalism. And not only that. In his writings he expresses regret that so many censors of *The Structure of Scientific Revolutions*, published as the last of the *Fundamentals* of the *Encyclopaedia of United Sciences* by the Chicago neo-positivists, neglected an important aspect of his thesis, that where he treated scientists as reasonable people capable of inspiring trust. On the other hand, it would not make much sense to ask whether theoretical changes in science are rational or not, where certain theories have prevailed instead of others and, most of all, as Pera posits in his book, whether "being rational means following the prevailing theory."

According to Pera Kuhn explained, more or less, how one theory or another prevailed, by introducing the concept of 'conversion' in which many authors of methodological knowledge were obliged to believe during their rapport with historical issues of science. This meant relating their existence among some modern and contemporary authors to the significance of given epistemological questions, without which certain problems might appear impossible to solve. In particular, for scholars of methodological knowledge the rationality of scientific theories had always been rooted in the properties of theories bound by rules, that is, by given requisites, without feeling the need to depend on results that scientific dialectics gradually imposed upon competing theories. The mistake of referring to rationality seemed to reside in the fact of discussing it all the time in the absence of a sole *rationis universalis*.

Pera's reasoning led him also to question himself about the veracity of scientific theories, in order to push investigation beyond gratuitous affirmation of a Cartesian kind, whereby scientists have always come to dispose of theories ideal for doubting whether they may turn out to be true. In some parts of his work Pera convinced the reader of the benefit of reasoning in terms of the usefulness of accepting the truth of a theory only following an ideal discussion concerning its existence. The result was that of finding oneself within the ambit of results derived from an uncertain debate regarding 'all the factors of scientific dialectics'. This way the 'game' might become real for all the players.

EPISTEMOLOGICAL KNOWLEDGE AND THE HISTORY OF SCIENCE

Making the rhetorical model capable of asserting the decline of method became the task of the philosopher of science towards the end the twentieth century, as Pera posited in a work which aroused considerable interest among methodologists of social sciences in the late 1990's. The same issue had been raised by his *Apologia del metodo* when the book on rhetoric was only being thought up. These were years when the philosophy of science paid for its crisis by leaving itself open to criticism of the transition between the idea of science as 'demonstration' and as 'argumentation'. From the 'point of view' of post-Popperian historians, to accuse the traditional vision of method, as intended by Bacon and Descartes, Popper and Lakatos, meant associating philosophical apologetics with criticism of the fundamentals of a considerable part of the epistemological knowledge produced until then, even if the reasons for referring to scientific method in apologetic terms, were limited to an attempt to establish *dialectics* as the sole meter by which to gauge historical aspects of science. This way, the circle seemed to close. Dialectics represented, in the end, the logic of the debate, meaning an exchange of arguments, all to be evaluated by a community sharing traditions and capable of relying on its own resources. What the historian of science needed to learn was that no referee would ever have interfered with what the debate succeeded in providing. Only dialectics might replace method. As Pera sought to assert "The overall pathway is, therefore, Darwinian (and Kuhnian): its direction is not predefined, but marked, from time to time, by the outcome of the debate. As in life, in science too one proceeds from (a previous situation) and because of (the reasons for changing that situation) one does not tend towards (a goal) and in order that one may (approach that goal), there is no future purpose capable of causing attraction, there are present reasons which compel."

Many authors had to work hard to acknowledge the overturned image of science that ensued at that precise time. Some of them showed varying degrees of agreement with Pera and his original thesis, as the author himself recalls in the *Preface* as the new edition of his *Apologia* of 1996. To this regard it is opportune to remember adhesions by philosophers, epistemologists and historians of science like Marcello Cini (physicist), Evandro Agazzi, Adelino Cattani, Pierluigi Barotta who, in works published at the end of the 1990's, sought to follow, selectively, the uneven trends of philosophy of science. Some of these scholars developed a 'reductionist' approach to knowledges which led them to believe that the history of science might now embrace institutions, methodologies and concepts reduced to minimal terms in order to explain facts. In this instance too, the work of the historian of science was bound to the epistemological knowledge of his/her own time, as far as the passage of ontological, and epistemological forms and of conceptual applications like the so-called 'Occam razor' or the 'principle of economy', in order to sanction the critical need to reduce the entities involved in explanations of phenomena including those proposed by philosophy. For example, historians of science, but also those involved in the cognitive sciences, etc. (see Evandro Agazzi, *The Problem of reductionism in Science,* 1991) took note of the fact that it was possible to propose studying the mind as a subject of physics, rather than remain anchored to studies of mind-body dualism. New *reasons* seemed to intervene in processes of dialectical comparison. These were the same ones which gave rise to new examinations of sure-concepts and all which was granted as known.

As regards the historians of science, it is worthwhile noting that one of Pera's important conclusions was that the new philosophers of science did not believe strongly in their ability to have an impact on scientific research. Their greatest ambition seemed to be that of producing a counter-methodological model in the wake of the *intra-moenia* revolutions or the 'revolutions of philosophy'. Unlike the discussions regarding the relationship between methodological knowledge and the history of science, reflections seem to have been different; in any case, full of other kinds of consequences. And not only. The latter may not be seen as the only ones capable of arousing the interests of methodologists of the social sciences, and, in the long run, of the new historians of scientific sociology, most of all where this discipline appeared to be still experiencing the primordial phase of its development. As we have already said, interests concerned mostly the claim of philosophers like Pera to follow an alternative route to method, by establishing a rhetorical model, resting on the new philosophers of science, who, compared to the rest, who proved better able than others to illustrate that the birth of professionalism philosophy had become an undertaking independent of and unrelated to science, as sustained by Paul Feyerabend in his *Science in a Free Society* published in the late 1970's. Moreover, the freedom of research itself, again in the opinion of the Austrian epistemologist needed to be guaranteed against all kinds of authority, including that of reason. Against all forms of dogmatism, including that of science, it needed to assert its freedom and its ability to give rise to alternatives and pluralism.

The fact is that epistemological knowledge was not capable of producing professional historians of science as methodological knowledge had long led people to believe. From this stemmed the refusal of any theory of method which sought to base scientific research upon more or less rigidly pre-established norms, as suggested by criticism of neo-positivism, also in the name of 'radical empiricism'. In this sense, not even epistemological knowledge would ever surpass its own origins.

It was the counter-methodological model which actually sprung from elaborations on the new philosophy of science, even if it failed, in Pera's opinion, to go beyond the idea of irrational/method duality, which still posed, he held, issues of a Cartesian kind. The fact is that, in the counter-methodological model of science one found a kind of cognitive production giving rise to assumption of positions regarding epistemic evaluations or claims of legitimising given types of epistemology or, ultimately, the epistemological knowledge produced up to that moment. This was to be one of the ultimate conquests capable of proving useful.

As we know, historians of science began to discuss more often about sociology, psychology and hermeneutics, despite their being deemed unfit to replace the alleged validity of rhetoric posited, as highlighted by Pera when discussing, as we have already seen, Kuhn and his ideas to this regard. In practice, as asserted by authors like Lakatos during their rational treatment-reconstruction of events regarding the historical subjects of science, Kuhn should have been considered an original product of conceptions of the 'irrational way by means of which rational authority changes." This was another achievement to register.

Anyway, according to Pera a considerable amount of the epistemological knowledge produced by the new philosophy of science might have been linked to historical subjects by means of the diffusion of irrational and 'propagandist' conceptions of scientific enterprise like those intuited by the Italian Ronchi, and Feyerabend-style anarchism, that is, availing of David Bloor's sociology of the 'strong programme' or some of Mary Hesse's interpretations, to arrive at the post-philosophies anticipated by Richard Rorty and which sought to replace the entire epistemological apparatus with 'normal conversation'. This modern attempt to substitute hermeneutics with epistemology too, gave rise to a series of epistemic evaluations, even when, in the case of philosophers like Gianni Vattimo, hermeneutic argumentation was 'interpreted' as 'provenance and distortion'.

In actual fact, all epistemological knowledge had, when all came to all, to examine itself by means of specular game of evaluation carried out by scholars involved in epistemology itself. This would have made the craft of the historian of science even more difficult than that of the methodologist, had Kuhn not intervened with his limitations to ulterior development, forcing specialists to replace the old Cartesian dilemma with his 'brainteasers', veritable *puzzles* for modern scientists. With Kuhn the problem of the institutionalization of the work of the historian of science, outside the forbidden gardens of 'professional' philosophy, became a pragmatic reality also within the precincts of university departments, where it was logical to seek changes in scientific matters without fear of being obliged to evaluate the most advanced products of epistemology in a counterproductive manner, as happened in the case of some philosophers of science.

To a certain extent, it seemed that the historian of science was destined to assimilate the saying whereby advances in science concerned changes brought about by scholars themselves who sought devices permitting them to interpret the many difficulties encountered, as Pera recalled in the final chapter of his *Scienza e retorica,* referring to a text successful in the 1950's and called *The Art of Scientific Investigation,* written by William Ian Beardmore Beveridge. It sufficed to agree on a theory and practice shared preferences and by availing of given techniques, render operative the rhetorical model, which provided no precision tools or even strategies to follow. From the 'point of view' of the historian of science, rhetoric would prevail- in the end- when seeking to meet demands to fill logical or conceptual, empirical gaps, with strategies arranged according to values (empirical and extra empirical) shared by interpretations, in hierarchical order.

But let us ask ourselves whether what holds for historians of science holds also for those who produce epistemological knowledge? Vice versa, did those who believed in the establishment of epistemology during the twentieth century succeed in following the route of professionalism without regretting bygone days and the pompous past of the many producers of scientism?

For some critics of the social sciences and Italian methodologists, the profession of historians of science succeeded in producing little more than a handful of 'narratives accounts' of some of the events and 'procedures' of the scientists of the past, like, scholarly interpretations of the course of ideas in modern scientific culture, were too close to disciplines such as sociology, as in the case of Mario D'addio's *Il caso di Galilei* (1993, reprinted in 1997), where the author examined the subtle interplay of historical, political and cultural, therefore, theological, exegetical, philosophical questions affecting Galileo's tormented life. And the D'addio's is not the only example of this kind of work. Another, though from a different angle, was the work of epistemologists of the historical development of knowledge, carried out within the fields of disciplines like physics and treated by historians of science where a series of geographical economical, technical, institutional as well as philosophical and cultural factors were brought into play. This was the case, for example, of Fritz Krafft writing on physics and its future, an essay translated by Franco Voltaggio and published in Italy in 1990 as *La fisica e il suo devinire*, for the methodological sciences and philosophy of language series directed by Francesco Barone and Vittorio Somenzi. The problem addressed by Fritz Krafft in his essay was self-understanding in physics caused by the discovery of an intrinsic historicity, depending on a complex 'theoretical space of experience', the quest for which defines the development of knowledge concerning anything that we may call 'physics'. Certainly, the mingling and discovery of so very dissimilar natural events made the different between a rigid way of practicing history and the professionalism of moderns and contemporaries. In most cases, similar works had the merit of increasing the already abundant wealth of monographs by scholars of the social and political sciences and by intellectuals who, in other instances, were able to follow the example of some rare European historians of science, who, devoted to endeavours of a retrospective nature were, to put it frankly, a long mark from the task of linking their investigations of the sources of scientific concepts with the possible epistemological choices of scientists.

With regard to this latter exigency, we should recall the works of some of Italy's best known "professional" historiographers like Vincenzo Cappelletti and Paolo Casini, who, later on, gave life to Pisa's 'Domus Galileiana' and, in the case of the former in particular, to the Società Italiana di Storia della Scienza. They also fostered 'openness' towards the studies on epistemology and the historiography of scientific subjects. Their contributions, with those of a few others, led to a debate on the object of the history of science in the thinking of the first 'professional' historians coincident with precarious developments within the historiography of the social sciences and, in any case of philosophical historiography, at a time when new interest was being shown in sociology. Moreover, surrounding the debate on the 'object' of the history of science and historiographical studies, Italy witnessed a sort of renaissance in the 1950's. This was favoured by contributions by several researchers including those who assembled around the chair and school of the University of Rome. These researchers succeeded in indicating differentials concerning the cognitive identity of science and historiographical issues that may be referred to them, as Vincenzo Cappelletti pointed out in the 1981 preface to the first book (*Automa e Vita. Saggio sulla fisiologia di P.-J. Barthez*) by Federico di Trocchio's a student of Vittorio Somenzi's who dedicated his premise and his work, instead, to showing how, from a certain 'point of view', an interest in concepts might give rise from the beginning to a selection of options of a theoretical nature capable of producing knowledge and, in any case, a specific approach to the various sciences (he dealt with biological sciences). As regards Somenzi's Roman School we should recall, among others, Roberto Cordeschi and Massimo Stanzione, the author of *Episemologie naturalizzate* (1990) and professor of Philosophy of Science at the University of Cassino.

Generally speaking, the questions raised by these scholars regarding the production of concepts and epistemological knowledge, were usefully dealt with in more recent discussions of the tasks and the profession of historians within the development of the scientific subjects, following a route quite unlike that of the previous decades. In practice, it impacted on many areas featuring studies and research on scientific historiography and epistemology. This meant that the separation between epistemological knowledge and encyclopaedic knowledge, present also in some 'classics' of the social sciences, appeared more evident.

There is no doubt that Kuhn himself found a new way of considering the history of science and see it as a privileged place for the confrontation of epistemological knowledge, regarding, for example, the philosophy of science itself where changes in the perception of the scientist and various shifts in meanings resembled conceptual revolutions recurring in the course of scientific subjects. It does not seem right, from a more and more modern angle, to deal with concepts without dealing with the revolutions of knowledges. The concept itself of all the epistemological knowledge produced, intended as a sequence of examples capable of providing an image of the scientific advancement due to conjectures and confutations, ceases to exist. But the same might be said of the notion of an epistemological knowledge believed to derive exclusively from the theories of the new philosophers of science, as Pera sought to convey when he appealed even to Kuhn, who disavowed this 'internal' genesis unless it was compared with 'external' factors, in other words, with all kinds of continuity-discontinuity present in the production of knowledge. The best analysts, or rather, the most advanced analysts of this process -to which every scientific community should submit due to its ability to generate consensus, rather than developing persuasive techniques- seemed to be the social scientists. They were the most critical, in the long run, when it came to processes of conceptual hetero-direction or scientific revolutions present in the debate and in the dynamics of science, opposed, at times, to those devoid of purpose of technology and/or technological knowledge. Pera provided the example, but without going into matters more deeply, of 'strong programme' sociologists, who, over time, had created an inversion of opinion regarding Kuhnian followers, by re-evaluating more than might have been imagined, Kuhn's tension towards the sociology of science, rehabilitating, finally, some developments of the latter. This justified, in part, the fact whereby conceptual problems were addressed by social scientists as fundamental after they had become of interest to historians of science. This was a problem Pera failed completely to explore when he discussed only the intensity of the relativism produced at a certain moment in the twentieth century, as well as the scarce propensity on the part of sociologists to critically discuss the philosophy of science, until Kuhn made things clear.

On these issues, an 'Italian' testimony of Robert King Merton was important. In an interview conducted by Massimiliano Bucchi which appeared some years ago in n. 4 of "Rassegna Italiana di Sociologia" Merton analysed his contribution to the sociology of science, answered questions and also stated that his way of speaking about the 'normative structure' of science, and, in any case, of sociology of science, had always dealt with attempts at conceptualisation taking place after historians of science, who were in the habit of doing so, described and foregrounded a minority of 'great men'. The problem was really that of focusing on concepts at the base of modern science as a social institution. Moreover, in the interview (of the 5th April 2001) Merton re-evaluated the Edinburgh school, mentioning authors like Donald McKenzie, David Bloor and Trevor Pinch that have reconsidered their positions and spoke at length about the studies of Karin Knorr Cetina.

On the same occasion, Merton told of how, with Paul Lazarsfeld, he had set up a post-doctorate programme to bring together about fifty young American scholars. One of those who participated was Thomas Samuel Kuhn who thus had the occasion of frequenting social scientists and discuss with them. So, according to Merton, Kuhn noticed that social scientists spoke about fundamental concepts while physicists with whom Kuhn usually discussed, who took those concepts for granted and concentrated on specific aspects. What Kuhn has expected to find was what he called the 'paradigm' and so the discovery of the laws changing ideas in science took place. It is important to notice, also, that what Kuhn formulated in terms of scientific change was for Merton a discovery in terms of serendipity. As we know Merton used serendipity to define the unexpected characterising scientific discovery and essential to it in that scientific discovery may occur when one fails to find something or finds something else.

We would like to think that Robert King Merton and Paul Felix Lazarsfeld belong to the history of science although they failed to convince Kuhn of the importance (or not) of carrying out an epistemic evaluation of the social sciences, also because a long time afterwards social scientists were to return, in any case, to discuss Kuhn's contribution to their disciplines without forgetting, obviously, the opposite. As far as this is concerned, it is the case to mention nos. 53/54 of "Sociologia e ricerca sociale" of October 1997 edited by Enzo Campelli which feature papers presented at a seminar of Italian researchers and scholars on T. S. Kuhn and how ideas of science change.

This collection of essays and studies reveals a renewed interest on the part of methodologists of social sciences in the theme of historicity of knowledge addressed by Kuhn in well-known terms of epistemological cognitive analysis and of that complex enterprise which he believed dealt with the construction of theories and which Kuhn called 'normal science'. As he specified in *The Function of Dogma in Scientific Research*, although the scientist usually does not struggle to invent new genres of fundamental theories, these theories emerge repeatedly from the continuous practice of research; a veritable warning, this, for philosophers like Pera. Again, the world of the history of science and a considerable number of fundamental topics attributable to the epistemological knowledge produced in a specific epoch, met.

Furthermore, it needs to be said that Kuhn himself, reported (in 1990) benefitted greatly from his academic experience at Princeton and from relations with Carl Gustav Hempel who, as we know, studied the formation of concepts and theories in the empirical sciences, also provoking with his essay *The Theoretician's Dilemma*, a whole series of ideas used later by Kuhn in research into the historic and 'evolutive' approach to the philosophy of science. Kuhn underlined the importance of having glimpsed in his continuous juxtaposing of philosophy and the history of science (the latter acknowledged by Hempel) analytical tools developed within the logic empiricism in such a way as to admit that the interest shown in philosophy contained a debt towards some of the difficulties encountered within that tradition, comparable, in any case, to those met when trying to develop a historical approach to science.

Considering the results, we might be led to believe that the first to show an interest in fundamental changes were those willing to deal with the production of knowledge at a time when the traditional concepts of some of the disciplines were being accused and/or discussed once more in the light of other developments. Furthermore, there was no want - besides the specific journal of 1997- particular occasions for debates in which Italian social scientists faced each other, with a view to taking stock of the problem of in-depth epistemology studies of relations between theory/research stemming from philosophical/hermeneutical and social issues leading also to assertion of the historicity of concepts and scientific theories.

To this regard we shall recall briefly the conference on Theory and Research held at The University of Cassino from the 21st to the 24th May 1997 and organized by the

Department of Philosophy and Social Sciences. On that occasion, Franco Ferrarotti and a researcher of the history of scientific subjects like Filippo Barbano spoke, the former on theses regarding theory and the formation of concepts, the latter about hypotheses tending towards new investigations of epistemological history relating to certain 'periods' described by post-positivist criticism of development and the history of science. In that ambit, the transition between Talcott Parsons's a-historicity of sociology and Robert King Merton's 'systematics of theory' represented an example of the pathway followed by the history of scientific subjects at a moment crucial to 'inner' development, including the limits of this trajectory of the 'official' history of sociological science, as Franco Ferrarotti pointed out in his paper without 'trespasses' or examples of an epistemological nature. In reality, especially in the opinion of some, epistemology, intended as a critical reflection on method was not to be hypostatized in relation either to its possible inter-disciplinary dimension but historicised, in a certain sense, that is, practiced in those areas and become research as '*pre-post* and *circum* science', as underlined by Ferrarotti in an interview with the writer of the present book regarding Italian sociology, published in n. 1, 2004 of "Sociologia," the above-cited four-monthly journal of the historical and social sciences.

According to Ferrarotti epistemology ran the risk of being left to its own devices, turning around in circles to become a new kind of scholastics. Even Kuhn's American experience was significant, as it showed that the history of science tended to concern itself also with the conceptual development of other fields of knowledge, where the latter were useful for the development of an approach, even when those concepts (which was not the case with Kuhn!) were treated in relation to their limitations and inadequacies, their epistemological vagueness and subordination to given schemes of thought, categories, etc. It was precisely in Kuhn's case it was possible to claim, with some contemporary researchers, including some Italians like, Paul Hoyniengen Huene, Stefano Gattei and others, that one of the most important changes in Kuhnian thinking after the publication of *Structure,* might have been the distant stance he assumed with regard to perceptive vision of scientific revolutions in favour of a more conceptual conception. Some of the reconstructions surrounding Kuhn's contribution to the philosophy of science delivered in years previous to the Italian debate -with which we shall deal in a moment - led, where the topics discussed were associated with doubts concerning the philosophical contribution made by *Structure,* to positions in favour of a theory of scientific development forged by historical events. For purely philosophical references, it suffices to recall the explanations advanced by Kuhn concerning the mysterious concept of 'incommensurability' and, in any case, relating to previous conceptions produced in 1962. As regards the second aspect, the approach assumed by studies and debates following the publication of *Structure* in a period of true 'socialization of the history of science' applies, as defined by the sociologist of science Peter Weingart at the beginning of the 1980's; this author is of great utility (see Carmela Agodi's essay in n. 53/54 of

'Sociologia e ricerca sociale', 1997) when locating Kuhn within a critical perspective and with regard to his 'impact' on the social sciences. Furthermore, it should be underlined that Kuhn's change of outlook marked the emergence of serious methodological problems relating to perceptions of the study of the history of science. This topic interested so many fundamental disciplines and led to discussions in other fields regarding the development of science and, as a result, the grand revolution of knowledge. The tasks of the historian of science soon took precedence over to philosophers of science to be discussed in the light of a different configuration of conceptual changes regarding knowledge and its links with the epistemological history deriving from 'internal' troubles to science and/or the development of the history of scientific ideas.

Antonio Fasanella expressed opinions regarding a number of these topics in an essay dedicated to Kuhn in the above-mentioned "Sociologia e ricerca sociale" journal. He referred, in particular, to some criticism by Feyerabend and, in part, by Paul Hoyniengen Huene, regarding Kuhn's 'naturalism' which opposed other considerations of the history of science, presented in favour of a different approach to the philosophy of science. It was only by discussing Kuhn that some scholars like Fasanella revealed the strategic character of a problem like that of identifying the most coherent and transparent tasks (like those indicated by Kuhn) of historians of science and their activity concerning sociology and the social sciences. While scholars and methodologists of the social sciences like Enzo Campelli and Alberto Marradi tended to discuss Kuhn mostly in relation to the conceptual debt (widely treated at a Conference in Hamburg by Stephen Toulmin in 1981) he owed to the Polish doctor and biologist Ludwik Fleck, author of *Genesis and development of a scientific fact* -a volume translated into Italian four years after the American edition published in 1979 with a preface by Kuhn- mainly from the 'point of view' of an epistemological construction which seemed to precede a given view of a quest for historiographical continuity to be implemented and applied to scientific subjects.

Having said this, Fleck was also acknowledged as a worthy forerunner of 'historical and comparative investigations' of changes which may affect ways of implementing certain phenomena, where the changes are closely related to concepts that change 'meaning'. Ludwick Fleck seemed to anticipate the concept of 'scientific revolution' without, however, contributing to changes in perspectives, as did Kuhn, who made a different use of his insights, obtaining great success among scholars of sociological subjects. His was a conceptualization which led scientists out of the gnoseological vacuum of an epistemological *imaginabilis*. The contents of science were increasingly affected, and could be explained by "the history of thought, from psychology and the sociology of thought," as Fleck announced in a theory not overly in keeping with the historiography of science advocated later by Thomas Samuel Kuhn.

The methodologist Gianni Statera in an essay of 1997 published in "Sociologia e ricerca sociale" revealed renewed interest in Kuhn's thinking, when discussing his success with sociologists; this argument was surrounded by misunderstandings (also of a 'rational' nature, one might say) and never completely taken for granted either in Italy or elsewhere, considering Kuhn's statements regarding his competence in the field of social sciences. "Sociologists -writes Statera- rather than contemporary physicists were close to experiences of conflicting paradigms, of scientific revolutions, of alternative *Gestalten* incommensurable to each other, to sudden collapses of 'world views' incompatible with other conflicting ones. The American scholar was attributed credit, however, for representing step-by-step processes by means of which both transformations of scientific knowledge and the formation of consensus around certain goals of the scientific community develop, even when similar results were not univocally acceptable, although agreement should stem from shared acceptance of given theoretical and conceptual rules, procedures and tools, aimed at permitting the accomplishment of certain operations.

In this sense, Kuhn revaluated the role of the methodologist of social sciences, where profession needed to be mediated by scientific rationality. Later on Statera admitted that, according to Kuhn, a scientific community might arrive at sharing the logic of scientific investigation, with its peculiarities within a well-defined framework, that–to put it more clearly- of the publication and control of procedures, the 'justification' subsequent to possible 'discovery', the potential repeatability of an experiment, the operational correctness of its theories and concepts and reiterated control of every generalization.

In practice, Thomas Samuel Kuhn also fully satisfied the logic of scientific inquiry. His teaching was quite clear even from the 'point of view' of the historian of scientific sociology who seeks, today, to establish the priority within research of the link between concepts produced by sociological disciplines and the 'revolution of knowledges' taking place. It might prove very useful to study some of the philosopher-producers and critics of science, in order to reach conclusions from them suited to historical epistemology. Historians of the scientific sociology would certainly be presented with a series of theoretical and methodological options from which to start and, maybe, make a choice between *orthodoxy* and heresy.

But to explain matters better, we shall make a brief comparison between our reflection and Statera's in an effort to bring reflection back to its essential state. Thomas Samuel Kuhn, insisted in his books on passing from a logical consideration of scientific theories to a purely historical study of their formation and affirmation. This is an important point of arrival which we cannot neglect when considering the present permanence of the revolution of knowledges. On the basis of Gianni Statera's affirmations one might be led to believe that this process rationally legitimises, from within, the quest for a scientific method, logic and epistemology.

Scientific theories, as we know, develops within given paradigms (though some today prefer the term 'tradition') which, once established, are capable of driving the activity of scientists without being questioned. Paradigms, as Gianni Statera asserted in his article, should be due to a theoretical-conceptual rationalization *ex post* of a disturbance occurring or implemented within a branch of scientific research. If the progress of research accumulates anomalies that fail to be explained on the basis of the given paradigm, after various attempts at modifying it, the paradigm is abandoned and changed, leading to what Kuhn defined a 'scientific revolution'. This was the case with transition to the new Copernican 'point of view'. In the long run, it is the scientific community which determines the boundaries between a kind of scientific orthodoxy and heresy.

In practice, scientists who advance new paradigms are considered 'heretics', but from among them an exception may produce new kind of orthodoxy. This is one of the reasons why scientific progress cannot be said to proceed along a methodically linear pathway, but only by means of radical changes in paradigm, attesting to the relativity of the validity of different theories and hypotheses. Such changes in position are generated by epistemological reflections on contributions to the history of scientific subjects like sociology as well as other disciplines due to the closeness of social scientists to the 'confines' of topics examined by the various scientific traditions. The boundaries of the epistemological problem in the social science still aroused, between 2000 and 2015, some interest on the part of several scholars specializing in the various disciplines and belonging to different areas of theoretical and critical development. Of special note was the Conference held in Rome in April 2010 on horizons and boundaries in epistemological research -*Orizzonti e confini della ricerca epistemological.* It was attended by authors like Luciano Zani, Paolo De Nardis, Paolo Parrini, Guglielmo Rinzivillo, Federico Di Trocchio, Paola Borgna, Luigi Frudà, Leonardo Cannavò and Roberta Lanfredini. The minutes of this Roman meeting are contained in a volume edited by me: *Una storia delle scienze per i nuovi saperi. Discussioni e ricerche, "ATTI"* del Convegno: *Orizzonti e confini della ricerca epistemologica,* published in Rome, in 2011.

TECHNOLOGICAL KNOWLEDGE AND THE HISTORY OF SCIENCE

Social scientists and contemporary sociologists seemed to assist helplessly at changes in scientific knowledge. This meant that some of them were starting, seriously, to ask themselves how to enable their disciplines to follow the conceptual development of the history of science which seemed to assume the guise of an epistemological history. Recent, re-born interest in the history of science, related to the development of the philosophy of science and, to some extent, to the development of the social sciences, produced studies by Bruno Beretta, Massimiano Bucchi, Elena Gagliasso, Giovanni Boniolo, Paolo Vidali and others which focused on new perspectives capable of arousing the interest of sociologists. Apart from these publications, these proved to be of particular interest to authors on the European scene of the sociology of scientific knowledge (Bruno Latour and others), including works translated into Italian and published in Italian magazines (like "Tecnoscienza"- an on-line journal with publications by the Association for the Social Studies of Science and Technology (STS) as of 2005). In future, the definition of a variety of areas and the field of the conceptual and epistemological interests of the social disciplines, in the light of logical and procedural developments in the history of science, might become a crucial task.

This already appeared to be, in part, one of the tasks to be carried out by the new generation of social scientists, charged with showing how the very existence of the history of science and of a conceptual, typological historiography, might inform the discipline regarding the new course science subjects were pursuing. Full assimilation by the social sciences of the epistemological topics proposed by historians of science and technology still seemed to suffer from the poor consideration in which technological subjects were held by authors involved in the dissemination of 'materials' produced by various contributions to the debate between science and philosophy, even before dealing with 'technology', in certain areas of development like Italy. It needs to be said, however, that some examinations of contemporary epistemology carried out in Italy, from a number of different angles –like those of Alessandro Pagnini, Paolo Parrini, Silvano Tagliagambe- were accompanied by more thorough reflections concerning upstream reception by the social sciences of the studies and contributions of methodological science as well as those of the so-called pragmatic and fallibilist models. This after dwelling on evolutionary scientific knowledge which, in some cases, led to models of 'technological use', so to speak. To tell the truth, very few attempts were made to follow an evolutionary approach posing epistemological questions to the social sciences like those advanced by scientific Marxism (see, Umberto Melotti, *L'esigenza di una nuova scienza unitaria dell'uomo e della sociatà* in "Annali do Sociologia/Soziologisches Jahrbuch," 1998-99). Instead, those following this approach preferred to propose a quest for the salient features of paradigmatic evolutionism 'concealed' by scientific twentieth-century historiography (see also some recent studies by Elena Gagliasso, 2003). More keenly motivated seemed to be studies and research by epistemologists and methodologists (not in the specific field of the 'technical' and 'technological') who paid attention to the production of conceptual materials as well as to cognitive activities and questions related to them and the sociological theories, scientific knowledge and philosophy of science of young Italian scholars and researchers, like Roberto Fideli, Eleonora Venneri, Agnese Vardenaga, Giampietro Gobo, Sergio Mauceri, Francesca Ridolfi, Sandro Landucci and many others.

A considerably large amount of the historiographical development found in former relations with the practical results of this kind of thinking -that is, the relationships between science and philosophy- did not seem to have exhausted all its potential regarding the current epistemological revolution activated by changes in the 'culture' and work of researchers. These scholars now began to reason, as we might say, inside and outside the laboratory, in the light of the fact that the scientific revolution had already taken place during a major technological revolution, derived also from a conceptual revolution between 'knowledge' and technical-scientific enterprise which had become operative. The neo-empirical conception of science itself, still considered in vogue among some methodologists and philosophers of science, had found antecedents in many philosophers, as the 'official' historian of neo-positivism Joergen Joergensen pointed out towards the middle of the twentieth century. This until scholars began to question the reductionism existing prior to foundational claims by scientists and in many western social scientists, including Max Weber, who, from the 'point of view' of a sociological history of philosophical subjects and not without serious theoretical difficulties, had come to conceive science as analysis of facts chosen from a practically infinite totality of facts, on the basis of interests and values which, when all came to all, science could not found.

In a certain sense, contemporary sociological criticism of the conception of neo-empiricist science marks the presence of philosophical issues which have remained unresolved and/or appear insoluble in the eyes of social scientists themselves, where the sociological discipline lacks a quest for epistemological-kind outcomes, and, most importantly, where spontaneous choices of a cultural (and political) nature are permitted and finally validate the pursuit of more objective laws and relationships between facts thus selected and analysed without, of course, the ancient claim of considering them universal laws.

The reflections of many historians of scientific knowledge pass through this view of conceptual horizon's shift beyond topics concerning the sure-foundation of science and beyond an exclusive practice of assertion and assiduous control of the truth of scientific theories and the knowledge produced. Many philosophers of science have tried to examine only the transition from empiricism in the science of authors like Rudolph Carna and Bertrand Russell to Popper and reception within different areas, of analytic philosophy as well as of some of the debates surrounding the guidelines of new epistemology and so-called 'epistemological relativism'. In their research they were able to ascertain the inability of scholars to practice a time leap and go back; this was evident with regard to all the examinations of a particular kind of scientific production carried out. In Italy, precisely, authors like Francesco Barone and Alberto Pasquinelli still tried to discuss results inherited from Carnap while discussing epistemology, the philosophy of science and the philosophy of language (1995).

We have mentioned changes in contemporary researchers' 'culture' but this does not mean that constant reference to some of the contributions made to scientific thought are to be deemed less relevant. A number of changes in cultural processes which gave rise to and, as we write, give rise to artefacts still being analysed, availing of approaches like that of Karl Popper, by historians of science who discuss natural-knowledge models, like Enrico Bellone, University of Padova (2000), resumed studies by Luca Cavalli Sforza and M. W Feldman.

We also have the impression that the major changes brought about by the revolution of scientific knowledges will lead, ultimately, to a technological use of concepts, causing a continuous redefinition of certain epistemological issues. The attraction felt by researchers engaged most in examinations of relationships between today's science and technology-two distinct dimensions by tradition- as emphasised by several historians of science, led to analysis of a set of processes capable of creating practices which generated scientific facts and/or technical artefacts in the strict sense, a result never independent of descriptions of pluralities of events and social practices stemming from concepts forming a weave of different processes. On this issue see studies on 'techno-science', where the applicative aspect tended, ultimately, to locate itself at conceptual and theoretical level, giving rise to inquiry into the merely 'technocratic model', whereby the proponents of the model itself began to question the very fact that 'techno-science' was always right. On the other hand, we should not forget that other proponents of these studies were led to consider as inconclusive results of research into technological sophistication or into areas where the theoretical aspect was governed by extreme norms which transformed the ambit of scientific practice into 'a chat room', as epistemologist Giovanni Boniolo of the University of Padua sustained when referring to 'bioethics'. Lastly, we should recall the transversal interest shown by many in studies of the 'impact' which technological knowledge may have, given that modern analysts of the human-machine like Prof. Alberto Baldissera of the University of Turin (1992) tried to redefine some concepts of contemporary sociology applied by those who dealt with risks posed by new technologies in the light of innovative developments in the cognitive and organizational sciences.

As someone noted, the cognitive sciences themselves, have helped to overcome the epistemological dilemma, and not only with regard to 'technical' issues but also those of an organizational nature, especially in the field of scientific knowledge, where the social sciences had only to address real paradoxes. With respect to these one may read a number of publications inspired, more or less, by a search for an interpretational key (see, for example, Natale Ammann, *Elementi di epistemologia sociologica* Milan, Franco Angeli, 2004).

In any case, the afore-mentioned 'technocratic model' would not be able to draw to itself the heirs of a paradigmatic shift as evident as that of so-called 'biotechnology', where the use of science was not faulted because of being subject to multinational corporations. This would be true in the case of sociologists of science as well as of philosophers of science, obliged to recur to presumptions of their very own critical existence.

On extreme example of the application of this whole series of arguments might be that of the study of heavenly bodies in the field of theoretical physics. It seems strange, but it is simply by using those concepts and, in particular, those concepts called 'neutron stars' and 'white dwarfs' (small hot, dense stars) that J. Robert Oppenheimer and his disciple Hartland Snyder formulated hypotheses regarding blacks holes as far back as

1939. To give another example of how epistemology reflects on scientific thinking, we might also recall conventionalist Jules-Henry Poincaré, who before Einstein published his famous article in the "Annals of physics" (1904-1905 circa) had abandoned the concept of absolute time acting on criticism of the concept of 'simultaneity of events' in different places, thus giving rise to a kind of qualitative philosophical anticipation the limited theory of relativity destined to become a conceptual revolution in the history of physics.

Examples of this kind might continue, especially when taking into consideration the course of the history of science currently considered most reliable within today's international paradigmatic discussion and debate. As astronomer and physicist James Jeans pointed out, it was when the universe begins to look more like a grand thought than a great machine, that the history of quantum physics was born. Following its developments, one should keep in mind that the name of the theory derived from the concept of 'quantum, how much' introduced by Max Planck in 1900 and taken up by Einstein in 1905.

The fact is that sociologists are dealing at present with an age of change not very unlike that prophesised by Austrian journalist Robert Jungk, the author of the history of atomic scientists entitled *Brighter than a Thousand Suns: A Personal History of the Atomic Scientists [*translated into Italian as *Gli apprendisti stregoni* meaning the sorcerer's apprentices] as a future or a future discipline which would seem to be independent without, however, forgetting humans as 'producers' of technological concepts. This, in short, was the significance of the research carried out by Jungk with Frenchmen Gaston Berger and Bertrand de Jouvenel and which helped define the terms of a system of human and social forecasting at Berlin's University of Technology. Even today's sociologists might be genuinely flabbergasted by how their discipline has changed, as were (for reasons related to practical applications of their ideas) scientists like Niels Bohr, Otto Han, Robert Oppenheimer, Enrico Fermi, Edward Teller and all the protagonists of Jungk's famous book who, in the shadow of the titanic figure of Albert Einstein, realized they could no longer control the forces they had managed to unleash.

Historians of physics have long been ruled by what historians of sociology may be obliged to adhere to in the near future, meaning the world of concepts and of contemporary technology which may well become the protagonist of yet another 'Copernican revolution' of scientific knowledge. This is, in part, one of the reasons why groups of scientists – like British researchers – are seeking, since the dawn of the twenty-first century, to rethink relations between science, technology and society and reconceptualise the rapport between science and the public, where technical and scientific knowledge may be investigated in areas subject to a particular kind of social evolution. With reference to this we might refer to Luciano Gallino's most recent studies regarding technology and public property in the age of democracy or to less recent works like that by Massimiliano Bucchi concerning the afore-mentioned 'biotechnologies', 1998. One should ask oneself, along with these researchers and scientists, what the public thinks of contemporary science and technology to the point of questioning the formation of public opinion regarding the construction of social studies and, in particular, results of studies conducted by sociologists of scientific knowledge. In the case of Italy, we recall, besides studies by Leonardo Cannavò and Gallino, those of Paola Borgna of the University of Turin (2001).

In all the above mentioned cases, the use of broad concepts referring to 'knowledges' implies reference to technology as a multidisciplinary area of research and development closely linked to productive processes. Its advancement also questions the fact that human intelligence is capable of governing, in the long run, the intelligence of machines, as the so-called 'futurologists' began to fear, often before some well-known sociologists did. In fact, unlike technique, which deals with the workings of machinery, technology means the development of machinery based on scientific principles. Technology was destined to become ubiquitous, like the knowledge produced in research centres, science academies and laboratories and a few other institutions whose goal it is to produce it deliberately, regardless of its actual consumption though not immune to possible manipulation.

Overall study of this process is important while of some interest are studies regarding its evaluation, an argument derived from the methodology of social research which has played a role recently, especially in areas of appraisal of / in the social sciences and discussions of the validity or otherwise of research. To this regard we recall the sociological work of the Genoese methodologist Mauro Palumbo, as well as that of other authors from varied backgrounds and universities, which has helped broaden discussions on the discipline and the different results obtained. Beyond epistemological investigations of forms of evaluation, tests regarding teaching and scientific policies in universities remain confined to specific areas.

The danger of subversion of functions that may be manipulated by human intervention, has, according to some, already occurred in the area of technocratic intervention in sectors of politics, as demonstrated, some time ago, by sociologist Jean Meynaud who examined relationships between politics and technology and, therefore, the influence of technocracy (as a kind of upper level bureaucratic power) in industrialized societies. Other more modern scholars like Jack Goody and his studies on the manipulation of the surrounding environment, approached various cases of ousting and manipulation in terms of the 'technologies of the intellect' as referred to writing.

Similarly, the 'futurologists' seemed to continually redefine concepts of scientific revolution in the light of technological revolution, where the terms of frequent technical change and changes in 'theoretical knowledge', provided a sure tool for measuring similarly great changes, more than studies of other aspects of the natural, cultural reality. While we are totally unable to foresee the progress of epistemological knowledge, for example, we are practically certain that the course of given human productive activities needs to be connected, increasingly, to scientific knowledge. This is the sense that we may claim that technical activities should be intended as conscious human activities, just as we intent to refer to activities acting as guidelines to human conceptual production. In the case of history too there exists a rational answer of the so-called 'futurologists', provided by the presentation of a method in social sciences comparable, according to authors like Jungk or Ossip K. Flechtheim, the author of *History and Futurology*, to field theory in physics and the theory of Gesthalt in psychology and social psychology.

In reality, operative upheavals starting from claims of dealing with future knowledge as true scientific knowledge, and, in part, the upheavals akin to many efforts by sociologists to conceptualise and forecast, have assumed the traits of a series of approximations of an epistemological nature sanctioning the creation of a kind of naivety conveyed to investigations in the field of technology. This took place in the distant past and, in many ways, it continues to take place not so far from our everyday lives. Many disciples of technological knowledge seem avail of the abstractions belonging to the science of the future and of those who believed in exploring new branches of knowledge aimed, as Flechtheim wrote in 1996 in a book dedicated to Erich Fromm, at predicting probable futures rather than planning a sole one.

Some rational anticipations and/or previsions proved of interest to sociologists as pathways of 'unfulfilled prophecies' which suddenly became the stuff of a history of the immediate future, at a time so unlike that when some historians of Italian sociology were considering producing concrete analyses concerning relations between society and technology. One such study was that relating to the sociology of science by authors Filippo Barbano, Franco Bonacina and Pietro Prini (for example in *Società e Tecnologia* in long-ago 1969). Between 1970 and 1980, volumes and research by many other sociologists were issued and included translations of books dedicated to the 'work of the scientist' and questions of interaction between the philosophical and social aspects of science and technology (like John Ziman whose book published in 1984 was translated in Italy by Libero Sosio in 1987 e published by Laterza).

Predictive science, in any case, launched some pragmatic-technical tools to avail of to study certain advancements and raise questions surrounding what Pierre Lévy called 'cognitive ecology' when referring to new relationships between IT technologies and the telecommunications (ICT). We also wish to recall the society portrayed by paladins like Daniel Bell or the 'spaces' which Jean-François Lyotard defined in 1979 as a veritable human condition dominated by technology (in particular by information technology) and by knowledge as goods subject to all the economic-social implications that might stem from it. This way, the relationship between scientific knowledge, the powers managed by humans and the places reserved for learning science, appeared as themes of considerable importance, reconnecting discourse with examinations of scientific knowledge detached or subordinable, anyway, to technological investment, in an era when technological innovation was translated into knowledge destined to be enhanced or deteriorate in relation to future prospects.

Questions of this kind and similar problems saw historians of science in recent times far more distant from sociological previsions than historians of 'futurology' and justifiably referring back to theoretical and epistemological implications, without forgetting the fact that studying the course of probable events might have aroused some interest among all those who showed an interest in philosophy, the human sciences or humanities. The fact that scientific sociology, like physics and chemistry, are born and develop continually around analyses of concepts did not exclude the fact that many historians of sociology were educated in concepts derived from philosophy or other sources. This also meant that they were led to admit that by following the development of the history of science, and later on, of the philosophy of science, they would surely have provided their studies with certain characteristics rather than others. It is today that the observational lens of historians of sociology need to be changed rather than use the so-called 'interpretational key' useful in certain situations, or, in the long run, to 'point of view', if we wish to continue to refer to a plurality of imaginable and possible choices to make among many others. Surely, we are now in an epoch completely unlike that foreseen by the Greek philosopher Pythagoras in the fifth century B.C. when he held that man should be the indisputable measure of all things.

This, even when revolutionary discoveries seem to be a matter of chance. The history of scientific subjects of the last centuries is full of examples testifying to the future of some experiments and proofs capable of causing change in scientific developments favouring 'creative' new propositions of theoretical constructs, reference schemata, the formulation of hypotheses and the production of new concepts. We cannot deny that the history of chemistry may be considered – and rightly so- more full of surprises than the history of sociology although the history of chemistry and of scientific sociology could never be confounded with the history of chemists or sociologists, etc. By a quirk of destiny certain stories have entered the domain of narration like that of the Swiss chemist Albert Hoffmann, who discovered LSD in 1943, when by accident a drop of the acid fell on his hand during a laboratory experiment and made him feel dizzy. Another more famous episode regards the discovery of Rontgen rays or X-rays; an even more famous one is that of Isaac Newton and the window of his College, beneath which stood an apple tree. This is a strange way of conceiving the development of science. It is an attempt to render divulgative, usual and for 'everyone', professions which are actually for the few and/or for those who intend availing of special lenses to scrutinize the world and interpret it.

It is also the case to point out, as some scholars have, that because of Kuhn that conceptual framework called 'paradigm' acts as a lens for those seeking to find their bearings during research, where the observation of reality is no mere matter of recording facts or objective data but a theoretical activity, which, in turn, acts as a selector of perspectives on the basis of more or less relevant concepts. Similarly, a simple strategy in/of different 'optics' prior to identification of a theory would not make much sense, except that of 'justifying' 'discovery', so to speak, in an era when contexts of the latter tend to replace those of the former, in extremely explicit terms of a contemporary reflection on science and/or sciences.

An answer that so many contemporary researchers and students seem to expect, also in the wake of the debate triggered by 'programmes' and sociologies of scientific knowledge up until deconstructionism, etc. seems to follow an epistemological-historical approach that should reveal the nature of new scientific changes as they occur or are examined, avoiding the age-old legacy of traditional analysis and most of all analyses by historians using non-sociological categories, that is, not drawn from the conceptual and epistemological background of sociology as a scientific subject which has come to maturation recently. The example of the sociology of science which investigates the conditions preceding the production of scientific events or, better still, scientific facts (*causality*) relating to operative practices and processes among actors and scientists in laboratories, the place where the latter act, developing technical documents that do not always correspond to the neutrality of a methodological programme, seems quite significant to us. It is in this sense that the history of sociological science might appear today as an exercise for sociologists and no longer for historians educated in concepts developed over the years and the centuries by the philosophy of 'philosophers', not by the philosophy of 'scientists'.

Obviously, it should be noted that the most relevant lesson many modern and contemporary historians of sociology and of the so-called scientific subjects remains, probably, that of having actually witnessed the scientific revolution of the whole of the twentieth century. Moreover, from the end of the nineteenth century to the beginning of the 1960's and beyond, science was submitted to a series of transformations not only concerning some of its technical aspects but also fundamental methods and objectives. Despite this, the consequent limit remained that of continuing to reason in some areas and in various fields of research and epistemology as if this miracle were as yet to take place.

In actual fact the twentieth century is undoubtedly the century of epistemology's confrontation with issues of technological and scientific development. This we have to say reflects the 'point of view' of some sociologists of scientific knowledge engaged in studying the field of transformation. From the 'point of view' of historians of science, the human being, according to a famous definition by Benjamin Franklin, is an animal which creates instruments and, as a result, unlike other animals, has always sought to produce his/her own took kit. This kind of discourse harkens back, instead, to that set of studies aimed at precisely defining, in terms of scientific rationalization, the various technical processes with which raw materials are transformed into finished products, studies otherwise known as 'technology', at least as far as a fairly straightforward definition of process and/or processes is concerned. It seems evident that the link between technology and the history of science might represent some sociologists' interest in the issue, recently, also with regard to epistemological studies of the production of 'technical knowledge and know-how'. Among the most important, recent approaches, we should remember those concerning technology as a 'social construct' and which saw the publication, at the end of the 1980's, of studies by authors like Wiebe E. Bijker, Thomas P. Huges and Trevor Pinch whose interest derived from a socio-scientific approach to the study of technology (see T. Pinch and W. E. Bijker, *The social construction of Facts and Artefacts: Or how the Sociology of Science and the Sociology of Technology Might benefit Each Other* in 'Social Studies of Science', 1984, 14). To these we might add a contribution like that of Harry Collins and Pinch himself translated into Italian as *Il Golem tecnologico* (2000), a continuation of a previous study by the same authors -issued in the United Kingdom in 1995- concerning al one should know about science. But the same may be said of studies by a number of authors who dealt with 'technological artefacts', or better still, the tortuous pathway of the technological development rooted in historical and social variations, observed in Italy by authors like Massimiliano Bucchi of *Scienza e Società* (2002, 2010) or Peppino Ortoleva, University of Turin 1995 and in 2002, in 'historical-media' terms, what is, by linking technologies to the history of the media, the evolution of culture with the development of capitalist institutions.

But the fact is that, originally, this whole series of problems implied not so much the skills of the social sciences and/or of social scientists, but the interest of historians of science, who, for many years, sought explain in their books the causes of the separation between science and technology even up to the present day. To illustrate these issues, it seems useful to take a glance at this science-technology separation manifested itself up until the mid-nineteenth century. Steam engines, for example, were already functioning for more than seventy years when scientists tried to devise a theory, established and specified only half a century later. Similarly, we note that the construction of mechanical and precision instruments preceded the theoretical work of nineteenth-century scholars of mechanics. In the same way the manufacture of mineral acids preceded Antoine-Laurent Lavoisier's chemical system. As the historians of science point out, it was not until the mid-nineteenth century that exchanges between these two fields of activity began to balance out. The chemical industry swiftly reaped the fruits of the scientific discoveries of the first half of the century; power generators and the electric telegraph appeared very shortly after the scientific work of Alessandro Volta, Humphry Davy, Andre-Marie Ampere and Michael Faraday. And so forth.

With regard to yet another issue, it seems useful to recall that from today's sociological 'point of view' and/or that of scientific knowledge, analysts are not encouraged to reflect on the relationships in question; in the same fashion historians of technology always refer back to September 1882 when Thomas A. Edison inaugurated the Pearl Street power station in New York.

From a certain angle, therefore, with a view to a more certain collocation of so-called 'technical knowledge', it would suffice to admit that during the nineteenth century scientific research began exert its influence on technology. It was only in the twentieth century that science became, in turn, a specific feature of technical progress. It was precisely from that moment historians of scientific subjects or sociologists, began to discover that the association not only of scientific knowledge but of scientific principles too to technology continued to increase and become more complex, giving rise to an entire system of technical sciences or 'technologies' ranging from electronics to automation, from disciplines concerning space travel to calculators and data-elaboration devices.

Recently, a number of historians of science like Paolo Rossi (1997) sought to retrace the historical course of scientific knowledge, by locating the latter in close association with its own capacity for change and growth in relation to the bond existing between certain speculative and technical-knowledge practices. The attempt aims at to providing a framework for the birth of a new science in Europe in consideration of the development of a 'technological knowledge' capable of imposing change on the theoretical vision itself of scientific subjects. Results different from those of Rossi (in *La memoria del sapere*, for example) are those we find in the works of those who seek to establish links between the development of their own disciplines (physics, astronomy, etc.) and the progress made by the sociology of science and technology.

In other words many scholars and historians of science continue to discuss the areas of applied research as if these were, by now, clearly defined and delimited by a sort of technical or 'technological' knowledge tending very often towards anticipation of methodological knowledge itself, as pointed out recently regarding the use of tools when choosing between 'qualitative' and 'quantitative' approaches in sociology. In this case, reference should be made to due recourse by scientists to an acclaimed form of rationality. Outside of this not even the most overt technologies could be outlined or devised in relation to the variety of guidelines capable of defining them.

Some approaches to the problem of the definition of guidelines have been identified by the Italian methodologist Gianni Statera in an essay of his regarding the myth of qualitative research which he published in 1992. In this paper the author posed several questions about the use of 'qualitative' techniques intended – in the long run- as mere techniques to avail of in sociological research of a so-called 'quantitative' and/or conventional kind, where the influence of these self-same 'techniques' were the outcome of 'objects' modified by the researcher or researchers. For Statera, the relationship between the use of 'technical' and scientific knowledge might be posited, in part, as a problematic aspect of choices of guidelines to within a perspective of the development of science and the result of historicisable change. Reference to this kind of discourse might also lead back to the epistemology of the social sciences which, as Davide Sparti from the University of Siena pointed out in the course of his ten-year study, should be intended as authentic reflection on scientific knowledge in the social sciences (see *Epistemologia delle scienze sociali*, chapter II, 2002).

Effectively, and this holds for Statera too, scientific sociology connoted itself thanks also to the use of a methodological practice capable of following, only on a temporary basis, the results of post-positivistic epistemology relating to technology, new 'qualitative' exigencies of sociology and so-called 'hermeneutic sociologies' (2000), on the basis of a series of series of steps capable of ensuring the rigour of procedures when making methodological choices. Epistemology, too, according to Gianni Statera, including its radical anarchist and relativistic tendencies, is taking form thanks to a need for justification and irreversible 'technical' issues regarding the coherence, publication and control of assertions made by research. These worries, even if they do not lead to a veritable 'opening' on the part of this researcher towards and/or evident interest in epistemological history, as Robert King Merton's Italian exegete, Filippo Barbano, pointed out after Merton's death (see *La società e la scelta dell'indagine scientifico-sociale,* 2001), are present in his recent articles, ranging from *Logica dell'indagine scientifico-sociale,* 1994 to *Il pensiero di Kuhn e la methodologia delle scienza sociali* (1997), an essay which not only reveals the bonds existing between unresolved 'paradigmatic conditions' of sociology and 'internal' reception of imports of Merton's sociology of science, but also confirms the advent of an epoch where sociologists may reason in new terms concerning research itself as well as the history of the discipline.

This means that even the concrete terms of a debate on scientific knowledge might be reviewed in relation to the broader 'confines' of new demands, regarding new research issues, as that found the few Italian researchers who have recently brought sociology back to 'infinite' and unsolved discussions still present on its horizon today: for example, those regarding contemporary epistemology, considered as driven by gnoseology, without renouncing completely some ontological assumptions, that is, those which treat 'historicity' as a metaphysical problem referred to the objects of knowledge as well as to some more or less rigid demarcations between the sciences of nature (see certain doubts expressed by Giovanni Boniolo, 2001) and the social sciences (two aspects discussed in interesting studies by Maurizio Bonolis from Rome's La Sapienza University published in "Sociologia e ricerca sociale" n. 67, 2002 and n. 78, 2005). It is quite significant that themes of this kind continue to circulate in an epoch when Bacon's Dedalus has grown so powerful, in an age of techno-science when an avalanche of 'technological knowledge' impacts on the human and social sciences and which, eventually, raise questions without knowingly providing a comprehensive answer regarding the epistemological status of sociology (see Bonolis's 2005 essay which returns to Enzo Campelli's article on 'medium range' explanation in sociology, that too published in "Sociologia e ricerca sociale ', n. 75 in 2004). The same happened, in part, to the origin of the scientific revolution, when certain issues became widespread as a form of thought. In 2016 Enzo Campelli devoted a volume called *Nullius in verba. Il metodo nella rivoluzione scientifica.*

At the end of the day, what happened recently to the 'soft' philosophy and philosophers of science, is happening to the humanities and social sciences, which will be led to permanently question themselves regarding definable or indefinable limits, insurmountable within the confines of the empirical world, since overcoming them may still be considered a problem for the advancement of scientific disciplines, like sociology. To this regard one of the concerns which Richard Rorty manifested in *Consequences of Pragmatism* in 1982 bears eloquent witness to this. He held that with the advent of a post-philosophical 'culture' all the problems of philosophy would cease definitely to be a source of concern including those of an epistemological nature (he implied).

The same cannot be said of relationships between science and the advent of 'technological know-how', where a narrowing of boundaries might determine a new philosophical and epistemological reaction the broadening of which could lead to recognition of increasingly more specialized versions of science and/or of particular areas of scientific research.

In a certain way, it is 'typical' for sociologists in Italy, that they, precisely, still struggle with two different ways of addressing research without having carried out proper examinations of theoretical assumptions regarding the investigation of knowledge, without deep reflection regarding the historical episteme of social knowledge. On many occasions they simply allow themselves be driven by methods which may be employed to conduct scientific research or simply 'perfect' specific tools to use in a method to acquire knowledge (techniques). The technical aspects of certain procedures of investigation involve, on their part, some methodological problems, common to all social sciences, and, partially, to the physical and mathematical sciences which may concern the logic of propositions, relations between theory and the collection of empirical data, between hypothesis and verification, between the subjective purposes of research and the 'objectivity' of research. Furthermore, it needs to be said that to follow the procedures of established protocols means dealing, ultimately, with epistemology, where the interests of historians of scientific sociology may–in the end- take into account the development of procedures of various conceptual and 'cognitive' aspects of the discipline and its evolution. If the methodology of the social sciences may discuss, without doubt, 'operative' trends, epistemology (rather than epistemologists in unison) seems to suggest taking some steps backwards and pose all those conceptual and 'cognitive' questions connected to research, including the difficulties concerning epistemological orientations.

This is why the concerns of philosophers like Rorty do not appear to be very well founded. The reason is simple: in the history of science, epistemological history, must take into due account the permanent empirical world, and, which is, as far as it is possible to admit, 'immanent' to the researcher and not discuss its total dissolution. It is pretty obvious that a 'post-scientific culture' would not eliminate all the epistemological problems in question.

A few decades ago two Italian researchers, Rosanna Memoli and Giancarlo Gasperoni, in an article which appeared in *Italian Sociology in the Nineties*, a monographic volume published in 1998 and edited by Giancarlo Rovati of the Italian Association of Sociology (AIS), pointed out how some Italian methodologists had evaluated, in a sense, the extent of the changes taking place in the human and social sciences assuming a certain kind of stance critical also of the reception of studies concerning the epistemological bases of cognitive processes in sociology. Rosanna Memoli has recently edited volumes questioning the role of scientists in an interdisciplinary perspective, including one with philosopher of science Elena Gagliasso, containing papers presented at a Conference held in the Philosophy Faculty on the occasion of the publication of a book by Marcello Cini, Italian physicist and epistemologist, who died in 2012.

The starting point, however, was that of having established a confrontation with a post-empirical epistemology capable of providing an arrival point for historians of scientific subjects confronted with the pluralism of new advocates of epistemology and/or of a new epistemology commensurate to an 'integrated methodology' to apply to social research, like the 'theoretical proposal' advanced by authors including Costantino Cipolla of the University of Bologna, editor of volumes on the epistemology of tolerance published in 1997. In the years that followed significant methodological work of scholars besides Costantino Cipolla, like Piergiorgio Corbetta, Leonardo Cannavò and Alberto Marradi was published. They are listed in the bibliography below. These methodological manuals favoured the debate on social research between 1999 and 2007 when scientific production also influenced university teaching. At that time reference to 'schools' seemed to vanish, although the academic parcelling sociology in the field remained.

The fact that changes taking place and posing important difficulties to epistemological frameworking, following the various attempts at new conceptualization within the various disciplines, is, to say the least, self-evident. In the same way, one sees that contemporary social researchers are evidently desirous to approach the history of science availing of concepts stemming from a 'unitary' consideration of science and/or of the different sciences (see Maurizio Bonolis's *Storicità e storia della sociologia,* 2007). Maurizio Bonolis's work extends over time and may be read in the pages of the "Sociologia and Ricerca Sociale" journal, where the topics range from realism to the action theory. Vol. 33, 99 was dedicated to Franco Leonardi, Italian epistemologist and sociologist, who died in 2012. There is also an interview conducted by the author of present book published in 2015 in a professional journal details of which appear in the bibliography.

The theory itself of methodological and epistemological pluralism might interest areas of discussion requesting freedom for scientists, the thesis sustained by that anarchist of knowledge Paul k. Feyerabend, is asserting itself more and more in order to continue within the ambit of the history of science, and along other routes, the methodological pluralism desired by authors like the Italian Alessandro Bruschi, who determined –in this sphere too - a sort of comeback of epistemological cognizance of technical or 'technological knowledge'. Strangely enough, the progress of technological knowledge corresponds to an increase in interest in the quality of sociological research, at least in academic circles regarding separate organizational and cultural issues. The problem of the quality of sociological knowledge was still central to a national Conference organized recently by Italian sociologists in Florence (10th – 12th October 2013). The discussions focused on the role of sociology in studies and analyses of society, its ability to develop interpretive categories capable of deciphering complexity and taking action regarding it. The Conference was dedicated to the memory of sociologists Antonio De Lillo and Guido Martinotti (who died in 2012). Typically, this type of Conference was modelled on the academic nature of sociology in Italy which always tends to create large containers for wide-ranging discussion, as was the case also with the very recent Congress held in Verona (10th – 12th November, 2016) which mustered interest around the issue of inequality, justice and fairness in a global context. Between one subtle discussion and another, there emerged the conviction that scientific and technological knowledge might join to complete the conceptual revolution of an epoch when the depositaries of the dominant science which always tends to favour themselves and thus increase the permanent aura of their sacrality. A certain form of debate on these issues is contemplated in Italian sociology of cultural processes, a discipline which, in recent years has been very successful regarding ways of dealing with increases in technological knowledge. Part of that success is due to advancements in communication sociology at Italian universities with the diffusion of the merits of a cultural project involving the new technologies. To this regard it is important to mention the cultural modernization project set up over a decade ago by Mario Morcellini and his students at Rome's *Sapienza* University and by scholars like Marino Livolsi and Graziella Mazzoli who study the relationships existing between communications and the new technologies. This interesting topic of cultural sociology deals with the sociology of artificial artistic and aesthetic experiences as we read in the studies of Massimo Negrotti. Other interests regard reflective and qualitative sociological research (Alberto Melucci, Laura Bovone etc.) while others dwell on the sociology of culture, as in the text *Il gioco della cultura* edited by Simonetta Piccone-Stella and Luca Salmieri. Finally, reference should also be made to studies seeking to provide a hermeneutics of social research and communication, like that of Paolo Montesperelli's (2014) recent contribution, listed in the bibliography.

References

AA.VV [Various authors], edited by Angelo Pagani, *Campi di applicazione della sociologia* [*Fields of application of sociology*], Bologna, Il Mulino, 1963.

AA.VV, AIS, *Anni in salita* [*Uphill years*], edited by Giandomenico Amendola, Milan, Franco Angeli, 2004.

AA.VV, AIS, *Giovani sociologi 2003* [*Young sociologists 2003*], Milan, Franco Angeli, 2004.

AA.VV, *Antologia di scienze sociali* [*Anthology of social sciences*], edited by Angelo Pagani, Bologna, Il Mulino, 1963.

AA.VV, *Atti del Congresso di studi metodologici* [*Minutes of the Congress on methodological studies*], Centre of Methodological Studies, Turin 17-20th December 1952, Turin, Edizioni Ramella, 1954.

AA.VV, *Atti del Convegno indetto a celebrazione del trentennio di attività del Centro di Studi Metodologici: Turin, 1946-1976* [*Minutes of the Convention held to celebrate the thirty years of activity of the Centre for Methodological Studies*: *Turin, 1946-1976*] organised by The Centre of Methodological Studies, Turin, published by Leprotto & Bella, 1977.

AA.VV, *Atti del Convegno sui problemi metodologici di storia della scienza* [*Minutes of the Convention on the methodological problems of the history of science*] (Turin, Centre of Methodological Studies, 29-30 March 1967), Florence, Barbèra, 1967.

AA.VV, *Atti del Primo Congresso Nazionale di Scienze Sociali* [*Minutes of the First National Congress of Social Sciences*], Bologna, 1958.

AA.VV, *Attualità del materialismo dialettico* [*The relevance of dialectical materialism*], Rome, Editori Riuniti, 1974.

AA.VV, *Bilancio della fenomenologia e dell'esistenzialismo* [*Balance of phenomenology and existentialism*], Padua, 1960.

AA.VV, *Crisi della ragione. Nuovi modelli del rapporto tra sapere e attività umane* [*Crisis of reason. New models of the relationship between knowledge and human activities*], edited by Aldo Gargani, Turin, Einaudi, 1979.

AA.VV, *Evandro Agazzi filosofo della scienza* [*Evandro Agazzi, philosopher of science*], in "Il Protagora," Year XXXVII July-December, vol. 14, Mimesis, 2010-2011.

AA.VV, Fenomenologia e sociologia [Phenomenology and sociology], in *Archivio di Filosofia*, [*Archives of philosophy*], Padua, 1951.

AA.VV, *Filosofia e sociologia* [*Philosophy and sociology*], Bologna, Il Mulino, 1954.

AA.VV, *Fondamenti logici della scienza* [*The logical foundations of science*], Turin, Francesco De Silva, 1947.

AA.VV, *Il problema della filosofia* [*The problem of philosophy*], Rome-Milan, 1952.

AA.VV, *Il problema delle scienze nella realtà contemporanea* [*The problem of science in the contemporary reality*], edited by Michele Massafra and Fabio Minazzi, Milan, Franco Angeli, 1985.

AA.VV, *Il sociologo e le sirene. La sfida dei metodi qualitativi* [*Sociology and the Sirens. The challenge of qualitative methods*] edited by C. Cipolla and A. De Lillo, Milan, Franco Angeli, 2004.

AA.VV, *L'identità sociale della sociologia in Italia* [*The social identity of sociology in Italy*] edited by Costantino Cipolla, Milan, Franco Angeli, 2012.

AA.VV, *L'immagine della scienza. Il dibattito sul significato dell'impresa scientifica nella cultura italiana* [*The image of science. The debate on the meaning of scientific undertaking in Italian culture*] edited by Giulio Giorello, Milan, Il Saggiatore, 1977.

AA.VV, *La filosofia dal '45 a oggi* [*Philosophy from 1945 to the present*] edited by Valerio Verra, with essays by N. Abbagnano, F. Barone, P. Filiasi Carcano, A. Pasquinelli, Turin, 1976.

AA.VV, *La filosofia della scienza oggi (Europe 1993*) [*The philosophy of science today (Europe 1993)*], edited by Fabio Minazzi, Naples, Italian Institute of philosophical studies, 1991.

AA.VV, *La filosofia italiana dal dopoguerra ad oggi* [*Italian philosophy from the postwar period to the present*] Bari, Laterza, 1985.

AA.VV, *La scienza e la critica del linguaggio* [*Science and the criticism of language*] Venice, Marsilio, 1980.

AA.VV, *La sociologia di Achille Ardigò* [*Achille Ardigò's sociology*], edited by Costantino Cipolla and Sebastiano Porcu, Milan, Franco Angeli, 1997.

AA.VV, *La storia della filosofia come sapere critico* [*The history of philosophy as critical knowledge*], Milan, Franco Angeli, 1984.

AA.VV, *Le più recenti epistemologie: Popper-Hempel* [*The latest epistemologies. Popper-Hempel*], Padua, Gregoriana, 1974.

AA.VV, *Max Weber:* Cinquant'anni dopo [Max Weber: fifty years later] in *Rassegna Italiana di Sociologia*, [*Italian Review of Sociology*], A. XI, n. 4, Oct.-Dec. 1970.

AA.VV, Metascienza [Meta-science] in *Sociologia e ricerca sociale*, [*Sociology and social research*] n. 13, April 1984.

AA.VV, *Neoempirismo logico, semiotica e filosofia analitica - Antologia di scritti* [*Logical neo-empirism, semiotics and analytical philosophy - An anthology of writings*], Brescia, Editore La Scuola, 1964-1984.

AA.VV, *Neopositivismo e unità nella scienza* [*Neopositivism and unity in science*], Milan, Bompiani, 1973.

AA.VV, *Omaggio a Husserl* [*Homage to Husserl*], edited by Enzo Paci, Milan, Il Saggiatore, 1960.

AA.VV, Paul Felix Lazarsfeld: un "classico" marginale [Paul Felix Lazarsfeld: a marginal "classic"] *Sociologia e ricerca sociale*, [*Sociology and social research*] n. 58-59, 1999.

AA.VV, Per una sociologia scientifica. In ricordo di Gianni Statera [For a scientifical sociology. In memory of Gianni Statera] special issue of *Sociologia e ricerca sociale*, [*Sociology and social research*], n. 66, 2001.

AA.VV, Problemi della concettualizzazione nelle scienze sociali [Problems of the conceptualisation in the social sciences] special issue of *Sociologia e ricerca sociale*, [*Sociology and social research*], 43, 1994.

AA.VV, *Prospettive della logica e della filosofia della scienza Atti del Convegno Triennale della Società Italiana di Logica e Filosofia delle scienze* [*Perspectives of logic and philosophy of science, minutes of the Triennial Convention of the Italian Society of Logic and Philosophy of Sciences*] Cesena and Urbino, 15-19 February 1999, edited by V. Fano, G. Tarozzi and M. Stanzione, Soveria Mannelli, Rubbettino, 2001.

AA.VV, *Ricerca sociologica e ruolo del sociologo* [*Sociological research and the role of the sociologist*] Centre of Methodological Studies, edited by Pietro Rossi, Bologna, Il Mulino, 1972.

AA.VV, *Rivoluzioni scientifiche* [*Scientific revolutions*] edited by Ian Hacking, Preface by Giulio Giorello, Rome-Bari, Laterza, 1984.

AA.VV, *Scienza e cultura. Teoria e programmi di ricerca sociale* [*Science and culture. Theory and Programmes of social research*] edited by Filippo Barbano, '*Atti*' del Convegno Italiano di Sociologia, Rome, October 1981, Milan, Franco Angeli, 1983.

AA.VV, *Scienza e filosofia. Saggi in onore di Ludovico Geymonat* [*Science and philosophy. Essays in honour of Ludovico Geymonat*] edited by Corrado Mangione, Milan, Garzanti, 1985.

AA.VV, *Scienza e storia - Analisi critica e problemi attuali* [*Science and history - Critical analysis and current problems*] edited by Silvano Tagliagambe and Antonio Di Meo, in 'Critica marxista', Rome, Editori Riuniti, 1980.

AA.VV, *Sociologia. I concetti base* [*Sociology. The basic concepts*] Bologna, Il Mulino, 2013.

AA.VV, *Storia delle scienze* [*The history of sciences*] edited by Nicola Abbagnano, Turin, UTET, 1965, 3 vol. (4 volumes).

AA.VV, *Sul marxismo e le scienze* [*On Marxism and the sciences*] in 'Critica Marxista', n. 6, 1972.

AA.VV, *Sulla filosofia italiana del Novecento* [*On Italian philosophy of the nineteenhundreds*] edited by B. Bonghi and F. Minazzi, Milan, Franco Angeli, 2008.

AA.VV, T. S. Kuhn: come mutano le idee sulla scienza [T.S. Kuhn. How ideas of science change] edited by Enzo Campelli, in *Sociologia e ricerca sociale*, [*Sociology and social research*], n. 53-54, 1997.

AA.VV, *Vittorio Somenzi. Antologia e testimonianze 1918-2003* [*Vittorio Somenzi. Anthology and testimonies*] Fondazione Banca Agricola Mantovana, Publi. Paolini, Mantua, 2011.

AA.VV, Wissenschaftliche Weltauffassung Der Wiener Kreis, 'Veröffentlichungen des Vereines Ernst Mach' [Scientific conception of the world The Vienna Circle, 'Publications of the Ernst Mach' Association] Wien, Artur Wolf Verlag, 1929; Italian translation edited by Alberto Pasquinelli, *La concezione scientifica del mondo. Il circolo di Vienna* [*The scientific concetpion of ther world. The Viennese circle*] Bari, Laterza, 1979.

AAVV, *Sociologia. Materiali sui fondamenti e le questioni di metodo* [*Sociology. Materials regarding fundaments and questions of method*] edited by Martino Ancona, Naples, Liguori, 1983.

Abbagnano, N., (edited by), *Storia delle scienze* [*History of science*] Turin, UTET, 1965, 3 volumes.

Abbagnano, N., *Dizionario di Filosofia* [*Dictionary of Philosophy*] Turin, UTET, 2001.

Abbagnano, N., *La fisica nuova. Fondamenti di una teoria della scienza* [*New Physics. Fundaments of the theory of science*] Naples, Guida, 1934.

Abbagnano, N., *Problemi di sociologia* [*Problems of sociology*] Turin, Taylor Publisher, 1959.

Abbagnano, N., *Scritti scelti* [*Selected writings*] Turin, Taylor, 1967.

Acquaviva, S. S., Franco Ferrarotti and Italian Sociology in "Social Research," vol. 33, n. 2, *Essays in Economic Theory in Honour of Adolph Lowe and Hans Neisser*, Summer 1966, pp. 332-353.

Acquaviva, S. S., *Il problema della logica nelle scienze umane* [*The problem of logic in the human sciences*] Venice, Marsilio, 1964.

Addario, N., Novità e paradossi del costruttivismo radicale: problemi di una epistemologia naturalizzata [Novelty and paradox in radical constructivism: problems of a naturalised epistemology] in *Teoria sociale* [*Social theory*], 2, 1993.

Agassi, J., *Epistemologia, Metafisica e storia della scienza* [*Epistemology, Metaphysics and the history of science*] Italian translation by Adelino Cattani, Introduction by Massimo Baldini, Rome, Armando Armando, 1978.

Agassi, J., The Nature of Scientific Problems and Their Roots in Metaphysics, in Science in Flux, Dordrecht, Kluwer, 1975; Italian transl, in *Le radici metafisiche delle teorie scientifiche* [*The metaphysicsl roots of scientific theories*], edited by Emanuele Riverso, Rome, Borla, 1983.

Agazzi E., (*Introduction*), *Il concetto di progresso nella scienza* [*The comncept of progress in science*] Milan, Feltrinelli, 1976.

Agazzi E., Filosofia della scienza [Philosophy of Science] in *Questioni di storiografia filosofica* [*Issues of philosophical historiography*] edited by A. Bausola, Brescia, Ed. La Scuola, 1978.

Agazzi E., Valore epistemologico della storia della scienza [Epistemological value of the history of science], *Civiltà delle Macchine*, [*The civilisation of machines*], n. 1, 1970.

Agazzi, E., (edited by), *La filosofia della scienza in Italia nel '900* [*The philosophy of science in Italy in the nineteen hundreds*] Milan, Franco Angeli, 1986.

Agazzi, E., (edited by), *Storia delle scienze* [*The history of science*] Rome, Città Nuova, 1984.

Agazzi, E., Minazzi, F., Geymonat., L. *Filosofia Scienza e Verità* [*The Philosophy of science and Truth*] Milan, Rusconi, 1989.

Agazzi, E., *Scientific Objectivity and Its Contexts*, Springer Verlag, 2014.

Agazzi, E., *Sulla filosofia della scienza* [*On the philosophy of science*] Milan, Mimesis, 2015.

Agazzi, E., *Temi e problemi di filosofia della fisica* [*Topics and issues of the philosophy of physics*] Rome, Abete, 1969.

Agazzi, E., Tra scientismo e scetticismo [Between scientism and scepticism] in AA. VV, *Il mondo incerto* [*The uncertain world*] edited by Marcello Pera, Rome-Bari, Laterza, 1994.

Agnoli, M. S., *Concetti e pratica nella ricerca sociale* [*Concepts and practice in social research*], Milan, Angeli, 1994.

Agnoli, M. S., Il legame micro-macro nella pratica della ricerca sociale [The micro-macro link in the practice of social research] in *Sociologia e ricerca sociale*, [*Sociology and social research*], n. 43, 1994.

Agodi, M. C., Boudon, l'individualismo metodologico e lo 'stato paradigmatico' della sociologia [Boudon, methodological individualism and the 'paradigmatic state' of sociology] in *Sociologia e ricerca sociale*, [*Sociology and social research*], V, n. 17-18, 1985, pp. 29-69.

Agodi, M. C., La sociologia della conoscenza ed il problema dei fondamenti delle scienze. La debolezza del programma 'forte' [The sociology of knowledge and the problem of the fundaments of the sciences. The weakness of the 'strong' programme] in *Sociologia*, 20, Jan.- April 1986.

Agodi, M. C., Lazarsfeld e la "natura" della classificazione nelle scienze sociali [Lazarsfeld and the "nature" of classification in the social sciences] in *Sociologia e ricerca sociale*, [*Sociology and social research*], XX, n. 58-59, 1999, pp. 117-149.

Agodi, M. C., T. S. Kuhn e la sociologia della conoscenza scientifica [T. S. Kuhn and the sociology of scientific knowledge] in *Sociologia e ricerca sociale*, [*Sociology and social research*], XVIII, n. 53-54, 1997, pp. 183-226.

AIS Associazione Italiana di Sociologia [The Italian Association of Sociology] edited by G. Amendola, *Anni in salita* [*Uphill years*] Milan, Franco Angeli, 2004.

Alberoni, F., *Classi e generazioni* [*Classes and generations*], Bologna, Il Mulino, 1970.

Alexander, J. C., La logica teorica oggi: il pensiero sociologico alla luce della 'Struttura' di Parsons [Theoretical logic today: sociological thinking in the light of Parsons's 'Structure'] translation by G. Nevola, in *Quaderni di Sociologia*, [*Sociology notebooks*], XXXIII, n. 9, 1987.

Aliotta, A., Il nuovo positivismo e lo sperimentalismo [New positivism and experimentalism] in *Opere Complete* [*Complete works*] Rome, Perrella, 1954, VII.

Aliotta, A., *Il nuovo positivismo* [*New positivism*] s.1.e., Gubbio, 1954.

Aliotta, A., La fenomenologia, il circolo di Vienna e il mio sperimentalismo [Phenomenology, the circle of Vienna and my experimentalism] in *Logos*, n. 1, 1941.

Aliotta, A., *La reazione idealistica contro la scienza* (1912) [*The idealist reaction against science*] full reprint edited by Cleto Carbonara, Naples, Libreria scientifica, 1970.

Amaturo, E., *Metodologia della ricerca sociale* [*Methodology of social research*] Turin, UTET, 2012.

Amendola G., - Ardigo' A. (edited by), *Ricerca sociologica informatica e società italiana* [*Sociological informatic research and Italian society*] Milan, Franco Angeli, 1986.

Ammassari, P., Dell'uso analitico dei dati empirici [Of the analytical use of empirical data] in *Rassegna Italiana di Sociologia*, [*Italian Review of Sociology*], 1,1962.

Ammassari, P., *Saggi metodologici* [*Methodological essays*] edited by Angelo Saporiti, Milan, Franco Angeli, 1995.

Ammassari, P., Validità e costruzione delle variabili: elementi per una riflessione [Validity and construction of variables: elements for reflection] in *Sociologia e ricerca sociale*, [*Sociology and social research*], 13, 1984.

Ammaturo, N., *Elementi di epistemologia sociologica* [*Elements of sociological epistemology*] Presentation by Vincenzo Cesareo, Milan, Franco Angeli, 2003.

Ancarani, V., *La scienza decostruita* [*Science deconstructed*] Introduction by Filippo Barbano, Milano, Franco Angeli, 1996.

Ancarani, V., *La scienza decostruita. Teorie sociologiche della conoscenza scientifica* [*Science deconstructed. Sociological theories of scientific knowledge*] Milan, Franco Angeli, 1996.

Ancarani, V., *Struttura e mutamenti nelle scienze. L'epistemologia storica di Gaston Bachelard* [Structure *and chenges in science. Gaston Bachelard's historical emistemology*] Milan, Angeli, 1981.

Antiseri, D., *Dal neopositivismo alla filosofia analitica* [*From neopositivism to analytical philosophy*] Rome, Abete, 1966.

Antiseri, D., *Epistemologia contemporanea e didattica della storia* [*Contemporary epèistemology and the didactics of history*] Rome, Armando, 1974.

Antiseri, D., *Il compito delle scienze sociali, teoriche* [*The task of the theoretical I social sciences]* n 'Il Mondo 3', 1, June 1995.

Antiseri, D., *Karl R. Popper. Epistemologia e società aperta* [K*arl R. Popper. Epistemology and open society*], Rome, Armando, 1972.

Antiseri, D., *La filosofia del linguaggio: metodi, problemi e teorie* [*The philosophy of language: methods, problems and theories*] Brescia, Morcelliana, 1972.

Antiseri, D., *Le ragioni del pensiero debole* [*The reasons of weak thought*] Rome, Borla, 1993.

Antiseri, D., Reale, G., *Storia della filosofia* [*History of philosophy*] vol. 11: Scienza, epistemologia e filosofi americani del XX secolo [Science, epistemology and American philosophers of the twentieth century] Milan, Bompiani, 2014.

Antiseri, D., *Sulla storiografia di Popper e Hempel* [*On the historiography of Popper and Hempel*] in "Proteus," II, 1970.

Antiseri, D., *Teoria unificata del metodo* [*Unified theory of method*] Padua, Liviana Editrice, 1981.

Ardigo', A., *Per una sociologia oltre il post-moderno* [*For a sociology beyond the post-modern*] Rome-Bari, Laterza, 1988.

Ayer, A. J., *Language, Truth and Logic*, London, Victor Gollancz, 1946, II edition; Italian translation edited by G. De Toni, *Linguaggio, verità e logica*, Milan, Feltrinelli, 1987.

Bachelard, G., *Le nouvel esprit scientifique* [*The new scientific spirit*] Paris, Alcan, 1934 (IV ediz. PUF, Paris, 1946); Italian translation edited by Ludovico Geymonat e Pietro Redondi, Il nuovo spirito scientifico, Bari-Rome, Laterza, 1978.

Baldini, M., *Congetture sull'epistemologia e sulla storia della scienza* [*Conjectures regarding epistemology and the history of science*] Rome, Armando Editor, 1986.

Baldini, M., *Epistemologia e storia della scienza* [*Epistemology and history of science*] Florence, Città di Vita, 1974.

Baldini, M., *Introduzione a Karl R. Popper* [*Introduction to Karl R. Popper*] Rome, Armando Editor, 2002.

Baldini, M., *Problemi e prospettive di storia della scienza* [*Problems and prospectives of the history of science*] Rome, Ed. Città Nuova, 1986.

Baldini, M., *Teoria e storia della scienza* [*Theory and history of science*] Rome, Armando, 1975.

Banfi, A., *La ricerca della realtà* [*The quest for reality*] Bologna, Il Mulino, 1996

Banfi, A., *Principi di una teoria della ragione* [*Principles of a theory of reason*], new edition, Florence, 1960.

Barbano, F., (edited by), *Scienza e cultura* [*Science and culture*] Milan, Franco Angeli, 1983.

Barbano, F., La metodologia della ricerca nella sua impostazione teoretica [The theoretical approach to the methodology of research] in *Sociologia*, July - September, 1958.

Barbano, F., Trasformazioni e tipi della teoria sociologica contemporanea, in *Questioni di Sociologia*, (n. 347), Brescia, La Scuola Editrice, 1966, vol. I. [Transformations and types of contemporary sociological theory, in *Sociology Issues*]

Barbano, F., *Introduction to the Italian edition of R. K. Merton, Science, Technology and Society in Seventeenth Century England*, New York, Harper Torchbooks, Harper & Row, 1970; Italian translation by Pasquale De Gaetano edited by Alberto Izzo, *Scienza, tecnologia e società nell'Inghilterra del XVII secolo*, Milan, Franco Angeli, 1975.

Barbano, F., *La sociologia della scienza in Italia: situazioni e problemi in Robert King Merton* - Jerry Gaston (edited by), *The sociology of science in Europe*, 1977; Italian translation *La sociologia della scienza in Europa*, Milan, Franco Angeli, 1980.

Barbano, F., Le teorie sociologiche tra storicità e scienza [The sociological theories between historicism and science] Introduction to italian edition by R. K. Merton, *Social Theory and Social Structure*, New York, The Free Press, 1968; v. *Teoria e struttura sociale*, Bologna, Il Mulino, 1983, vol. I.

Barbano, F., Sola, G., *Sociologia e scienze sociali in Italia dal 1861 al 1890* [*Sociology and social sciences in Italy from 1861 to 1890*] Milan, Franco Angeli, 1985.

Barbano, F., *La sociologia in Italia. Storia, temi e problemi 1945-60* [*Sociology in Italy. History, topics and problems 1945-60*], Rome, Carocci, 1998.

Barbano, F., Teoria e ricerca: storia epistemologica e circolarità ermeneutica [Theory and research: epistemological history and hermeneutic circularity] in *Sociologia*, n. 2-3, 1998.

Barbano, F., La società e la scelta dell'indagine scientifico-sociale [Society and the scientific-social investigational choice] in *Sociologia e ricerca sociale*, [*Sociology and social research*], n. 66, 2001.

Barbano, F., *La sociologia in Italia. Le trasformazioni degli anni '70* [*Sociology in Italy. The tyransformations of the 1970's*] Milan, Franco Angeli, 2003.

Barnes, B., *Conoscenza scientifica e teoria sociologica* [*Scientific knowledge and sociological theory*], Naples, Liguori, 1979.

Barnes, B., *T. S. Kuhn: la dimensione sociale della scienza* [*T. S. Kuhn: the social dimension of science*] Bologna, Il Mulino, 1985.

Barnes, S. B. T., *Kuhn and Social Science*, London, Macmillan, 1982; Italian translation by Luca Ciancio, *T. S. Kuhn, la dimensione sociale della scienza*, Bologna, Il Mulino,1985. [*T. S. Kuhn, the social dimension of science*]

Barnes, S. B., *Scientific Knowledge and Sociological Theory*, London, Routledge and Kegan Paul, 1974; Italian translation *Conoscenza scientifica e teoria sociologica*, Naples, Liguori, 1979.

Barone, F., *Il neopositivismo logico* [*Logical neopositivism*], Bari, Laterza, 1986, 2 volumes.

Barone, F., Il trascendentalismo kantiano e l'epistemologia [Kantian transcendentalism and epistemology] in *Filosofia*, [*Philosophy*], 1, 1973.

Barone, F., *Introduzione a Otto Neurath, Il Circolo di Vienna e l'avvenire dell'empirismo logico* [*Introduction to Otto Neurath, The Circle of Vienna and the future of logical empiricism*], Rome, 1977.

Barone, F., La contemporanea discussione metodologica e la storiografia della scienza [The contemporary methodological discussion and the historiography of science] in *Physis*, XXII, 1980.

Barone, F., Metodologia e storiografia della scienza nel Novecento [Methodology and historiography of science in the twentieth century] in Francesco Barone, *Immagini filosofiche della scienza* [*Philosophical images of science*], Bari, Laterza, 1985.

Barone, F., Neopositivismo e filosofia analitica [Neopositivism and analytical philosophy] in *Grande Antologia Filosofica* [*Great Philosophical Anthology*] , Milan, Marzorati, 1978.

Barone, F., Vailati e l'analisi del linguaggio [Vailati and the analysis of language] in *Rivista critica di storia della filosofia*, XXXVIII, 3, 1963. [*Critical review of the history of philosophy*]

Battistelli F., *Breve storia del pensiero sociologico* [*Brief history of sociological thought*], Rome, Lithos, 2017.

Battisti, F. M., *Sociologia matematica* [*Mathematical sociology*] Gangemi Editor, 1982.

Bausola, A., (edited by), *Questioni di storiografia filosofica* [*Questions of philosophical historiography*] Brescia, Ed. La Scuola, 1978.

Bausola, A., *Indagini di storia della filosofia: da Leibniz a Moore* [*Exploration of the history of philosophy: from Leipniz to Moore*] Milan, 1969.

Bellone, E., *Il mondo di carta* [*The paper world*], Milan, Mondadori, 1976.

Bellone, E., *Il sogno di Galileo. Oggetti e immagini della ragione* [*Galileo's dream: objects and images of reason*] Bologna, Il Mulino, 1980.

Bellone, E., *La scienza negata. Il caso italiano* [*Science denied. The Italian case*] Turin, Codice edizioni, 2005.

Bellone, E., La storia della scienza in Ludovico Geymonat, Storia del pensiero filosofico e scientifico [The history of science in Ludovico Geymonat, History of philosophical and scientific thought] vol. X, *Il Novecento* [*The twentieth century*], Milan, Garzanti, 1996.

Bellone, E., *Saggio naturalistico sulla conoscenza* [*Naturalistic essay on knowledge*] Turin, Boringhieri, 1992.

Beretta, M., *Storia materiale della scienza* [*The material history of science*], Milan, Bruno Mondadori, 2002.

Bernardini, S., *Critica della ragione filosofica* [*Criticism of philosophical reason*] Soveria Mannelli, Rubbettino, 2006.

Bernardini, S., Il secondo Methodenstreit [The second Methodenstreit] in AA.VV, *Sociologia. Materiali e fondamenti sulle questioni di metodo* [*Sociology. Materials and fundaments of methodological issues*], V, edited by Martino Ancona, Naples, Liguori, 1983.

Bernardini, S., *Logica della conoscenza scientifica* [*The logic of scientific knowledge*] Naples, Liguori,1980.

Berti, E., *L'unità del sapere in Aristotele* [*Unity of knowledge in Aristotle*] Padua, Cedam, 1965.

Bettin G. L., Turi P., edited by *La sociologia di Luciano Cavalli* [*Lucian Cavalli's sociology*] Florence, Florence University Press, 2007 (2008).

Bianco, F., Comprensione, spiegazione, interpretazione [Understanding, explanation, interpretation] in Pietro Rossi (edited by), *Max Weber e l'analisi del mondo moderno* [*Max Weber and the analysis of the modern world*] Turin, Einaudi, 1981.

Bianco, F., Di Bernardo, G., (edited by), *Episteme e azione* [*Episteme and action*] Milan, Franco Angeli, 1991.

Bianco, F., *Introduzione a Dilthey* [*Introduction to Dilthey*] Bari, Laterza, 1986.

Bianco, F., *Le basi teoriche dell'opera di Max Weber* [*The theoretical bases of Max Weber's work*] Rome-Bari, Laterza, 1997.

Bianco, M., *La costruzione della scienza* [*The construction of science*] Milan, Franco Angeli, 1984.

Bisogno, P., Forti, A., *Ricerca scientifica e bisogni umani* [*Scientific research and human needs*] Milan, Franco Angeli, 1979.

Bisogno, P., *Introduzione alla politica della scienza* [*Introduction to the politics of science*] Milan, Franco Angeli, 1977 (1979).

Bisogno, P., *Memoria del futuro. Scritti sulle immagini della scienza* [*Memory of the future. Writings on the images of science*] Milan, Franco Angeli, 1991.

Bisogno, P., Tonini, V., *Razionalità e storicità* [*Rationality and historicity*], Rome, La Goliardica Editrice Universitaria di Rome, 1983.

Bisogno, P., Verso l'unificazione del sapere [Towards the unification of knowledge] in *Prometheus*, 1987, n. 3, pp. 135-138.

Bloor, D. *Knowledge and Social Imagery*, Routledge and Kegan Paul, London, 1976, second edition 1991; Italian translation by Giovanna Bettini, Introduction to

Alessandro Dal Lago, La dimensione sociale della conoscenza [*The social dimension of knowledge*] Milan, Raffaello Cortina,1994.

Bloor, D., *La dimensione sociale della conoscenza* [*The social dimension of knowledge*] Milano, Cortina, 1991.

Boniolo, G., *Filosofia della scienza* [*Philosophy of science*] in AGG. TI. a [update of] Ludovico Geymonat, Storia della filosofia e del pensiero scientifico [History of philosophy and scientific thought] Milan, Garzanti, 1996.

Boniolo, G., Dorato, M. *La legge di natura. Analisi storico-critica di un concetto* [*The law of nature. Historical-critical analysis of a concept*], Milan, McGraw-Hill, 2001.

Boniolo, G., *Metodo e rappresentazioni del mondo* [*Method and representations of the world*], Milan, Bruno Mondadori, 1999.

Bonolis, M., Analiticità e realismo nella teoria sociologica contemporanea [Analyticity and realism in contemporary sociological theory], in *Sociologia e ricerca sociale*, [*Sociology and social research*], XXIX, n. 85, 2008, pp. 22-52.

Bonolis, M., Conseguenzialismo metodologico e teoria dell'azione [Methodological conseguensialism and action theory], in *Sociologia e ricerca sociale*, [*Sociology and social research*], XXXI, n. 92, 2010, pp. 5-33.

Bonolis, M., Galileo e Popper (Ontologia, ontologismo, filosofia dei valori) [Galileo and Popper (Ontology, ontology, philosophy of values)], in *Sociologia e ricerca sociale*, [*Sociology and social research*], n. 67, 2002.

Bonolis, M., Il problema metafisico della storicità [The metaphysical problem of historicity], in *Sociologia e ricerca sociale*, [*Sociology and social research*], n.78, 2005.

Bonolis, M., Per Franco Leonardi [For Franco Leonardi], in *Sociologia e ricerca sociale*, [*Sociology and social research*], vol. XXXIII, 99, 2012, pp. 5-8.

Bonolis, M., Per una critica della "razionalità assiologica," [For a critique of "axiological rationality"], in *Sociologia e ricerca sociale*, [*Sociology and social research*], vol. XXXIV, 103, 2014, pp. 5-35.

Bonolis, M., *Storicità e storia della sociologia* [*Historicity and history of sociology*], Milan, Franco Angeli, 2007.

Borgna, P., Immagini della scienza per una teoria delle scienze, in Una storia delle scienze per i nuovi saperi. Discussioni e ricerche [Images of science for a theory of science, in A history of science for new knowledge. Discussions and research] *Minutes of the Conference: Research Horizons and epistemological boundaries, Rome 8 April 2010*, edited by Guglielmo Rinzivillo, Rome, Casa Editrice La Sapienza, 2011.

Borgna, P., *Le immagini pubbliche della scienza. Gli italiani e la ricerca scientifica e tecnologica* [*Public images of science. Italians and scientific and technological research*], Preface by Luciano Gallino, Turin, Edizioni di Comunità, 2001.

Borutti, S., *Filosofia delle scienze umane: le categorie dell'antropologia e della sociologia* [*Philosophy of the human sciences: the categories of anthropology and sociology*], Milan, Bruno Mondadori, 1999.

Borutti, S., *Teoria e interpretazione: per un'epistemologia delle scienze umane* [*Theory and interpretation: for an epistemology of the human sciences*], Milan, Guerini e Associati, 1991.

Boudon, R., *Le juste et le vrai. Études sur l'objectivité des valeurs et de la connaissance* [*The right and the true. Studies on the objectivity of values and knowledge*], Paris, Fayard, 1995; Italian translation *Il vero e il giusto. Saggio sull'obbiettività dei valori e della conoscenza*. Bologna, Il Mulino, 1997.

Bovone, L., *Tra riflessività e ascolto. L'attualità della sociologia* [*Between reflexivity and listening. The topicality of sociology*], Rome, Editor Armando, 2010.

Braga, G., Livelli di concettualizzazione in sociologia [Levels of conceptualisation in sociology] in *Studi di Sociologia*, I, 4, 1963.

Bridgman, P. W., The Way Things Are, Harvard University Press, 1959; Italian translation edited by Bruno Cermignani, *La critica operazionale della scienza* [*Operational criticism of science*], Turin, Boringhieri, 1989.

Bronowski, J., Progresso scientifico e contesto culturale [Scientific progress and cultural context], in *Civiltà delle Macchine* [*The civilisation of machines*], n. 3, 1964.

Bruschi, A., *Conoscenza e metodo: Introduzione alla metodologia delle scienze sociali* [*Knowledge and method: Introduction to the methodology of social sciences*], Milan, Bruno Mondadori, 1990.

Bruschi, A., *La competenza metodologica. Logiche e strategie nella ricerca sociale* [*Methodological competence. Logics and strategies in social research*], Rome, La Nuova Italia Scientifica, 1996.

Bruschi, A., Logica e metodologia [Logic and methodology], in *Sociologia e ricerca sociale*, [*Sociology and social research*], n. 35, 1991.

Bucca, G., La sociologia della conoscenza tra 'razionalismo critico' e 'teoria critica della società' [The sociology of knowledge between 'critical rationalism' and 'critical theory of society'], in *Sapienza*, 28, 1975.

Bucchi, M. *Science in Society. An Introduction to social studies of science*, London & New York, Routledge, 2004.

Bucchi, M. *Scienza e società* [*Science and society*], Milan, Raffaello Cortina Editore, 2010.

Burgalassi, M. M., *Itinerari di una scienza* [*Itineraries of a science*] , Preface by Filippo Barbano, Milan, Franco Angeli, 1996.

Buzzoni, M., *Filosofia della scienza* [*Philosophy of science*], Brescia, La Scuola, 2008.

Callon, M., *La Science et ses réseaux. Genèse et circulation des faits scientifiques* [*Science and its networks. Genesis and circulation of scientific facts*], Paris, La Découverte; Unesco; Strasbourg, Conseil de l'Europe, Textes à l'appui. Anthropologie des sciences et des techniques [Supporting texts. Anthropology of science and technology], 1989; (texts by Michel Callon, John Law, Andrew Pickering, Peter Groenvegen and Arie Rip).

Campelli, E., *Da un luogo comune. Elementi di metodologia nelle scienze sociali* [*From a common place. Elements of methodology in the social sciences*], Rome, Carocci, 1999 (2009).

Campelli, E., Elogio della vaghezza. Riflessione quasi epistemologica sul tempo presente [In praise of vagueness. An almost epistemological reflection on the present time], in *Sociologia e ricerca sociale*, [*Sociology and social research*], 89, 2009, pp. 21-58.

Campelli, E., *Il metodo e il suo contrario. Sul recupero della problematica del metodo in Sociologia* [*Method and its opposite. On the recovery of the problematic of method in Sociology*], Milan, Franco Angeli, 1991.

Campelli, E., *Nullius in verba. Il metodo nella rivoluzione scientifica* [*Nullius in verba Method in the scientific revolution*], Milan, Franco Angeli, 2016.

Campelli, E., Per una spiegazione di medio raggio. Ancora sul problema della spiegazione in sociologia [For a medium-range explanation. On the issue of explanation in sociology again] in *Sociologia e ricerca sociale*, [*Sociology and social research*], n. 75, 2004.

Cannavo', L., (edited by), *Professione scienziato* [*Profession, scientist*], Milan, Franco Angeli, 1989.

Cannavo', L., (edited by), *Studi sociali della scienza, aspetti e problemi* [*Social studies of science, aspects and problems*], Rome, La Goliardica, 1989.

Cannavo', L., Alle origini dell'analisi operazionale in sociologia [At the origins of operational analysis in sociology], in *Sociologia e ricerca sociale*, [*Sociology and social research*], X, April 1983.

Cannavo', L., Definizioni operative, concettualizzazione e logica della ricerca [Operative definitions, conceptualisation and the logic of research], in *Sociologia e ricerca sociale*, [*Sociology and social research*], XIII, 1984; also in A. Marradi (edited by), *Metascienza. Riflessioni sulla conoscenza scientifica* [*Metascience. Reflections on scientific knowledge*], Rome, EuRome - Editrice Universitaria di Roma - La Goliardica, 1991.

Cannavo', L., Dentro la ricerca: ethos ed etica della scienza a confronto [Within research: the ethos and the ethics of science compared], in *Sociologia e ricerca sociale*, [*Sociology and social research*], n. 32, 1990.

Cannavo', L., Il ciclo della ricerca sociale e la metodologia critica [The cycle of social research and critical methodology], in *Ricerca sociale*, edited by Leonardo Cannavò and Luigi Frudà, Rome, Carocci, 2007.

Cannavo', L., Il processo di operazionalizzazione nella ricerca sociale [The process of operationalisation in social research], in *Ricerca sociale*, edited by Leonardo Cannavò and Luigi Frudà, Rome, Carocci, 2007.

Cannavo', L., Qualità e quantità: tra metodologia sociologica e sociologia della scienza [Quality and quantity: between sociological methodology and sociology of science], in *Sociologia e ricerca sociale*, [*Sociology and social research*], n. 28, 1989.

Cannavo', L., *Sociologia della conoscenza scientifica. Dal paradigma organizzativo ai programmi cognitivi e comunicativi* [*The sociology of scientific knowledge. From the organisational paradigm to cognitive and communicative programmes*], Rome, La Goliardica Editrice, 1984.

Cannavo', L., Sociological Models of Scientific Knowledge in *International Sociology*, XII, n. 4, 1997.

Capecchi, V,. Causalità e correlazione nella problematica sociologica [Causality and correlation in sociological issues], in *Studi di Sociologia*, II, 1964.

Capecchi, V., Appunti per una riflessione sulla metodologia della ricerca in sociologia [Notes for a reflection on the methodology of research in sociology], in *Quaderni di Sociologia*, [*Sociology notebooks*], n. 4-5, 1985.

Capecchi, V., *Per una storia della ricerca azione in Italia* [*For a history of action research in Italy*], in "Inchiesta," Year XXXVI, n. 151, January-March, 2006.

Capecchi, V., Struttura e tecniche della ricerca [Research Structure and techniques], in P. Rossi (edited by), *Ricerca sociologica e ruolo del sociologo* [*Sociological research and the role of the sociologist*], Centre of Methodological Studies, Bologna, Il Mulino, 1972.

Cappelletti, V., *La scienza tra storia e società* [*Science between history and society*], Rome, Edizioni Studium, 1978.

Cappelletti, V., *La struttura del conoscere secondo Ludwig Wittgenstein* [*The structure of knowing according to Ludwig Wittgenstein*], in "La Nuova Critica," VII-VIII, 1958-1959.

Cappelletti, V., *Le scienze umane nella cultura e nelle società odierne* [*The human sciences in today's culture and society*], Rome, Studium, 1993.

Cappelletti, V.,*'Epistemologia'* in *Enciclopedia del Novecento* [*'Epistemology' in the Encyclopedia of the twentieth century*], Istituto della Enciclopedia Italiana fondata da Giovanni Treccani, Rome, 1977, II.

Cardano M., Miceli, R., *Il linguaggio delle variabili*, Turin, Rosemberg & Sellier, 1991.

Carnap, R., Die physikalische Sprache als Universalsprache der Wissenschaft [The language of physics as the universal language of science] in *Erkenntnis*, II, 1931; Italian translation Il linguaggio della fisica come linguaggio universale della scienza in R. Carnap, *La filosofia della scienza* [The language of physics as the universal language of science in R. Carnap, *The philosophy of science*], Brescia, 1964.

Casini, P., *Alle origini del Novecento. "Leonardo" 1903-1907* [*At the origins of the twentieth century. "Leonardo" 1903-1907*], Bologna, Il Mulino, 2002.

Casini, P., Scienza e filosofia in Italia nel primo Novecento [Science and philosophy in Italy in the early twentieth century], in *Terzoprogramma*, 4, 1966.

Casini, P., Teoria e storia delle rivoluzioni scientifiche secondo Thomas Kuhn [The theory and history of scientific revolutions according to Thomas Kuhn], in *Rivista di Filosofia*, LXI, 1970.

Cassano, F., Il gioco della scienza [The game of science], in *Rassegna Italiana di Sociologia*, [*Italian Review of Sociology*], n. 1, 1989.

Cassano, F., *La certezza infondata. Previsione ed eventi nelle scienze sociali* [*The unfounded certainty. Forecasting and events in the social sciences*], Bari, Dedalo, 1983.

Castellana, M. G., Bachelard: bilancio critico di un'epistemologia [Bachelard: a critical summing up of an epistemology], in *Il Protagora*, Lecce, 5, 1984.

Cavallaro, R., *Paradigmi e sociologia. Un dibattito insensato?* [*Paradigms and sociology. A meaningless debate?*], Catania, ISVI, 1984.

Cavalli A., Moscati, R., Per capire la società che cambia: scritti in ricordo di Antonio De Lillo [To understand tchanging society: written in memory of Antonio De Lillo], in *Quaderni di Sociologia*, [*Sociology notebooks*], Vol. LVII, 2013, 62.

Cavalli, A., La funzione dei tipi ideali e il rapporto fra conoscenza storica e sociologica [The function of ideal types and the relationship between historical and sociological knowledge], in Pietro Rossi (edited by), *Max Weber e l'analisi del mondo moderno* [*Max Weber and the analysis of the modern world*], Turin, Einaudi, 1981.

Cavalli, A., La sociologia e le altre scienze sociali: prospettive di integrazione interdisciplinare, in *Ricerca sociologica e ruolo del sociologo* [Sociology and other social sciences: prospects of interdisciplinary integration, in *Sociological research and the role of the sociologist*], edited by Pietro Rossi, Bologna, Il Mulino, 1972.

Centrone, M., *Che cos'è la scienza. Introduzione all'epistemologia contemporanea* [*What science is. An itroduction to contemporary epistemology*] Bari, Dedalo, 1993.

Ceri, P., Borgna P., *La tecnologia per il XXI secolo* [*Technology for the 21st century*], (edited by), Turin, Einaudi, 1998.

Ceri, P., I quattro volti dell'anti-sociologia [The four faces of anti-sociology], in *Quaderni di Sociologia*, [*Sociology notebooks*], n. 4-5, 1985.

Ceruti, M., *Il vincolo e la possibilità* [*Constraint and possibility*], Milan, Raffaello Cortina Editor, 2009.

Cesareo V., (edited by), *L'icona tecnologica* [*The technological icon*], Turin, Edizioni della Fondazione Agnelli, 1989.

Cesareo V., *Sociologia, teorie e problemi* [*Sociology, theories and issues*], Milan, Vita e Pensiero, 1993 (2006).

Chiodi, P., *Esistenzialismo e fenomenologia* [*Existentialism and phenomenology*], Milan, Comunità, 1963.

Chiodi, P., Grmeck, M. D., Somenzi, V., (edited by), *La scoperta scientifica. Aspetti logici, psicologici e sociali* [*Scientific discovery. Logical, psychological and social aspects*], Rome, Armando, 1984.

Chiodi, P., Moritz Schlick, in *Rivista di Filosofia*, XLV, 1954.

Chiodi, P., Sulla possibilità della sociologia come scienza [On the possibility of sociology as a science], in AA.VV, *Filosofia e Sociologia* [*Philosophy and sociology*], Bologna, Il Mulino, 1954.

Cini, M., *Trentatré variazioni su un tema. Soggetti dentro e fuori la scienza* [*Thirty-three variations on a theme. Subjects inside and outside of science*], Rome, Editori Riuniti, 1990.

Cini, M., *Un paradiso perduto* [*A paradise lost*], Milan, Feltrinelli, 1994.

Cipolla, C., Cipriani, R., Colasanto, M., D'alessandro, L., *Achille Ardigò e la sociologia, Introduction* [*Achille Ardigò and sociology, Introduction*], by Vincenzo Cesareo, Milan, Franco Angeli, 2010.

Cipolla, C., *Dopo Popper. Saggio sulla grandezza e limiti dell'epistemologia popperiana e sul suo superamento* [*After Popper. Essay on the greatness and limits of Popperian epistemology and its outstripping*], Rome, Borla, 1990.

Cipolla, C., *Epistemologia della tolleranza* [*The epistemology of tolerance*], Milan, Franco Angeli, 1997, v. volumes.

Cipolla, C., *Teoria della metodologia sociologica* [*Theory of sociological methodology*], Milan, Franco Angeli, 1988 (1995).

Cipriani R., (edited by), *Nuovo manuale di sociologia* [*New manual of sociology*], Santarcangelo di Romagna (RN), Maggioli Editor, 2016.

Cipriani, R., *L'analisi qualitativa. Teorie, metodi, applicazioni* [*Qualitative analysis. Theories, methods, applications*], Rome, Armando, 2008.

Collins, H. M., The Sociology of Scientific Knowledge: Studies of Contemporary Science, in *Annual Review of Sociology*-Vol. 9, 1983, pp. 265-285.

Coniglione, F., *La scienza impossibile. Dal Popperismo alla critica del razionalismo* [*The impossible science. From Popperism to the critique of rationalism*], Bologna, Il Mulino, 1980.

Coniglione, F., *Nello specchio della scienza. Ricerca scientifica e politiche nella società della conoscenza* [*In the looking-glass of science. Scientific research and policies in the society of knowledge*], Milan, Bruno Mondadori, 2009.

Coniglione, F., *Popper addio, Dalla crisi dell'epistemologia alla fine del Logos occidentale* [*Popper adieu, From the crisis of epistemology to the end of the Western Logos*] Acireale-Rome, Bonanno Editor, 2008.

Corbetta, P., *Metodologia e tecniche della ricerca sociale*, [*The Methodology and techniques of social research*], Bologna, Il Mulino, 1999.

Corradi, C., Notes on qualitative sociology in Italy in *Qualitative Sociology*, Springer, vol. 11, Issue I, March 1988, pp. 77-87.

Cotesta, V., *Linguaggio e verità. Studi di sociologia della conoscenza* [*Language and truth. Studies in the sociology of knowledge*], Armando, Rome, 1999.

Cotesta, V., *Realismo versus costruttivismo. Un vecchio dibattito dalle nuove implicazioni* in *Costruzionismo e scienze sociali* [*Realism versus constructivism. An old debate with new implications in Constructionism and social sciences*], edited by Ambrogio Santambrogio, Perugia, Morlacchi, 2010.

Crespi, F., *La prospettiva sociologica nel sapere contemporaneo* [*The sociological perspective in contemporary knowledge*], Milan, Giuffré, 1961.

Crespi, F., *La sociologia come scienza e teoria dell'azione sociale* [*Sociology as a science and theory of social action*], Milan, Giuffré, 1964.

Crespi, F., *Le vie della sociologia* [*The ways of sociology*], Bologna, Il Mulino, 1993.

D'addio, M., *Il caso Galilei. Processo scienza verità* [*The Galilei case. Trial science truth*], Rome, Studium Edition, 1993 – reprint 1997.

Dal Lago, A., (with Giglioli, P. P.), *Etnometodologia* [*Ethno-methodology*], Bologna, Il Mulino, 1983.

Dal Lago, A., La scienza come forma di vita? [Science as a form of life?], in *Rassegna Italiana di Sociologia*, XXIII, n. 2, April-June 1982.

Dal Lago, A., *Oltre il metodo. Interpretazione e scienze sociali*, [*Beyond method. Interpretation and social sciences*], Milan, Unicopli, 1989.

Dal Pra, M. (edited by), *Il pensiero di John Dewey* [*The thinking of John Dewey*], Milan, Fratelli Bocca Editori, 1952.

Dal Pra, M., L'empirismo logico dal 1930 ad oggi [Logical empiricism from 1930 to today], in *Rivista critica di storia della filosofia*, VII, 1952.

Dal Pra, M., Minazzi, F. *Ragione e storia* [*Reason and history*], Milan, Rusconi, 1992.

Dal Pra, M., *Studi sul pragmatismo italiano* [*Studies of Italian pragmatism*], Naples, 1984.

Dalla Chiara, M. L., (edited by), *Italian Studies in the Philosophy of Science*, Dordrecht, 1981.

Davidson, D., On the Very Idea of the Conceptual Scheme, in *Proceedings of the American Philosophical Association*, 47, 1974; Italian translation Sull'idea stessa di schema concettuale, in R. Egidi, (edited by), *La svolta relativistica nell' epistemologia contemporanea*, Milan, Franco Angeli, 1988. Also reported in *Verità e interpretazione*, Bologna, Il Mulino, 1994.

De Giovanni, B., *G.* Semerari, la lotta per la scienza [G. Semerari, the struggle for science], in *Filosofia* (Review), 17, 2, 1966.

De Lillo, A., (edited by), *Il mondo della ricerca qualitativa* [*The world of qualitative research*], Turin, UTET, 2010.

De Masi D., (edited by), La società post-industriale [Post-industrial society], in *L'avvento post-industriale* [*The post-industrial advent*], Milan, Franco Angeli, 1985.

De Nardis, P., Individualismo metodologico, semantica e logica dell'indagine [Methodological individualism, semantics and logic of investigation], in *Sociologia e ricerca sociale*, [*Sociology and social research*], 43, 1994.

De Nardis, P., *L'equivoco sistema* [*The equivocal system*], Milan, Franco Angeli, 1988.

De Nardis, P., *Sociologia del limite* [*The sociology of the limit*], Rome, Meltemi, 1999.

De Nardis, P., *Teoria sociale e analisi socio-istituzionale* [*Social theory and socio-institutional analysis],* Rome, Carucci, 1978.

De Nardis, P., Una nuova rappresentazione per la sociologia scientifica in AA. VV, *Le nuove frontiere della sociologia* [*A new representation for scientific sociology* in AA.VV, *The new frontiers of sociology*], edited by Paolo De Nardis, Rome, Carocci, 1998.

Dei F., Simonicca, A., (edited by), *Ragione e forme di vita* [*Reason and forms of life*], Milan, Franco Angeli, 1990.

Dewey, J. *Logic, the Theory of Inquiry*, New York, Hanry Holt and Co., 1938. Italian translation by Aldo Visalberghi, *Logica, teoria dell'indagine*, Turin, Einaudi, 1970.

Di Bernardo, G., Bianco, F., (edited by), *Episteme e azione* [*Episteme and action*] Milan, Franco Angeli, 1991.

Di Bernardo, G., *L'indagine del mondo sociale* [*Invetigation of the social world*], Milan, Franco Angeli, 1979.

Di Bernardo, G., La fondazione epistemologica delle scienze sociali in F. Minazzi edited by *Filosofia, scienza, bioetica. Studi internazionali in onore di Evandro Agazzi* [The epistemological foundation of the social sciences in F. Minazzi edited by *Philosophy, science, bioethics. International studies in honour of Evandro Agazzi*], Rome, Presidenza del Consiglio dei Ministri, 2007, pp. 155-170.

Di Bernardo, G., *Le regole dell'azione sociale* [*The rules of social action*], Turin, Il Saggiatore, 1983.

Di Bernardo, G., *Linguaggio, scienza, sociologia* [*Language, science, sociology*], in "Sociologia," III, 3, 1969.

Di Nuoscio, E., *Epistemologia del dialogo* [*The epistemology of dialogue*], Rome, Carocci, 2011.

Di Nuoscio, E., *Il cammino della scienza* [*The pathweay of science*], Milan, Mondadori, 2008.

Di Nuoscio, E., *Il mestiere dello scienziato sociale. Un'introduzione all'epistemologia delle scienze* [*The social scientist's trade. An introduction to the epistemology of science*] Naples, Liguori, 2006.

Digilio M., Un nuovo Methodenstreit: Popper-Albert contro Adorno-Habermas [A new Methodenstreit: Popper-Albert against Adorno-Habermas], in *La Critica Sociologica*, [*Sociological criticism*], n. 8, Winter 1968-1969.

Dolby, R. G., A Sociology of knowledge in natural science, *Science studies*, 1971, pp. 3-21.

Donati, P., Leggere la società. Problemi teorici nella sociologia italiana [Reading society. Theoretical problems in Italian sociology], in *Studi di Sociologia*, April- September, 1985.

Donati, P., *Teoria relazionale della società: i concetti base* [*The relational theory of society: the basic concepts*] Milan, Franco Angeli, 2009.

Egidi, R. Relativismo epistemologico e linguaggi della scienza [Epistemological relativism and languages of science], in *Paradigmi*, 2, 1984.

Egidi, R., Anti-individualismo epistemologico e scienze sociali [Epistemological anti-individualism and the social sciences], in *Sociologia e ricerca sociale*, [*Sociology and social research*], 43, 1994.

Egidi, R., *Il linguaggio delle teorie scientifiche* [*The language of scientific theories*], Naples, Guida, 1979.

Egidi, R., *La svolta relativistica nell'epistemologia contemporanea. Essays by Davidson, Feyerabend, Hacking, Hübner, Krüger, Laudan, Putnam, Radnitzky, Rorty, Stegmüller* [*The relativistic turning point in contemporary epistemology. Essays by Davidson, Feyerabend, Hacking, Hübner, Krüger, Laudan, Putnam, Radnitzky, Rorty, Stegmüller*], Milan, Franco Angeli, 1992.

Egidi, R., *Ontologia e conoscenza matematica* [*Ontology and mathematical knowledge*], Florence, Sansoni, 1963.

Egidi, R., *Un saggio su G. Frege* [*An essay on G. Frege*], Florence, Sansoni, 1963.

Enriques F., De Santillana, G., *Compendio di storia del pensiero scientifico fino ai tempi moderni* [*Compendium of the history of scientific thought up to modern times*], Bologna, Zanichelli, 1937; facsimile reprint, 1982.

Enriques, F., *Il significato della storia del pensiero scientifico* [*The meaning of the history of scientific thought*], Bologna, Zanichelli, 1936.

Enriques, F., *Problemi della scienza* [*Problems of science*], Bologna, Zanichelli, 1906, 1909[2]; rist. an., ivi 1985.

Fasanella, A., *Anarchismo metodologico e scienze sociali* [*Methodological anarchism and social sciences*], Preface by Gianni Statera, Milan, Franco Angeli, 1987.

Fasanella, A., *Concettualizzazione e spiegazione sociologica. Il modello nomologico-inferenziale e la sua applicabilità alle scienze sociali*, Milan, Franco Angeli, 1993. [*Conceptualization and sociological explanation. The nomological-inferential model and its applicability to the social sciences*]

Fasanella, A., La generalizzazione in sociologia: una rilevante questione metodologica [Generalization in sociology: an important methodological issue], in P. De Nardis (edited by), *Le nuove frontiere della sociologia* [*The new frontiers of sociology*], Rome, Carocci, 1998.

Federici, M. C., *Dove fondano le libertà dell'uomo: l'approccio sociologico di Vilfredo Pareto con alcune riflessioni su L. Von Mises* [*Where human liberties are founded: Vilfredo Pareto's sociological approach, with some reflections on L. Von Mises*], Roma, Borla, 1991 (2007).

Ferarrotti F., *Il* sistema concettuale in Max Weber [The conceptual system in Max Weber], in *Quaderni di Sociologia*, [*Sociology notebooks*], vol. VIII, October-December, 1964.

Ferrarotti, F. Alcuni termini attuali della ragione sociologica [Some present-day terms of sociological reason], in *Rassegna Italiana di Sociologia*, [*Italian Review of Sociology*], II, 1961.

Ferrarotti, F. *Introduzione a Gianni Statera, La conoscenza sociologica* [*Introduction to Gianni Statera, Sociological knowledge*], Rome, Carucci, 1968.

Ferrarotti, F., *I miei anni con Adriano Olivetti a Ivrea e dintorni, da New York a Matsuyama* [*My years with Adriano Olivetti in Ivrea and thereabouts, from New York to Matsuyama*], Solfanelli, 2016.

Ferrarotti, F., Introduzione alla sociologia come scienza [Introduction to sociology as a science], I e II Part, in *Quaderni di Sociologia*, [*Sociology notebooks*], n. 1, Summer-Autumn, 1951; III Parte, in *Quaderni di Sociologia*, [*Sociology notebooks*], n. 5, Summer, 1952.

Ferrarotti, F., *La concreta utopia di Adriano Olivetti* [*Adriano Olivetti's concrete utopia*], Edizioni Dehoniane Bologna, 2013.

Ferrarotti, F., *La sociologia alla riscoperta della qualità* [*Sociology and the rediscovery of quality*], Bari, Laterza, 1989.

Ferrarotti, F., *La sociologia come partecipazione e altri saggi* [*Sociology as participation and other essays*], Turin, Taylor, 1961.

Ferrarotti, F., *Metodologia sociologica e ricerca storica*, in *La morte della filosofia dopo Hegel* [*Sociological methodology and historical research, in The death of philosophy after Hegel*] published by l'Istituto di filosofia dell'Università di Rome, Asti, Arethusa, 1958.

Ferrarotti, F., Primi appunti sul concetto di scienza sociale [Preliminary notes on the concept of social sciences] in *"Atti" del Primo Congresso dell'Associazione Italiana di Scienze Sociali* [*Minutes of the first congress of the Italian Association of Social Sciences*] Bologna, Il Mulino, 1958.

Ferrarotti, F., Sociologia e filosofia, oggi [Sociology and philosophy, today], in *Paradigmi*, n. 13, Jan-April 1987.

Ferrarotti, F., *Trattato di Sociologia* [*Tractactus of Sociology*] Turin, UTET, 1968, 2001.

Ferrarotti, F.. Nota sull'attuale evoluzione delle scienze e dei loro rapporti [Note on the current evolution of the sciences and their relationships], in *La Critica Sociologica*, [Sociological criticism], n. 50, Summer 1979.

Feyerabend, P. K., *Against Method. Outline of an Anarchist Theory of Knowledge*, London, New Left Books, 1975; Italian translation of Libero Sosio, *Contro il metodo. Abbozzo di una teoria anarchica della conoscenza*, Preface by Giulio Giorello, Milan, Feltrinelli, 1979. See also edition in the Series 'Interventi', *Contro il Metodo*, Milan, Lampugnani Nigri, 1973.

Feyerabend, P. K., *Problems of Empirism* I, II, in *Beyond the Edge of Certainty* (1965) e in *The Nature and Function of Scientific Theory* (1969), edited by R. Colodny, Englewood Cliffs N. J., Prentice-Hall; Italian translation *I problemi dell'empirismo*, Milan, Lampugnani Nigri, 1971.

Feyerabend, P. K., *Science in a Free Society*, London, New Left Books, 1978; Italian translation of the revised German edition 1980, by Libero Sosio, *La scienza in una società libera*, Milan, Feltrinelli, 1981.

Filiasi Carcano, P., *Antimetafisica e sperimentalismo* [*Antimetaphysics and experimentalism*], Rome, 1941.

Filiasi Carcano, P., *Introduzione alla lettura delle "Ricerche filosofiche" di Wittgenstein* [*Introduction to the reading of Wittgenstein's "Philosophical Research"*] Rome Bulzoni, 1976.

Filiasi Carcano, P., *La metodologia nel rinnovarsi del pensiero contemporaneo* [*Methodology in the renewal of contemporary thought*] Naples, 1957, VII.

Filiasi Carcano, P., *La sociologia della conoscenza in Fenomenologia e sociologia* [*The sociology of knowledge in Phenomenology and sociology*] Archivio di Filosofia, Padua, Giuffré, 1951.

Fistetti, F., *Neurath contro Popper.Otto Neurath riscoperto* [*Neurath against Popper.Otto Neurath rediscovered*], with Preface by R. Haller, Bari, Dedalo, 1985.

Fistetti, F., Semerari, F., *La malinconia di Hume: Sul pensiero di Giuseppe Semerari* [*The melancholy of Hume: On the thinking of Giuseppe Semerari*] Milan, Guerini Studio, 2007.

Formaggio, D., Premise, in *Fenomenologia e scienze dell'uomo* [*Phenomenology and the sciences of man*], Quaderni del seminario di filosofia delle scienze dell'uomo, number 1, Padua, 1982.

Forti, U., *Il metodo storico nell'insegnamento della fisica* [*The historical method in the teaching of physics*], Rome, 1928.

Forti, U., *Storia della scienza* [*History of science*], Rome, dall'Oglio, 1969, various volumes.

Fruda', L., Alle origini dell'empirismo logico e del razionalismo critico: il problema del metodo scientifico [The origins of logical empiricism and critical rationalism: the problem of the scientific method] in *Studi di Sociologia*, XXVI, n. 2, 1988.

Fruda', L., *Metodo, logica e scienze sociali* [*Method, logic and social sciences*], Rome, EuRome - Editrice Universitaria di Roma - La Goliardica, 1993.

Gadamer, H. G., Die phamomenologiche Bewegung [The phenomenological movement] in *Philosophische Rundschau*, 11, 1963; Italian translation *Il movimento fenomenologico*, Rome-Bari, Laterza, 1994.

Gagliasso, E., Della Rocca, M. Memoli, R., *Per una scienza critica. Marcello Cini e il presente: filosofia, storia e politiche della ricerca* [*For a critical science. Marcello Cini and the present: philosophy, history and research policies*], ETS, 2015.

Gagliasso, E., Memoli, R., Pontecorvo, M. E., *Scienza e scienziati: colloqui interdisciplinari* [*Science and scientists: interdisciplinary interchanges*], Milan, Franco Angeli, 2011.

Galavotti, M. C., Introduzione a C.G. Hempel, *Aspetti della spiegazione scientifica* [*Introduction to C. G. Hempel, Aspects of the scientific explanation*], (1965), Milan, Il Saggiatore, 1986.

Galavotti, M. C., *Spiegazioni probabilistiche: un dibattito aperto* [*Probabilistic explanations: an open debate*], Bologna, Clueb, 1984.

Gallino, L. (edited by), *Percorsi della sociologia italiana* [*Pathways of Italian sociology*], Milan, Franco Angeli, 1992.

Gallino, L., 'Sociologia della scienza' in *Enciclopedia Italiana* ['Sociology of science' in *the Italian Encyclopedia*], V, Appendix, Istituto dell' Enciclopedia Italiana, Rome, 1995.

Gallino, L., *Crisi della sociologia, ricerca sociologica, ruolo del sociologo* in *Ricerca sociologica e ruolo del sociologo* [*The crisis of sociology, sociological research, the role of the sociologist in Sociological Research the Role of the Sociologist*] edited by Pietro Rossi, Bologna, Il Mulino, 1972.

Gallino, L., *Dizionario di Sociologia* [*Dictionary of Sociology*], Turin, Utet, 2014.

Gallino, L., entry in '*Scienza, Sociologia della'* in *Dizionario di Sociologia* [*Science, Sociology in the Dictionary of Sociology*], Milan, Tea, 1993.

Gallino, L., *L'incerta alleanza. Modelli di relazione tra scienze umane e scienze naturali*, [*The uncertain alliance. Models of relationship between the huiamn and natural sciences*], Turin, Einaudi, 1992.

Gallino, L., *La sociologia, concetti fondamentali* [*Sociology, fundamental concepts*], Turin, UTET, 1989.

Gallino, L., *La sociologia. Indirizzi, specializzazioni, rapporti con altre scienze* [*Sociology. Directions, specializations, relationships with other sciences*] Turin, Einaudi, 1989.

Gallino, L., Metodologia neopositivistica e teoria sociologica [Neopositivistic methodology and sociological theory], in *Quaderni di Sociologia*, Turin, Taylor Publisher, n.1, Jan-March, 1973; also reported in AA.VV, *Sociologia. Materiali sui fondamenti e le questioni di metodo*, edited by Martino Ancona, Naples, Liguori, 1983.

Gallino, L., *Tecnologia e democrazia* [*Technology and democracy*], Turin, Einaudi, 2007.

Gallo M., *La fatica della ragione. Il problema dell'oggettività nel pensiero di K.R. Popper* [*The fatigue of reason. The problem of objectivity in the thinking of K.R. Popper*], Naples, Loffredo Publisher, 1990.

Gargani, A., (edited by), *La crisi della ragione* [*The crisis of reason*] Turin, Einaudi, 1979.

Gargani, A., *Il sapere senza fondamenti* [*Knowledge without bases*], Turin, Einaudi, 1975 (Mimesis, Milan, 2009).

Gargani, A., *Introduzione a Wittgenstein* [*Introdution to Wittgenstein*], Rome- Bari, Laterza, 1973.

Gargani, A., *Stili di analisi. L'unità perduta del metodo filosofico* [*Styles of analysis. The lost unity of the philosophical method*], Milan, Feltrinelli, 1993.

Gennaro G., (edited by), *Studi in Onore di Franco Leonardi* [*Studies in honour of Franco Leonardi*], Milan, Franco Angeli, 2007.

Geymonat, L., Analisi critica del recente indirizzo di logica formale del Carnap [Critical analysis of the recent formal logic of Carnap], in *Memorie dell'Accademia delle Scienze di Turin*, n. 70, 1942.

Geymonat, L., Bicchieri, C., *Il primo Schlick. Recenti critiche epistemologiche al modello neopositivistico* [*The early Schlick. Recent epistemological critiques of the neo-positivist model*], Milan, 1978.

Geymonat, L., Critica della scienza e lotta politica [Criticism of science and political struggle] in *Scientia*, 67, 8, 1973.

Geymonat, L., F. Enriques e la storia della scienza [F. Enriques and the history of science] in *"Atti" del Convegno Internazionale su Federigo Enriques*, Rome, Accademia dei Lincei, 1972.

Geymonat, L., *Filosofia e filosofia della scienza* [*Philosophy and the philosophy of science*], Milan, Feltrinelli, 1960.

Geymonat, L., G. Giorello, *Le ragioni della scienza* [*The reasons of science*], Rome-Bari, Laterza, 1986.

Geymonat, L., *Il problema della conoscenza nel positivismo* [*The issue of knowledge in positivism*] Turin, Bocca, 1931.

Geymonat, L., *Moritz Schlick fra realismo e neo-positivismo* [*Moritz Schlick between realism and neo-positivism*], Introduction to M. Schilick, *Tra realismo e neo-positivismo* [between realism and neo-positivism], Bologna, Il Mulino, 1974; Italian translation of essays printed by Gerald & Co di Vienna (1948), in "Erkenntnis" (1932, 1934), in *The Philosophical Review* (1936).

Geymonat, L., Neopositivist Methodology and Dialectical Materialism, in *Science and Society*, New York, XXXVII, 1973.

Geymonat, L., *Riflessioni critiche su Kuhn e Popper* [*Critical reflections on Kuhn and Popper*], Bari, Dedalo, 1983.

Geymonat, L., *Saggi di filosofia neorazionalistica* [*Essays on neo-nationalistic philosophy*], Turin, Einaudi, 1953.

Geymonat, L., *Scienza e realismo* [*Science and realism*], Milan, Feltrinelli, 1977; new extended edition, Milan, Feltrinelli, 1982.

Geymonat, L., *Studi per un nuovo razionalismo* [*Studies for a new rationalism*], Turin, Chiantore, 1945.

Gianquinto, A., *La filosofia analitica. L'involuzione della riflessione sulla scienza* [*Analytical philosophy. The involution of the reflection on science*], Milan, Feltrinelli, 1961.

Giorello G. (edited by), *Introduzione alla filosofia della scienza* [*Introduction to the philosophy of science*], Milan, Bompiani, 1999.

Giorello, G., Filosofia della scienza e storia della scienza nella cultura di lingua inglese [Philosophy of science and history of science in English-language culture], in Ludovico Geymonat, *Storia del pensiero filosofico e scientifico* [*History of philosophical and scientific thought*], Milan, Garzanti, 1976, 7 volumes, vol. VII.

Giorello, G., Programmi di ricerca, razionalità e progresso [Research programmes, rationality and progress], in *Quaderni*, Fondazione Feltrinelli, n. 2, 1978.

Gubert R., Tomasi, L., *Teoria sociologica ed investigazione empirica* [*Sociological theory and empirical inquiry*], Milan, Franco Angeli, 1995.

Guidicini, P., Castrignano, M., *L'utilizzo del dato qualitativo nella ricerca sociologica* [*The use of the qualitative datum in sociological research*] Milan, Franco Angeli, 1997.

Hacking, I., *Representing and Intervening*, Cambridge University Press, 1983; Italian translation di E. Prodi, *Conoscere e sperimentare*, Rome- Bari, Laterza, 1987.

Hacking, I., *The social construction of what?*, Harvard University Press, Cambridge, Massachusetts and London, England, 1999.

Hanson, N. R., *Patterns of Discovery. An Inquiry into the Conceptual Foundation in Science*, Cambridge University Press, 1958; Italian translation *I modelli della scoperta scientifica. Ricerca sui fondamenti concettuali della scienza*, Milan, Feltrinelli, 1978.

Hempel, C. G., *Aspects of Scientific Explanation and Other Essays of Philosophy of Science*, New York, Free Press, 1965; Italian translation *Aspetti della spiegazione scientifica*, Milan, Il Saggiatore, 1986.

Hempel, C. G., *Come lavora uno storico* [*How a histoprian works*], edited by Dario Antiseri, Rome, 1977.

Hempel, C. G., *Fondamentals of concept formation in empirical science*, Chicago, University of Chicago Press, 1952, Italian translation di Alberto Pasquinelli, *La formazione dei concetti e delle teorie nella scienza empirica*, Milan, Feltrinelli, 1976.

Hempel, C. G., *Philosophy of Natural Science*, New Jersey, Prentice-Hall, Inc., Englewood Cliffs, 1966; Italian translation by A. Berra, *Filosofia delle scienze naturali*, Bologna, Il Mulino, 1980[2].

Izzo, A, *La sociologia della conoscenza* [*The sociology of knowledge*], Rome, Armando, 1966.

Izzo, A., *Il condizionamento sociale del pensiero* [*The social conditioning of thought*], Turin, Loescher, 1970.

Izzo, A., *L'invincibile perplessità. Fondamenti, storia e problemi della sociologia della conoscenza* [*Invincible perplexity. Fundaments, history and problems of the sociology of knowledge*] Rome, Armando, 1999.

Izzo, A., Le critiche di Alfred Schütz a Max Weber e il lavoro del sociologo oggi [Alfred Schütz critiques of Max Weber and the work of the sociologist today] in *Fenomenologia e Società*, n. 17, 1982.

Izzo, A., *Ricerca di una sociologia critica* [*The quest for a critical sociology*], Naples, Liguori, 1975.

Izzo, A., *Storia del pensiero sociologico* [*The history of sociological thinking*], Bologna, Il Mulino, 2005, 3 volumes.

Jedlowski, P., *In un passaggio d'epoca. Esercizi di teoria sociale* [*In a period of epocal. Exercises in social theory*], transition.Naples -Salerno, Orthotes, 2012.

Knorr-Cetina, K. D., *The Manufacture of Knowledge. An Essay on the Constructivist and Contextual Nature of Science*, Oxford, Pergamon, 1981.

Koyré, A., *Filosofia e storia delle scienze* [Philosophy and the history of science], Milan, Mimesis, 2003.

Koyré, A., Storia della scienza - La genesi della cultura [The history of science-The genesis of culture], in *Civiltà delle Macchine*, [*The civilisation of machines*], n. 1-2, 1971.

Kuhn, T. S. *Dogma contro critica* [*Dogma against criticism*], edited by Stefano Gattei, Milan, Raffaello Cortina, 2000.

Kuhn, T. S., *La struttura delle rivoluzioni scientifiche* [*The structure of scientific revolutions*], Einaudi, 1978.

Kuhn, T. S., *La tensione essenziale* [*The essential tension*], Torino, Einaudi, 1985.

Kuhn, T. S., Sneed, J. D., Stegmüller, W., *Paradigmi e rivoluzioni nella scienza* [*Paradigms and the revolutions of science*], Rome, Armando Armando, 1983.

Kuhn, T. S., *The Essential Tension*, Chicago, The University of Chicago, 1977; Italian translation di Mario Vadacchino, *La tensione essenziale. Cambiamenti e continuità nella scienza* [*The Essential Tension. Changes and continuity in science*], Turin, Einaudi, 1985; extended edition by Claudio Bartocci and Giulio Giorello, Turin, Einaudi, 2006.

Kuhn, T. S., *The Structure of Scientific Revolutions*, Chicago, University of Chicago Press, 1962; second edition, 1970, with added of "Postscript-1969"; Italian translation di Adriano Carugo, *La struttura delle rivoluzioni scientifiche*, Turin, Einaudi, 1995.

Lakatos I., Musgrave A. (edited by), *Criticism and the Growth of Knowledge*, Cambridge, Cambridge University Press, 1970; Italian version edited by Giulio Giorello, *Critica e crescita della conoscenza*, Milan, Feltrinelli, 1993.

Lakatos, I., *History of Science and Its Rational Reconstructions*, in R. C. Buck, R. S. Cohen, (edited by), Philosophy of Science Association – 1970, Dordrecht, Reidel,

1971; Italian translation La storia della scienza e le sue ricostruzioni razionali, in I. Lakatos, *La metodologia dei programmi di ricerca scientifici. Scritti filosofici* [*The methodology of scientific research programmes. Philosophical writings*], I, Milan, Il Saggiatore, 1985.

Lakatos, I., *The Methodology of Scientific Research Programmes*, Philosophical Papers Volume I, Cambridge, Cambridge University Press, 1978; Italian translation edited by Marcello D'Agostino, *La metodologia dei programmi di ricerca scientifici. Scritti filosofici* I, Milan, Il Saggiatore, 1985, revised and extended by Matteo Motterlini, Premessa di Giulio Giorello, Milan, Il Saggiatore, 1996.

Landucci, S., *La rivoluzione pre-assertoria: Koiré, Fleck, Kuhn* [*The pre-assertory revolution: Koiré, Fleck, Kuhn*], Acireale-Rome, Bonanno, 2004.

Lanfredini, R., *Filosofia della conoscenza* [*The philosophy of knowledge*], Milan, Mondadori Education, 2007.

Lanfredini, R., *Oggetti e paradigmi* [*Objects and paradigms*], Rome, Teoria, 1988.

Lanfredini, R., Teorie della razionalità e conoscenza scientifica, in *Paradigmi*, 16, 1999.

Latour, B., *Science in Action. How to Follow Scientists and Engineers through Society*, Cambridge Mass, Harvard University Press, 1987; Italian translation *La scienza in Azione. Introduction to alla sociologia della scienza*, Turin, Edizioni di Comunità, 1998.

Latour, B., Woolgar, S., *Laboratory Life: the Social Construction of Scientific Facts*, London, Sage, 1980.

Laudan, L., *Beyond Positivism and Relativism: Theory, Method, and Evidence*, CO, Boulder, Westview Press, 1996.

Laudan, L., *Progress and its Problems. Towards a Theory of Scientific Growth*, University of California, 1977; Italian translation edited by Emanuele Riverso, *Il Progresso scientifico. Prospettive per una teoria*, Armando Armando, Rome, 1979.

Laudan, L., *Science and Values*, University of California Press, 1984; Italian translation by E. Prodi, *La scienza e i valori*, Rome-Bari, Laterza, 1987.

Lazarsfeld, P. F., *Saggi storici e metodologici* [*Historical and methodological essays*], edited by Carmelo Lombardo, Rome, Edizioni EUCOS, 2001.

Lentini, L., *Il paradigma del sapere. Conoscenza e teorie della conoscenza nell'epistemologia contemporanea* [*The paradigm of knowledge. Knowledge and theories of knowledge in contemporary epistemology*] Milan, Franco Angeli, 1990.

Lentini, O., *Saperi sociali, ricerca sociale* [*Social knowledge, social research*], Milan, Franco Angeli, 2003.

Leonardi, F., *Di che parla il sociologo. Problemi di epistemologia delle scienze sociali* [*What the sociologist talks about. Problems of epistemology of social sciences*], Milan, Angeli, 2007.

Leonardi, F., *Scritti metodologici 1960-1994* [*Writings on methodology 1960-1994*], Milan, Angeli, 2007.

Lombardi, F., *Dopo lo storicismo* [*After historicism*], Florence, Sansoni, 1970, 2a ediz.

Lombardi, F., *Filosofia e scienze* [Philosophy amnd science], in *Civiltà delle Macchine*, [*The civilisation of machines*], n. 4, 1965.

Lombardi, F., *Filosofia e società* [*Philosophy and society*], Florence, Sansoni, 1967.

Lombardi, F., Filosofia ed esperienza [Philosophy and experience], in *Società Italiana per il Progresso delle Scienze, 'Atti' della XXIV Riunione* (Palermo, 12-18 0ctober 1935), edited by L. Silla, Rome, SIPS, 1936, vol. V.

Lombardi, F., *La Filosofia di fronte alle scienze* [*Philosophy in the face of science*], Bari, 1962.

Lombardo, C., Conseguenze metodologiche del "programma forte": verso una sociologia debole [Methodological consequences of the "strong programme": towards a weak sociology] in *Sociologia e ricerca sociale*, [*Sociology and social research*], n. 35, 1991.

Lombardo, C., La formazione e la definizione dei concetti sociologici [The formatiomn and definition of sociological concepts] in AA.VV, Paul Felix Lazarsfeld: un "classico" marginale [Paul Felix Lazarsfeld: a marginal "classic"] in *Sociologia e ricerca sociale*, [*Sociology and social research*], n. 58-59, 1999.

Marconi, D., *Filosofia e scienza cognitiva* [*Philosophy and cognitive science*], Rome-Bari, Laterza, 2001.

Marconi, D., *Il mito del linguaggio scientifico. Studio su Wittgenstein* [*The myth of scientific language. A study of Wittgenstein*], Milan, Mursia, 1971.

Marconi, D., *L'eredità di Wittgenstein* [*The legacy of Wittgenstein*] Bari, Laterza, 1987.

Marletti, C., Bruzzone, E., (edited by), *Teoria, società e storia. Scritti in onore di Filippo Barbano* [*Theory, society and history. Writings in honour of Filippo Barbano*] Milan, Franco Angeli, 2000.

Marotta, M., Gregor, A. J., Sociology in Italy in *The sociological quarterly. Official Journal of the Midwest Sociological Society*, vol. 2, Issue 3, July 1961, pp. 215-221.

Marradi, A., Boudon: un sociologo che ha sbagliato specializzazione [Boudon: a sociologist who mistook specialisation] in *Rassegna Italiaa di Sociologia*, XXIII (3), 1982, pp. 445-54.

Marradi, A., Classificazioni, tipologie, tassonomie in *Enciclopedia delle scienze sociali* [Classifications, typologies, taxonomies in *Encyclopedia of social sciences*], Vol. II, Rome, Treccani, 1992; see also Classification, Typology, Taxonomy, in *Quality & Quantity*, XXIV, 2, 1990.

Marradi, A., Classificazioni, tipologie, tassonomie in *Enciclopedia delle Scienze Sociali* [Classifications, typologies, taxonomies in *Encyclopedia of social sciences*], Rome, Istituto della Enciclopedia Italiana, vol. II, 1993, pp. 22-30.

Marradi, A., *Concetti e metodi per la ricerca sociale* [*Concepts and methods for social research*], Florence, Giuntina, 1980.

Marradi, A., Filosofia, storiografia, sociologia della scienza: una proposta di risistemazione del campo semantico [Philosophy, historiography, sociology of science: a proposal for the reorganization of the semantic field], in *Laboratorio di Scienze dell'uomo*, III, 3-4, 1987.

Marradi, A., Linguaggio scientifico o torre di Babele? [Scientific language or tower of Babel ?], a in *Rivista Italiana di Scienza politica*, n. 1, XVII, 1987.

Marradi, A., Metodo come arte [Method as an art], in *Quaderni di Sociologia*, [*Sociology notebooks*], XL (10), 1996, pp. 71-92.

Marradi, A., *Metodologia delle scienze sociali* [*Methodology of the social sciences*], edited by Rita Pavsic and Maria Concetta Pitrone, Bologna, Il Mulino, 2007.

Marradi, A., Referenti, pensiero e linguaggio: una questione rilevante per gli indicatori [Referents, thought and language: a relevant issue about indicators] in *Sociologia e ricerca sociale*, [*Sociology and social research*], XV (43), 1994, pp. 137-207.

Marradi, A., Teoria: una tipologia dei significati, 1984, in A. Marradi (edited by), *Metascienza. Riflessioni sulla conoscenza scientifica* [Theory: a typology of meanings, 1984, in A. Marradi (edited by), *Metascienza. Reflections on scientific knowledge*] Rome, EuRome - Editrice Universitaria di Roma - La Goliardica, 1991.

Martinelli, A., *La modernizzazione* [*Modernisation*], Rome-Bari, Laterza, 2015.

Martinotti, G., *Il condizionamento della ricerca: destinazione, utilizzazione e strutture istituzionali* in *Ricerca sociologica e ruolo del sociologo* [*The conditioning of research: destination, use and institutional structures in sociological research and the role of the sociologist*], edited by Pietro Rossi, Bologna, Il Mulino, 1972.

Melandri, E., Gli inediti di Husserl [Husserl's unpublished works] in *Il Mulino*, VIII, 1959.

Melandri, E., *Logica ed esperienza in Husserl* [*Logic and experience in Husserl*], Bologna, 1960.

Melandri, E., Significato e verità nelle scienze sociali [Meaning and truth in the social sciences], Introduction to the Italian edition of A. Schütz, *Der sinnhafte Aufbau der sozialen Welt* [*The meaningful construction of the social world*] Wien, Springer-Verlag, 1960; Italian translation by Franco Bassani, *La fenomenologia del mondo sociale* [*The phenomenology of the social world*], Bologna, Il Mulino, 1974.

Melucci A., (edited by), *Verso una sociologia riflessiva. Ricerca qualitativa e cultura* [*Towards a reflective sociology. Qualitative research and culture*], Bologna, Il Mulino, 1998.

Memoli, R., (edited by), *Intersezioni tra discipline. Elaborare concetti per la ricerca sociale* [*Intersections between disciplines. Developing concepts for social research*], Milan, Franco Angeli, 2014.

Merton, R. K., *La sociologia della scienza. Indagini teoriche ed empiriche* [*The sociology of science. Theoretical and empirical inquiry*], Milano, Angeli, 1981.

Merton, R. K., *The sociology of science: Theoretical and Empirical Investigations*, Chicago, The University of Chicago Press; Italian translation *La sociologia della scienza. Indagini teoriche ed empiriche*, edited by Norman W. Storer, Preface by Filippo Barbano, Milan, Franco Angeli, 1981.

Minazzi, F., (edited by), *L'oggettività della conoscenza scientifica* [*The objectivity of scientific knowledge*], Milan, Franco Angeli, 1996.

Minazzi, F., (edited by), *Scienza e storia. Contributi per uno storicismo scientifico* [*Science and history. Contributions for a scientific historicism*], Verona, Bertani Publisher, 1985.

Minazzi, F., Epistemologia, criticismo e storicità [Epistemology, criticism and historicity], in L. Geymonat, G. Giorello, *Le ragioni della scienza* [*The Reasons of science*], Rome-Bari, Laterza, 1986.

Minazzi, F., *Filosofia, scienza e vita civile nel pensiero di Ludovico Geymonat* [*Philosophy, science and civilian life in the thinking of Ludovico Geymonat*] Minutes of the Conference in Milan on 29 November 2001, Naples, La Città del Sole, 2003.

Minazzi, F., *Giovanni Vailati. Epistemologo e maestro* [*Giovanni Vailati. Epistemologist and teacher*], Milan, Mimesis, 2011.

Minazzi, F., *Ludovico Geymonat epistemologo* [*Ludovico Geymonat epistemologist*], Milan-Udine, Mimesis,2010.

Mondadori, M., *Carnap e la significanza empirica* [*Carnap and the empirical meaning*] Bologna, 1969.

Mondadori, M., Struttura delle teorie scientifiche e progresso [Structure of scientific theories and progress], in *Quaderni*, Fondazione Feltrinelli, n. 2, 1978.

Mongardini, C., *Epistemologia e sociologia: temi e tendenze della sociologia contemporanea* [*Epistemology and sociology: the themes and trends of contemporary sociology*], Milan, Franco Angeli, 1985 (1991).

Mongardini, C., Tabboni, S., (edited by), *L'opera di R. K. Merton e la sociologia contemporanea* [*The work of R. K. Merton and contemporary sociology*], Genova, Egic, 1989.

Montaleone, C., *Scienze umane e metodologia. Weber, Popper, Durkheim* [*Human sciences and methodology. Weber, Popper, Durkheim*], Milan, Cisalpino- La Goliardica, 1975.

Montesperelli, P., *Comunicare e interpretare: introduzione all'ermeneutica nella ricerca sociale* [*Communicating and interpreting: introduction to hermeneutics in social research*], Milan, Egea, 2014.

Moore, G. E., *The refutation of idealism*, in "Mind," XI, 1903, New Series; Italian translation *La confutazione dell'idealismo*, in A. Pasquinelli (edited by), *Il neoempirismo*, Turin, UTET, 1969.

Mulkay, M. J., *Science and the Sociology of Knowledge*, Allen & Unwin Ltd, 1979; Italian translation *La scienza e la sociologia della conoscenza*, Milan, Edizioni di Comunità, 1981.

Mulkay, M., *Science and the sociology of knowledge*, Allen & Unwin, 1979.

Musgrave, A., *Senso comune, scienza e scetticismo* it. trasl. by Pier Daniele Napolitani of [*Common Sense, Science and Scepticism*], Cambridge University Press, 1993; now Milan, Raffaello Cortina Editor, 2000.

Musso, P., *Forme dell'epistemologia contemporanea* [*Forms of contemporary epistemology*], Rome, Urbaniana University Press, 2004.

Muzzetto, L., *Fenomenologia e etnometodologia. Percorsi della teoria dell'azione* [*Phenomenology and ethnomethodology. Pathways of action theory*], Milan, Franco Angeli, 1997.

Nagel, E., *The Structure of Science*, New York, Harcourt, Brace & World, 1961; Italian translation by Carla Sbrogi and Aurelia Monti, *La struttura della scienza. Problemi di logica della spiegazione scientifica* [*The structure of science. Problems regarding the logic of scientific explanation*] Milan, Feltrinelli, 1968.

Negrotti, M., La certezza sociologica [Sociological certainty] in *Sociologia*, III, 2, 1969.

Negrotti, M., *Per una teoria dell'artificiale* [*For a theory of the artificial*], Milan, Franco Angeli, 1993.

Neurath, O., *Le développement du Cercle de Vienne et l'avenir de l'empirisme logique* [*The development of the Circle of Vienna and the future of logical empiricism*], Paris, Hermann, 1935; Italian translation with Preface by Francesco Barone, *Il Circolo di Vienna e l'avvenire dell'empirismo logico* [*The Circle of Vienna and the future of logical empiricism*] Rome, Armando, 1977.

Nuzzaci, F., Il neorazionalismo di Ludovico Geymonat [Ludovico Geymonat's neo-rationalism], in *Il Protagora*, n. 77-78, 1971.

Nuzzaci, F., *Karl Popper. Un epistemologo fallibilista* [*Karl Popper. A fallibilistic epistemologist*], Naples, Glaux, 1975.

Paci, E., *Funzione delle scienze e significato dell'uomo* [*The function of the sciences and the meaning of man*] Milan, Il Saggiatore, 1964, 1970.

Paci, E., *Introduzione a Neopositivismo e unità nella scienza* [*Introduction to Neopositivism and unity in science*], Milan, Bompiani, 1958.

Paci, E., La fenomenologia come scienza del mondo e della vita [Phenomenology as a science of the world and of life], in *Aut-Aut*, n. 59, 1960.

Paci, E., *La filosofia contemporanea* [*Contemporary philosophy*], Milan, Garzanti, 1957.

Paci, E., Per lo studio della logica di Husserl [For the study of Husserl's logic], in *Aut-Aut*, n. 94, 1966.

Pagnini, A. (edited by), *Realismo/antirealismo. Aspetti del dibattito epistemologico contemporaneo* [*Realism / anti-realism. Aspects of the contemporary epistemological debate*], Florence, La Nuova Italia, 1995.

Palumbo, M. *Il processo di valutazione. Decidere, programmare, valutare* [*The evaluation process. Decide, plan, evaluate*], with a contribution by Nicoletta Stame, Milan, Franco Angeli, 2015 (new edition).

Parrini, P., *Conoscenza e realtà. Saggio di filosofia positiva* [*Knowledge and reality. An essay on positive philosophy*] Rome-Bari, Laterza, 1995.

Parrini, P., *Empirismo logico e convenzionalismo* [*Logical empirism and conventionalism*], Milan, Franco Angeli, 1983.

Parrini, P., *Filosofia e scienza nell'Italia del Novecento* [*Philosophy and science in twentieth-century Italy*], Milan, Guerini e Associati, 2004.

Parrini, P., *L'empirismo logico. Aspetti storici e prospettive teoriche* [*Logical empiricism. Historical aspects and theoretical perspectives*], Rome Carocci, 2002.

Parrini, P., *Linguaggio e teoria. Due saggi di analisi filosofica* [*Language and theory. Two essays on philosophical analysis*], Florence, 1976.

Parrini, P., *Una filosofia senza dogmi. Materiali per un bilancio dell'empirismo contemporaneo* [*A philosophy without dogmas. Materials for a balance of contemporary empiricism*], Bologna, Il Mulino, 1980.

Parsons, T., *The Structure of Social Action*, New York, McGraw-Hill, 1937; Italian translation *La struttura dell'azione sociale*, Bologna, Il Mulino, 1962; new edition1986.

Pasquinelli, A., (edited by), *Il neoempirismo* [*Neo-empiricism*], Turin, UTET, 1978.

Pasquinelli, A., (edited by), *L'eredità di Rudolf Carnap. Epistemologia, filosofia delle scienze, filosofia del linguaggio* [*The legacy of Rudolf Carnap. Epistemology, the philosophy of science, the philosophy of language*], Bologna, Clueb, 1995.

Pasquinelli, A., Filosofia e scienza in Vailati e Peirce [Philosophy and science in Vailati and Peirce] in *Rivista critica di storia della filosofia*, XXXVIII, 3, 1963.

Pasquinelli, A., *La sociologia del positivismo* [*The sociology of positivism*], Bologna, Zanichelli, 1953.

Pasquinelli, A., La Sociologia nel Circolo di Vienna [Sociology in the Circle of Vienna] in *Quaderni di Sociologia*, [*Sociology notebooks*], n. 3, 1952.

Pasquinelli, A., *Linguaggio, scienza e filosofia* [*Language, science and philosophy*], Bologna, Il Mulino, 1961 (1967).

Pasquinelli, A., *Nuovi principi di epistemologia* [*New principles of epistemology*], Bologna, Clueb, 1987.

Pasquinelli, A., Sociologia empirica, storiografia, storicismo [Empirical sociology, historiography, historicism], in *Il Mulino*, Bologna, 1954.

Pellizzi, C. Alcuni appunti di epistemologia e di etica sociologica [Some notes on epistemology and sociological ethics], in *Rassegna Italiana di Sociologia*, [*Italian Review of Sociology*], IV, 1963.

Pellizzi, C., 'Realtà', 'Segno' e altre note epistemologiche ['Reality', 'Sign' and other epistemological notes], in *Sociologia*, III, 3, 1969.

Pera, M., *Apologia del metodo* [*An apology of method*], Bari, Laterza, 1982.

Pera, M., *Induzione e metodo scientifico* [*Induction and scientific method*], Pisa, Editrice Tecnico Scientifica, 1978.

Pera, M., *Popper e la scienza su palafitte* [*Popper and science on stilts*], Rome-Bari, Laterza, 1981.

Pera, M., Preface (1982) B. Russell, *The Scientific Outlook*, London, George Allen & Unwin Ltd, 1931; Italian translation by Emilio A.G. Loliva, *La visione scientifica del mondo* [*The scientific view of the world*], Rome- Bari, Laterza, 2000.

Pera, M., *Scienza e retorica* [Science and rhetoric], Bari, Laterza, 1991.

Petruzzellis, N., *La crisi dello scientismo. Riflessioni su K.R. Popper, il neoempirismo e il razionalismo critico* [*The crisis of scientism. Reflections on K. R. Popper, neo-empirism and critical rationalism*], Milan, Nuovo Ist. Editoriale, 1984.

Phillips, D. L., *Wittgenstein and Scientific Knowledge. A Sociological Perspective*, London, the Macmillan Press, 1977; Italian translation *Wittgenstein e la conoscenza scientifica. Un approccio sociologico* [*Wittgenstein and Scientific Knowledge. A Sociological Approach*], Bologna, Il Mulino, 1981.

Picardi, E., *La chimica dei concetti* [*The chemistry of concepts*], Bologna, Il Mulino, 1994.

Piccone Stella S., Salmieri, L., *Il gioco della cultura* [*The game of culture*], Rome, Carocci editor, 2012.

Pickering, A., *La scienza come pratica e cultura* [*Science as practice and culture*], Milano, Edizioni Comunità, 2001.

Pickering, A., *Science as practice and culture*, it. transl., Milan, Community Editions, 2001.

Pinto, D., edited by *Contemporary Italian Sociology. A Reader,* Paperback, October, 1981.

Pizzorno, A., Una crisi che non importa superare, in *Ricerca sociologica e ruolo del sociologo* [A crisis not worth overcoming, in *Sociological research and the role of the sociologist*] edited by Pietro Rossi, Bologna, Il Mulino, 1972.

Poggi G., Ryan, C., I presupposti antropologici della teoria volontaristica dell'azione sociale [The anthropological presuppositions of the voluntarist theory of social action], in *Rassegna Italiana di Sociologia* [*Italian Review of Sociology*], VIII, 1967.

Poggi, S., *Husserl e la fenomenologia* [*Husserl and phenomenology*], Florence, Sansoni, 1973.

Poggi, S., Il *positivismo* [*Positivism*], Roma-Bari, Laterza, 1987.

Poggi, S., *Il realismo della ragione. Kant dai lumi alla filosofia contemporanea* [*The realism of reason. Kant from illuminism to contemporary philosophy*], Milan, Mimesis, 2012.

Poggi, S., *Introduzione al positivismo* [*Introduction to positivism*], Bari, Latreza, 1987.

Poggi, S., *Positivismo* [*Positivism*], Rome - Bari, Laterza, 1987.

Pontecorvo, C., (edited by), *Concetti e conoscenza* [*Concepts and knowledge*], Turin, Loescher, 1983.

Popper, K. R, Die Logik der Sozialwissenschaften [The logic of the social sciences], Köln und Opladen, 1962; Italian translation *La logica delle scienze sociali* [The logic of the social sciences], in AA. VV, *Dialettica e positivismo in sociologia* [*Dialectics and positivism in sociology*], Turin, Einaudi, 1972.

Popper, K. R., *Conjectures and Refutations. The Growth of Scientific Knowledge*, London, Routledge & Kegan Paul, 1963; it. trans. to Giuliano Pancaldi, *Congetture e confutazioni. Lo sviluppo della conoscenza scientifica* [*Conjectures and Refutations. The Development of Scientific Knowledge*], Bologna, Il Mulino, 1985.

Popper, K. R., Logik der Forschung [The logic of research], Wien, Julius Springer, 1935; English edit., *The Logic of Scientific Discovery*, London, Hutchinson, 1959; Italian translation by Mario Trinchero, *Logica della scoperta scientifica* [*The Logic of Scientific Discovery*], Turin, Einaudi, 1970.

Popper, K. R., *Objective Knowledge*, Oxford, Clarendon Press, 1972 (ed. riv. 1981); Italian translation and Introduction to A. Rossi, *Conoscenza oggettiva*, Rome, Armando, 1975.

Popper, K. R., Science and Philosophy, in *Contemporary British Philosophy*, III Serie, edited by H. D. Lewis, 1956; Italian translation di Mario Trinchero, *Scienza e Filosofia. Cinque saggi* [*Science and philosophy*. Five essays], Turin, Einaudi, 1969; see 2000 edition with introductory note by Roberta Lanfredini.

Possenti, V., *Epistemologia e scienze umane* [*Epistemology and human sciences*], Massimo, 2012.

Potter, T. M., Ross, D., *The Cambridge History of Science*, vol. VII – *The Modern Social Sciences,* US, Cambridge University Press, 2003.

Preti, G., *Lezioni di Filosofia della scienza (1965-1966)* [*Lessons in Philosophy of* Science *(1965-1966)*,], edited by Fabio Minazzi, Milan, Franco Angeli, 1989.

Preti, G., *Linguaggio comune e linguaggi scientifici* [*Common language and scientific languages*] Rome-Milan, Bocca, 1953.

Preti, G., *Praxis ed empirismo* [*Praxis and empiricism*], Preface by Salvatore Veca with an essay of Fabio Minazzi, Milan, Bruno Mondadori edition, 2007.

Preti, G., *Saggi filosofici, vol. I: Empirismo logico, epistemologia e logica* [*Philosophical essays, vol. I: Logical empirism, epistemology and logic*], Florence, La Nuova Italia, 1976.

Preti, G., *Scetticismo e conoscenza* [*Skepticism and knowledge*], Catania, CUECM, 1993.

Prodi, E., *Quale metodo per la scienza* [*What method for science*], Milan, Franco Angeli, 1993.

Putnam H., *Reason, Truth and History*, Cambridge, Cambridge University Press, 1981; Italian translation edited by A. N. Radicati di Brozolo, *Ragione, verità e storia*, Milan, Il Saggiatore, 1989.

Quine, W. V. O., Two Dogmas of Empirism, in *Philosophical Review*, LX, 1951, published with variants in W. V. O. Quine, *From a Logical Point of Wiew*, Cambridge, Harvard University Press, 2a ediz. 1961; Italian translation *Il problema del significato* [*The issue of meaning*], Rome, Ubaldini, 1966.

Quine, W. V., *Ontological relativity and Other Essays*, New York and London, Columbia University Press, 1969; Italian translation *La relatività ontologica e altri saggi*, Rome, Armando, 1986.

Radnitzky, G., L'oggettivismo e la seduzione del relativismo epistemologico [Objectivism and the seduction of epistemological relativism], in R. Egidi (edited by), *La svolta relativistica nell'epistemologia contemporanea* [*The relativistic turning point in contemporary epistemology*], Milan, Franco Angeli, 1988.

Redondi P. (edited by), *La verità degli eretici. Critica e storia della conoscenza* [*The truth of heretics. Critique and history of knowledge*], Milan, Il Saggiatore, 1978.

Redondi, P., *Epistemologia e storia della scienza* [*Epistemology and history of science*], Milan, Feltrinelli, 1978.

Reichenbach, H. *Modern Philosophy of Science*, London, 1959; Italian translation L'analisi filosofica della conoscenza scientifica [The philosophical analysis of scientific knowledge], Padua, 1968 and with the title *La nuova filosofia della scienza* [*The new philosophy of science*], Milan, 1968.

Reichenbach, H., *The Rise of Scientific Philosophy*, Berkley-Los Angeles, University of California Press, 1951; Italian translation edited by Domenico Parisi e Alberto Pasquinelli, *La nascita della filosofia scientifica* [*The birth of scientific philosophy*], Bologna, Il Mulino, 1961- 2003.

Remotti, F., Filosofia e scienze sociali [Philosophy and social sciences], in P. Rossi (edited by), *La filosofia* [*Philosophy*], Turin, UTET, 1995, vol. II, pp. 269-318.

Restaino, F., Di alcune tesi storico-metodologiche vailatiane nelle due Prolusioni del 1896 e 1897 [Some Vailatical historical-methodological theses in the two Prolusions of 1896 and 1897], in *Rivista critica di storia della filosofia*, XXXVIII, 3, 1963.

Ricolfi, L., *La ricerca qualitativa*, Rome, Carocci, 1997.

Rinzivillo, G., *La scienza e l'oggetto. Ricerche sulla fondazione della sociologia in Italia* [*Science and its object. Research concerning the foundation of sociology in Italy*], Acireale, Bonanno Publisher, 1994; next edition Milan, Franco Angeli, 2010, 2012 II edition.

Rinzivillo, G., Genesi e Prassi della Sociologia in *Italia –Sviluppo e origine della teoria in Alessandro Groppali* [Genesis and Practices of Sociology in *Italy - Development and origin of the theory in Alessandro Groppali*], Preface by Filippo Barbano, Rome, Seam, 2000.

Rinzivillo, G., Storia della scienza e storici della sociologia. Modelli e proposte per nuove interpretazioni sistematiche [History of science and historians of sociology.

Models and proposals for new systematic interpretations], *Sociologia*, n. 1, 2003, pagg. 43-50.

Rinzivillo, G. Filippo Barbano: storia della scienza e storici della sociologia [Filippo Barbano: history of science and historians of sociology], in *La Critica Sociologica*, [*Sociological criticism*], n. 151, Autunno 2004.

Rinzivillo, G., Dove va la sociologia italiana: intervista a Franco Ferrarotti [Where Italian sociology is going: an interview with Franco Ferrarotti] in *Sociologia*, n. 1, 2004, pages 85-88.

Rinzivillo, G., Alberto Izzo. Dalla storia alla sociologia della conoscenza [Alberto Izzo. From history to the sociology of knowledge], in *Sociologia*, n. 1, 2006, pages. 121-125.

Rinzivillo, G., *Sociologi in Italia e storia delle scienze* [Sociologists in Italy and the history of science], in *Sociologia - La società in Rete, Rivista quadrimestrale di Sociologia*, N. 3 - I year, 2006, pages 50-52.

Rinzivillo G., (edited by), *Tra i concetti e le regole. Contesti filosofici (e non) delle scienze sociali nel '900* [*Amid concepts and rules. Philosophical (and non) contexts of the social sciences in the 20th century*], Introduction to, edition on line www.scriptaweb.it, Naples, 2006; idem new edition, Rome, Tipografia Salesiana Pio XI, 2008.

Rinzivillo, G., *Tra i concetti e le regole. Contesti filosofici (e non) delle scienze sociali nel '900* [*Among the concepts and rules. Philosophical contexts (and not) of the social sciences in the 20th century*], Naples, ScriptaWeb, 2009.

Rinzivillo G., *La scienza e l'oggetto. Autocritica del sapere strategico* [*Science and the object. Self-criticism of strategic knowledge*], Milan, Franco Angeli, 2010, 2012.

Rinzivillo, G., *Epistemologia e sociologia italiana* [*Epistemology and Italian sociology*], Milano, Franco Angeli, 2010.

Rinzivillo G., Una storia delle scienze per i nuovi saperi. Discussioni e ricerche [A history of the sciences for new knowledge. Discussion and research] *ATTI* del Convegno: *Orizzonti e confini della ricerca epistemologica* [*Minutes of the Meeting on The horizons and boundaries of epistemological research*] Rome 8 April 2010, edited by Guglielmo Rinzivillo, Rome, Casa Editrice La Sapienza, 2011.

Rinzivillo, G., *Una epistemologia senza storia* [*An epistemology without a history*], Rome, Nuova Cultura, 2013.

Rinzivillo G., Intervista a Franco Leonardi [Interview with Franco Leonardi], in "Sociologia – La Società in… Rete," *Rivista professionale di Sociologia e Società*, single volume, Year VII, 2015, pages 18-26.

Riverso, E., *Esperienza e riflessione. Le tappe della filosofia e della scienza nella cultura occidentale* [*Experience and reflection. The stages of philosophy and science in Western culture*], Rome, Borla, 1985.

Riverso, E., I problemi della ricerca scientifica nella prospettiva di Karl R. Popper [The problems of scientific research according to Karl R. Popper] in *Rassegna di scienze filosofiche*, Year XIV, numbers 3,4, July-December 1961.

Riverso, E., *Il pensiero di B. Russell. Esposizione storico-critica* [*The thinking of B. Russell. A historical-critical exposition*] Naples, 1958; III edition, 1972.

Riverso, E., *Il pensiero di Ludovico Wittgenstein* [*The thinking of Ludwig Wittgenstein*], Naples, Edizioni Scientifiche Italiane, 1964.

Riverso, E., Il principio di verificabilità [The principle of verifiability], in *Logos*, I, 1969.

Rodolfi, F., Il dibattito sul pensiero di K. Popper [The debate on the thinking of K. Popper] in *Sociologia*, n. 1, 1997.

Rodolfi, F., Rivendicando il ruolo della tipologia [Vindicating the role of typology], in *Quaderni di Sociologia*, [*Sociology notebooks*], XLII (19), 1999, pp. 102-36.

Rodolfi, F., *Singole teorie o programmi di ricerca? Le immagini della scienza di Popper e Lakàtos* [*Individual theories or research programmes? Images of the science of Popper and Lakàtos*], Milan, Franco Angeli, 2001.

Rorty, R., *Consequences of Pragmatism*, Minneapolis, Minnesota University Press, 1982; Italian translation *Conseguenze del Pragmatismo*, Milan, Feltrinelli, 1986.

Rorty, R., *Philosophy and the Mirror of Nature*, Princeton, Princeton University Press, 1979; Italian translation and introductory note by Diego Marconi e Gianni Vattimo, *La filosofia e lo specchio della natura,* Milan, Bompiani, 1986.

Rossi Landi, F., (edited by), *Il metodo della filosofia* [*The philosophical method*], Bari, Laterza, 1957; reprinted 1967.

Rossi Landi, F., (edited by), *Il pensiero americano contemporaneo*, *I, Filosofia, epistemologia, logica*; *II, Scienze sociali* [*Contemporary American thinking, I, Philosophy, epistemology, logic; II, Social Sciences*], Milan, Edizioni di Comunità, 1958.

Rossi, A., Il “Methodenstreit” tra la scuola analitica di Popper e la scuola dialettica di Francoforte [The "Methodenstreit" between the analytical school of Popper and the dialectical school of Frankfurt], in *La nuova critica*, XXIII, 1969-70.

Rossi, A., *Popper e la filosofia della scienza* [*Popper and the philosophy of science*], Florence, Sansoni, 1975.

Rossi, P., (edited by), *Lo storicismo tedesco* [*German historicism*], Turin, UTET, 1977.

Rossi, P., *I ragni e le formiche. Un'apologia della storia della scienza* [*Spiders and ants. An apology of the history of science*], Bologna, Il Mulino, 1986.

Rossi, P., Karl Popper e la critica neopositivistica allo storicismo [Karl Popper and neopositivist criticism of historicism], in *Rivista di filosofia*, XLVIII, 1957.

Rossi, P., *La nascita della scienza moderna in Europa* [*The birth of modern science in Europe*], Rome-Bari, Laterza, 2005.

Rossi, P., *La scienza e la filosofia dei moderni: aspetti della rivoluzione scientifica* [*The science and philosophy of moderns: aspects of the scientific revolution*], Turin, Bollati Boringhieri, 1989.

Rossi, P., La struttura della scienza e le scienze sociali [The structure of science and the social sciences,], in *Quaderni di Sociologia*, [*Sociology notebooks*], XI, 3, 1962.

Rossi, P., *Le immagini della scienza* [*Images of science*], Rome, Editori Riuniti, 1977.

Rossi, P., *Lo storicismo tedesco contemporaneo* [*Contemporary German historicism*], Turin, Einaudi, 1979.

Rossi, P., *Ludwig Fleck e una rivoluzione immaginaria*] [*Ludwig Fleck and a revolution of the imaginary*], Introduction to Italian edition of *Enstehung und Entwicklung einer wissenschaftlichen Tatsache* [*Origin and development of a scientific fact*], Frankfurt, Suhrkamp, 1980; Italian translation by Maria Leonardi and Stefano Poggi, *Genesi e sviluppo di un fatto scientifico* [*Genesis and development of a scientific fact*], Bologna, Il Mulino, 1983.

Rossi, P., Mori, M., Trinchero, M., (edited by), *Il problema della spiegazione sociologica* [*The problem of sociological explanation*], Turin, Loescher, 1983.

Rossi, P., *Positivismo e società industriale*, Torino, Loescher, 1975. [*Positivism and industrial society*]

Rossi, P., Spiegazione e comprensione da Dilthey a Max Weber [Explanation and understanding from Dilthey to Max Weber], in *Rivista di Filosofia*, LXXV, 1984.

Rossi, P., *Storia e Filosofia. Saggi di storiografia filosofica* [*History and Philosophy. Essays on philosophical historiography*], Turin, Einaudi, 1969 - 1975- 2002.

Roversi, A., (edited by), *Max Weber. Saggi sulla dottrina della scienza* [*Max Weber. Essays on the doctrine of science*], Bari, De Donato, 1980.

Rusconi, G. E., Il concetto di società complessa [The concept of a complex society], in *Quaderni di Sociologia*, [*Sociology notebooks*], 1979, n. 2-3, pp. 261-272.

Russell, B. A. W., *Philosophical Essays*, London, George Allen and Unwin, 1910; Italian translation edited by Bruno Widmar, *Filosofia e scienza* [*Philosophy and science*] Rome, Newton Compton, 1972, 1981.

Russo, L., Santoni, E., *Ingegni minuti. Una storia della scienza in Italia* [*Minute intellects. A history of science in Italy*], Milan, Feltrinelli, 2010.

Rutigliano E., Tosini, D., Sulla crisi di orientamento della sociologia [On the crisis of the orientation of sociology], in *Quaderni di Sociologia*, [*Sociology notebooks*], v. 46, n. 2, 2002, pp. 91-112.

Rutigliano, E., Cambi, F., Quaglioni, D., *L'università a Trento, 1962-2002* [*The university in Trent 1962-2002*] Trento, Università degli Studi di Trento, 2004.

Rutigliano, E., L'influenza della teoria critica sulla sociologia italiana [The influence of critical theory on Italian sociology] in *La Critica Sociologica*, [*Sociological criticism*], n. 66, 1983.

Rutigliano, E., Max Weber nella recezione marxista [Max Weber in Marxist reception] in *La Critica Sociologica*, [*Sociological criticism*], n. 75-76, 1983.

Rutigliano, E., *Sociologi. Uomini e problemi* [*Sociologists. Men and issues*], Milan, Franco Angeli, 1990.

Rutigliano, E., *Teoria o critica. Saggio sul marxismo di Adorno* [*Theory or criticism. Essay on Marxism by Adorno*], Bari, Dedalo, 1977.

Sandri, G., Introduction to all'edizione italiana di Leszek Nowak, *U podstaw marksowskiej metodologii nauk* [*At the core of the Marxian methodology of science*], Warszawa, Panstwowe Wydawnictwo Naukowe, 1971; Italian translation di Riccardo Casimiro Levanski edited by Leonardo Benvenuti, *La scienza come idealizzazione: i fondamenti della metodologia marxiana* [*Science as idealisation: the fundaments of Marxian methodology*], Bologna, Il Mulino, 1977.

Santucci A., (edited by), *Scienza e filosofia nella cultura positivistica* [*Science and philosophy in positivist culture*], Milan, Feltrinelli, 1982.

Santucci A., *Pragmatismo e filosofia europea in Antonio Santucci, Storia del pragmatismo* [*Pragmatism and European Philosophy in Antonio Santucci, the history of Pragmatism*], Rome-Bari, Laterza, 1992.

Santucci, A., *Il pragmatismo in Italia* [*Pragmatism in Italy*], Bologna, Il Mulino, 1963.

Santucci, A., Introduction to a Karl Mannheim, *Ideologie und Utopie* [*Ideology and utopia*], Frankfurt, Schulte-Bumke, 1929; Italian translation the English edition of 1936, Bologna, Il Mulino, 1957.

Santuccio, M. Scienza [Science], in F. De Marchi and A. Ellena, B. Catarinussi, *Dizionario di Sociologia* [*Dictionary of sociology*], Rome, Edizioni Paoline, 1987.

Santuccio, M., Contributo ad una teoria della scienza come professione [Contribution to a theory of science as a profession], in *Sociologia dell'organizzazione*, 1973, n. 1, pp. 137-165.

Santuccio, M., *Scienza e società complessa* [*Science and complex society*], Padua, CSSR, 1991.

Santuccio, M., *Scienza e società* [*Science and society*], Milan, Franco Angeli, 1987 (1989).

Sartori, G., (edited by), *Social Science Concepts: A Systematic Analysis*, London, Sage, 1984.

Scaglia, A., *25 anni dell'Associazione Italiana di Sociologia* [*25 years of the Italian Association of Sociology*], Dipartimento di Sociologia e Ricerca Sociale, Università degli Studi di Trento, Quaderno 39, 2007.

Scarantino, L. M., *Giulio Preti. La costruzione della filosofia come scienza sociale* [*Giulio Preti. The construction of philosophy as a social science*], Preface by Paolo Parrini, Milan, Bruno Mondadori, 2007.

Schizzerotto, A., Preface in G. Bertin, *Valutazione e sapere sociologico* [*Assessment and sociological knowledge*], Milan, Franco Angeli, 1995.

Sciacca, M. F., Problema della scienza e scientism [The issue of science and scientism], in *Giornale della Metafisica*, 8, 5, 1953.

Sciarra, E., *Epistemologia e società. Popper, Wittgenstein, Gadamer* [*Epistemology and society. Popper, Wittgenstein, Gadamer*], Pescara, Tipolitografia Sigraf, 2008.

Sciarra, E., *L'epistemologia delle scienze storico-sociali nel pensiero di Karl Popper* [*The epistemology of historical-social sciences in the thinking of Karl Popper*], Chieti, 1981.

Sciolla, L., (edited by), *Processi e trasformazioni sociali* [*Social processes and transformations*], Rome-Bari, Laterza, 2009.

Semerari, G., La filosofia come scienza rigorosa e la critica al dogmatismo, in AA.VV, *Omaggio a Husserl* [Philosophy as a rigorous science and the criticism of dogmatism, in AA.VV [various authors], *Homage to Husserl*] edited by Enzo Paci, Milan, Il Saggiatore, 1960.

Semerari, G., *La lotta per la scienza* [*The struggle for science*], reprinted Milan, Guerini e Associati, 2013.

Semerari, G., *La scienza come problema. Dai modelli teorici alla produzione di tecnologie: una ricerca interdisciplinare* [*Science as a problem. From theoretical models to the production of technologies: interdisciplinary research*] Bari, De Donato, 1980.

Semerari, G., *Scienza nuova e ragione* [*New science and reason*], Bari, Lacaita, 1961.

Shapin S., *La rivoluzione scientifica* [*The scientific revolution*], Torino, Einaudi, 2003.

Shapin, S., *The Scientific Revolution*, it. transl., Turin, Einaudi, 2003.

Sini, C., Il mito dei miti. Confini problematici dell'epistemologia feyerabendiana [The myth of myths. Problems of the Feyerabendian epistemology], in *Atque*, [And yet] 10, 1994-1995.

Sini, C., *Il pragmatismo americano* [*American pragmatism*], Bari, Laterza, 1972.

Sini, C., *Introduction to alla fenomenologia come scienza* [*Introduction to phenomenology as a science*], Milan, 1965.

Sini, C., *Per una propedeutica alla fenomenologia come scienza rigorosa* [*For a propaedeutics to phenomenology as a rigorous science*] in 'Aut-Aut', n. 84, 1964.

Somenzi, V. Presentation to P. Frank, *Modern Science and Its Philosophy*, New York, Collier Books, 1961; Italian translation by Giuseppe Picca edited by Maria Carla Galavotti, *La scienza moderna e la sua filosofia* [*Modern science and its philosophy*], Bologna, Il Mulino, 1973.

Somenzi, V., Epistemologia evoluzionistica e creatività scientifica [Evolutionary epistemology and scientific creativity], in *La Nuova Critica*, [*New criticism*], n. 45, 1978.

Somenzi, V., Evoluzionismo ed epistemologia, in *Nuova Civiltà delle Macchine*, n. 4, 1984. [Evolutionism and epistemology, in *the New Civilization of Machines*]

Somenzi, V., Introduction to Percy Williams Bridgman, *The logic of Modern Physics*, London, Macmillan, 1927; Italian translation *La logica della fisica moderna*, Turin, Boringhieri, 1965, 1984.

Somenzi, V., La problematica delle scienze nel pensiero di Paolo Filiasi Carcano [The problematic of the sciences in the thinkimng of Paolo Filiasi Carcano], in *Incontri Culturali*, [*Cultural encounters*], n. 1-2, 1979.

Somenzi, V., Naturale e artificiale [Natural and artificial,], in *Società Italiana per il Progresso delle Scienze, 'Atti' della LXII Riunione* [*Italian Society for the Progress of Sciences, Minutes of the LXII Meeting*], (Urbino, 1995), *L'uomo tra natura e cultura* [*Man between nature and culture*], Rome, SIPS, 1995.

Somenzi, V., Scritti italiani di filosofia della scienza [Italian writings on the philosophy of science], in *Rivista critica di storia della filosofia* [*Critical review of the history of philosophy*], 6, 1954.

Sormano A., *Sociologia e sociologi in Italia*, Milan, Franco Angeli, 1988.

Sparti, D., *Epistemologia delle scienze sociali*, Bologna, Il Mulino, 2007.

Spirito, U., Il positivismo non è finito [Positivism is not over], in *Giornale critico della filosofia italiana*, [*The critical journal of Italian philosophy*], 58, 1979.

Spirito, U., Significato della fenomenologia [The meaning of phenomenology], in *Archivio di Filosofia*, [*Archives of philosophy*], 1951.

Spranzi, M., La sociologia e la retorica della scienza, in *Introduction to alla filosofia della scienza* [Sociology and the rhetoric of science, in *Introduction to the philosophy of science*], edited by G. Giorello, Milan, Bompiani, 1999.

Statera, G., (edited by), *La sociologia della scienza* [*The sociology of science*], Naples, Liguori, 1978.

Statera, G., 'Scienza e Società' ['Science and Society'] in *Enciclopedia delle Scienze Sociali Giovanni Treccani*, Rome, 1997, VII. [*Encyclopedia of Social Sciences Giovanni Treccani*]

Statera, G., Cannavo', L., (edited by), Sociologia della scienza e politiche della ricerca [The sociology of science and research policies], in *Sociologia e ricerca sociale*, [*Sociology and social research*], 24, 1987.

Statera, G., Contro la "nouvelle-vague" anti-metodologica [Against the anti-methodological "nouvelle-vague"], (1984), in Alberto Marradi (edited by), *Metascienza. Riflessioni sulla conoscenza scientifica* [*Metascience. Reflections on scientific knowledge*], Rome, EuRome - Editrice Universitaria di Roma - La Goliardica, 1991; published in Nel farsi della ricerca [In doing research], special number of *Sociologia e ricerca sociale*, [*Sociology and social research*], XXV, 2004.

Statera, G., Epistemologia e metodo scientifico, [Epistemology and scientific method] in *La cultura*, [*Culture*], V, 1967.

Statera, G., Il mito della ricerca qualitativa [The qualitative-research myth] in *Sociologia e ricerca sociale*, [*Sociology and social research*], XXXIX, 1992; published in Nel farsi della ricerca [In doing research], special number of *Sociologia e ricerca sociale*, [*Sociology and social research*], XXV, 2004.

Statera, G., Individualismo metodologico, ermeneutica, scienza sociale [Methodological individualism, hermeneutics, social science], in *Sociologia e ricerca sociale*, [*Sociology and social research*], n. 43, 1994.

Statera, G., *La conoscenza sociologica. Problemi e metodo* [*Sociological knowledge. Problems and method*], Naples, Liguori, 1974; new edition revised and expanded.

Statera, G., La sociologia della scienza in R. K. Merton [The sociology of science in R. K. Merton], in *La Critica Sociologica*, [*Sociological criticism*], 3, 1967.

Statera, G., La sociologia scientifica e le cosiddette 'sociologie ermeneutiche' [Scientific sociology and the so-called 'hermeneutical sociologies'], (2001) in published in Nel farsi della ricerca [In doing research], special number of *Sociologia e ricerca sociale*, [*Sociology and social research*], XXV, 2004.

Statera, G., Le epistemologie post-empiriste e la logica dell'indagine sociologica [Post-empirical epistemologies and the logic of sociological inquiry], in *Sociologia e ricerca sociale*, [*Sociology and social research*], n. 33, 1990; published in Nel farsi della ricerca [In doing research], special number of *Sociologia e ricerca sociale*, [*Sociology and social research*], XXV, 2004.

Statera, G., *Logica dell'indagine scientifico-sociale* [*The logic of scientific-social inquiry*], Milan, Angeli, 1994.

Statera, G., *Logica linguaggio e sociologia. Studio su Otto Neurath e il Neopositivismo* [*Logic language and sociology.A study on Otto Neurath and neopositivism*], Turin, Taylor, 1967.

Statera, G., *Manuale di sociologia scientifica*, Rome, SEAM, 1997. [*Handbook of scientific sociology*]

Statera, G., Phillips, D., *Introduzione alla sociologia della scienza*, Rome, Elia, 1977. [*Introduction to the sociology of science*]

Statera, G., Problemi metodologici delle scienze sociali [Methodological problems of the social sciences], in *Ricerche Metodologiche*, [*Methodological research*], II, n. 3-4, 1966.

Tagliagambe, S., *Filosofia della scienza* [*The philosophy of science*], (in collaboration with G. Boniolo, M.L. Dalla Chiara, G. Giorello, C. Sinigaglia), Milan, Raffaello Cortina, 2002.

Tagliagambe, S., *L'epistemologia contemporanea* [*Contemporary epistemology*], Rome, Editori Riuniti, 1991.

Tommasi, L., (edited by), *Sociology: The "New" Language*, Milan, Franco Angeli, 2012.

Tonini, V., *Dalla tecnologia alla sociologia* [From technology to sociology], in *Civiltà delle Macchine*, [*The civilisation of machines*], n. 5, 1967.

Toscano, M. A., *Introduzione alla sociologia* [*Introduction to sociology*], Milan, Franco Angeli, 2011 (2006).

Toscano, M. A., *Marx e Weber. Strategie della possibilità* [*Marx and Weber. Strategies of possibility*], Naples, Guida, 1988.

Toulmin, S. E., Filosofia e teoria della scienza, storia e sociologia della scienza [Philosophy and theory of science, history and sociology of science], in AA.VV, *Metascienza* [*Meta-science*], *Sociologia e ricerca sociale*, [*Sociology and social research*], V, n.13, April 1984.

Toulmin, S. E., The Philosophy of Science: An Introduction to, London, Hutchinson, 1953; partial Italian translation *Che cosa è la filosofia della scienza* [*What the philosophy of science is*], Rome, Ubaldini, 1968.

Trinchero, M., (edited by), *Il neopositivismo logico. Antologia* [*Logical neo-positivism. An anthology*], Turin, Loescher, 1978.

Trinchero, M., *La filosofia dell'aritmetica di G. Frege* [*G. Frege philosophy of arithmetic*], Turin, Giappichelli, 1967.

Vailati, G., *Gli strumenti della conoscenza* [The tools of knolwedge], edited by Mario Calderoni, Lanciano, Carabba, 1916.

Vailati, G., *Scritti di metodologia scientifica* [*Writings on scientific methodology*], edited by Michele Federico Sciacca, Florence, Principato, 1968.

Vailati, G., *Scritti* [*Writings*], edited by Mario Quaranta, Sala Bolognese, Forni edition, 1987, 3 volumes.

Vailati, G., *Sull'importanza delle ricerche relative alla storia delle scienze* [*On the importance of research related to the history of science*], reprinted in F. Rossi Landi (edited by), *Il metodo della filosofia*, Bari, Laterza, 1967.

Vanni Rovighi, S., *Gnoseologia* [*Gnoseology*], Brescia, Morcelliana, 1979.

Vassallo, N., *Filosofie delle scienze* [*Philosophies of science*], Turin, Einaudi, 2003.

Vassallo, N., *Frege on Thinking and Its Epistemic Significance*, with P. Garavaso, Lexington Books–Rowman & Littlefield, Lanham, MD, Usa, 2014.

Vassallo, N., *Teoria della conoscenza* [*Theory of knowledge*], Roma-Bari, Laterza, 2003.

Venneri, E., *La teoria sociologica tra realismo e strumentalismo* [*Sociological theory between realism and instrumentalism*], Soveria Mannelli, Rubbettino publisher, 1998.

Verdino, A., *Epistemologia della causalità* [*Epistemology of causality*], Rome, Bulzoni, 1972.

Vergati, S., *Sociologia delle comunità scientifiche. Sistema normativo e valori organizzativi* [*The sociology of scientific communities. The regulatory system and organizational values*], Rome, Carucci, 1981.

Viale, R., *Metodo e società nella scienza. Fattori metodologici, sociali e cognitivi delle decisioni scientifiche* [*Method and society in science. The methodological, social and cognitive factors of scientific decisions*], Milan, Angeli, 1991.

Viale, R., Oltre l'individualismo metodologico: il cognitivismo metodologico [Beyond methodological individualism: methodological cognitivism], in *Sociologia e ricerca sociale*, [*Sociology and social research*], 43, 1994.

Visalberghi, A., *Esperienza e valutazione* [*Experience and assessment*], Turin, Taylor, 1958 (1967).

Visalberghi, A., Il filosofo dello spirito scientifico [The philosopher of the scientific spirit] *Introduction to J. Dewey, Logic, the Theory of Inquiry* (1938); Italian translation 1949, new edition 1973, *Logica, teoria dell'indagine* [*Logic, theory of inquiry*], Turin, Einaudi, 1973, 2 volumes.

Voltaggio, F., *Bernard Bolzano e la dottrina della scienza* [*Bernard Bolzano and the doctrine of science*], Milan, Edizioni di Comunità, 1974.

Voltaggio, F., *Fondamenti della logica di Husserl* [*The fundaments of Husserl's logic*], Milan, Comunità, 1965.

Watkins, J., Science and Scepticism; Italian translation by G. Rigamonti, *Certezza e verità: per una epistemologia postpopperiana* [*Certainty and truth: for a post-Popperian epistemology*], Rome-Bari, Laterza, 1986.

Watkins, J., Tipi ideali e spiegazioni storiche [Ideal types and historical explanations], in *Controcorrente*, V, 1973.

Weber, M., L''oggettività' conoscitiva della scienza sociale e della politica sociale [The cognitive objectivity of social science and social policy] original title: Die 'Objektivität' sozialwissenschaftlicher und sozialpolitischer Erkenntnis [The 'objectivity' of sociological and socio-political knowledge, (1904), in *Gesammelte Aufsätze zur Wissenschaftslehre* [*Collected essays on teaching science*] Tübingen, Mohr, 1922 – second edition 1951; Italian translation and Introduction to Pietro Rossi, Il metodo delle scienze storico-sociali [The method of historical-social sciences], Turin, Einaudi, 1981. See also in the expanded edition *Saggi sul metodo delle scienze storico sociali* [*Essays on the method of historical social sciences*] edited by P. Rossi, Milan, Comunità, 2001.

Whitley, R. D. Le scienze come organizzazioni professionali [The science sas professional organisations] in *Sociologia e ricerca sociale*, [*Sociology and social research*], I, 2-3, 1980.

Widmar, B., *Introduzione alla filosofia della scienza*, Bari, Levante Editori, 1993. [*Introduction to the philosophy of science*]

Winch, P., *The Idea of a Social Science*, London, Routledge & Kegan Paul, 1958; Italian translation Il concetto di scienza sociale e le sue relazioni con la filosofia [The concept of social science and its relationships with philosophy] Milan, Il Saggiatore, 1972.

Wittgenstein, L. *Logisch-philosophicus Adhandlung* [*Logico-philosophial treatise*] London, Routledge & Kegan Paul, 1922; Italian translation di Amedeo G. Conte, *Tractatus logico-philosophicus*, Turin, Einaudi, 1989; see also new ed. with added

Notebooks 1914-1916, Oxford, Basil Blackwell, 1961, Italian translation *Quaderni 1914-1916*, Turin, Einaudi, 2004.

Wittgenstein, L. *Philosophische Bemerkungen* [*Philosophical remarks*], Oxford, Basil Blackwell, 1964; Italian translation edited by Marino Rosso, *Osservazioni filosofiche* [*Philosophical Observations*] , Turin, Einaudi, 1976; see also new edition 1999.

Wolff K. H., Wolf K. H., Some Aspects of Modern Sociology in *The Southwestern Social Science Quarterly*, vol. 21, n. 1, June 1940, pp. 42-52.

About the Author

Guglielmo Rinzivillo is a researcher and aggregate professor at the Department of Social Sciences and Economics at the *Sapienza* University of Rome. His interests are the history of sociological thought and the history of science. Among his latest monographs are *Karl Marx, dialettica e memoria* (Rome, 2013), *Natura, cultura e induzione nell'età delle scienze* (Rome, 2015) and *Scienza e valori in Karl Mannheim* (Rome, 2016). He is a Member of the European Society of the History of Science.

INDEX

A

B

C

D

O

P

Q

R

S

T

U

V

Advances in Sociology Research. Volume 25

Editor: Jared A. Jaworski

Series: Advances in Sociology Research

Book Description: *Advances in Sociology Research. Volume 25* opens with a chapter underscoring the importance of this age-old cultural practice of herding in relation to Basotho males and its implications to the non-formal education (NFE) policy and practice.

Hardcover ISBN: 978-1-53613-464-3
Retail Price: $250

Italy: Social, Environmental and Agricultural Issues

Editors: Gina Pittman

Series: Countries and Cultures of the World

Book Description: Italy's peculiar cultural variability is the result of population movements and invasions from the Mediterranean Sea, which took place in both historical and prehistoric times. Due to many interesting places such as the city of Rome, Central Italy was endured by migrants, who have contributed to Italy's culture, linguistics and also genetics.

Hardcover ISBN: 978-1-63484-408-6
Retail Price: $110